New Directions in Homotopy Theory

CONTEMPORARY MATHEMATICS

707

New Directions in Homotopy Theory

Second Mid-Atlantic Topology Conference
March 12–13, 2016
Johns Hopkins University, Baltimore, Maryland

Nitya Kitchloo
Mona Merling
Jack Morava
Emily Riehl
W. Stephen Wilson
Editors

AMS AMERICAN MATHEMATICAL SOCIETY

2010 *Mathematics Subject Classification.* Primary 14Fxx, 18Dxx, 19Exx, 55Mxx, 55Nxx, 55Pxx, 55Rxx.

Library of Congress Cataloging-in-Publication Data

Names: Mid-Atlantic Topology Conference (2nd : 2016 : Baltimore, Md.) | Kitchloo, Nitya, 1972– editor.

Title: New directions in homotopy theory : Second Mid-Atlantic Topology Conference, March 12–13, 2016, Johns Hopkins University, Baltimore, Maryland / Nitya Kitchloo [and four others], editors.

Description: Providence, Rhode Island : American Mathematical Society, [2018] | Series: Contemporary mathematics ; volume 707 | Includes bibliographical references.

Identifiers: LCCN 2017051421

Subjects: LCSH: Homotopy theory–Congresses. | Topology–Congresses. | AMS: Algebraic geometry – (Co)homology theory – (Co)homology theory. msc | Category theory; homological algebra – Categories with structure – Categories with structure. msc | K-theory – K-theory in geometry – K-theory in geometry. msc | Algebraic topology – Classical topics – Classical topics. msc | Algebraic topology – Homology and cohomology theories – Homology and cohomology theories. msc | Algebraic topology – Homotopy theory – Homotopy theory. msc | Algebraic topology – Fiber spaces and bundles – Fiber spaces and bundles. msc

Classification: LCC QA612.7 .M53 2016 | DDC 514/.24–dc23 LC record available at https://lccn.loc.gov/2017051421

Contemporary Mathematics ISSN: 0271-4132 (print); ISSN: 1098-3627 (online)

DOI: `http://dx.doi.org/10.1090/conm/707`

∞ The paper used in this book is acid-free and falls within the guidelines
established to ensure permanence and durability.
Visit the AMS home page at `http://www.ams.org/`

10 9 8 7 6 5 4 3 2 1 23 22 21 20 19 18

Contents

Preface

Introduction

This volume has grown out of the Second Mid-Atlantic Topology Conference, which was held in the Spring of 2016 in Baltimore[1]. The conference was motivated by the organizers' sense that algebraic topology, geometry, and category theory are in the midst of a generational shift of language, techniques, and paradigms, rooted in new thinking about higher homotopy-theoretic structures which profoundly alters our understanding of the nature of geometry and its applications across mathematics, reaching from data analysis to physics.

These fields are in ferment, with a forefront defined by a younger generation who bring to bear new perspectives from algebra on classical questions of pure and applied mathematics. The conference attempted to sample some of these developments, and to showcase some of the work of the research cohort with applications in fields such as

• arithmetic geometry (new aproaches to motives, e.g. via $\mathbb{A}^1$-homotopy theory; new techiques in K-theory via Hochschild and cyclic homology; homotopy-theoretic methods in automorphic forms, with applications to classical homotopy theory),

• geometric quantum field theory, e.g. through various cobordism hypotheses (via chiral homology and such, not necessarily E_∞ ring spectra), and

• category theory, e.g. through Voevodsky's univalent foundations project and related work on homotopy type theory by researchers drawn from algebraic topology, category theory, logic, and computer science (with more distantly related applications to topological data analysis).

Contents

Emmy Noether shifted the focus of algebraic topology from numerical invariants of space (such as Betti numbers) to the algebraic objects (e.g. modules of geometric cycles) underlying them. More recently, interest has broadened from topological (homeomorphism) invariants to homotopy theory in general, with an attendant expansion of potential applications. This has extended even further, to issues of logic and foundations, so that homotopy types, rather than sets, are becoming central, fundamental objects of mathematical discourse.

[1]Financial support was provided by the Johns Hopkins University Department of Mathematics and the NSF, via grant DMS 1619569. The conference web page is at `http://www.math.jhu.edu/~eriehl/matc2016/` with a list of talk abstracts at

`http://www.math.jhu.edu/~eriehl/matc2016/abstracts.pdf`

As Yuri Manin puts it:

> I am pretty strongly convinced that there is an ongoing reversal in the collective consciousness of mathematicians: the homotopical picture of the world becomes the basic intuition, and if you want to get a discrete set, then you pass to the set of connected components of a space defined only up to homotopy. ... Cantor's problems of the infinite recede to the background: from the very start, our images are so infinite that if you want to make something finite out of them, you must divide them by another infinity. — Yuri Manin, "We do not choose mathematics as our profession. It chooses us: Interview with Yuri Manin" by Mikhail Gelfand

The papers in this volume are all concerned with homotopy theory in this global sense. Tabuada, for example, surveys recent developments in algebraic geometers' theories of motives — for example, generalizations to categories of (not necessarily commutative) differential graded algebras. Heller and Ormsby, working with more classical motives defined over the field of real numbers, show that their associated spectra of complex points behave well, when regarded as genuinely $\mathbb{Z}/2$-equivariant spectra, while Kass and Wickelgren show that related questions about spectra associated to arithmetic objects (i.e. defined over number fields) conceal further mysteries.

Delicate questions of equivariant homotopy theory lurk behind another subset of papers, centered around more general categorical questions. Lind and Malkiewich on one hand, and Ponto and Shulman on another, bring categories of diagrams (or, parametrized families) of spectra, and their invariants, under firm control. Szymik extends (and clarifies) the notion of categorical center to the homotopy-theoretic contexts, while Zakharevich exhibits a closed structure on the multicategory of Waldhausen categories. Such work takes us "behind the homotopy groups", enabling us to work more directly with the geometric questions that engaged us in the first place.

A final circle of papers is concerned with hard computational questions about just how well constructions from algebraic geometry (moduli spaces of elliptic curves or formal groups) model phenomena in chromatic homotopy theory. McTague shows that the spectrum of topological modular forms is not easily represented geometrically in terms of string bordism, and Peterson pins down some of the subtleties in the theory of orientations for higher analogs of MString. Kitchloo, Lorman, and Wilson, working at the oddly difficult prime two, pioneer the study of higher $\mathbb{Z}/2$-equivariant chromatic theory.

Acknowledgements

The editors wish to thank the conference speakers, the authors of the papers in this volume, and the very conscientious referees for their support. This conference, and these proceedings, owe their existence to the energy and foresight of Nitu Kitchloo, and the other members of the editorial committee appreciate his continued leadership. We also wish to thank the Mathematics Department at

Johns Hopkins and the National Science Foundation for generous financial support, and the editorial staff at the American Mathematical Society for their help in the preparation of this volume.

Nitya Kitchloo
Mona Merling
Jack Morava
Emily Riehl
W. Stephen Wilson
Editors

Contemporary Mathematics
Volume **707**, 2018
http://dx.doi.org/10.1090/conm/707/14250

The stable Galois correspondence for real closed fields

J. Heller and K. Ormsby

For Ivo and Irving.

ABSTRACT. In previous work [**7**], the authors constructed and studied a lift of the Galois correspondence to stable homotopy categories. In particular, if L/k is a finite Galois extension of fields with Galois group G, there is a functor $c^*_{L/k} : \mathrm{SH}^G \to \mathrm{SH}_k$ from the G-equivariant stable homotopy category to the stable motivic homotopy category over k such that $c^*_{L/k}(G/H_+) = \mathrm{Spec}(L^H)_+$. The main theorem of [**7**] %citeHO:E2M says that when k is a real closed field and $L = k[i]$, the restriction of $c^*_{L/k}$ to the η-complete subcategory is full and faithful. Here we "uncomplete" this theorem so that it applies to $c^*_{L/k}$ itself. Our main tools are Bachmann's theorem on the $(2, \eta)$-periodic stable motivic homotopy category and an isomorphism range for the map $\pi^{\mathbb{R}}_\star \mathbb{S}_{\mathbb{R}} \to \pi^{C_2}_\star \mathbb{S}_{C_2}$ induced by C_2-equivariant Betti realization.

1. Introduction

In [**9**], Levine showed that the "constant" functor $c^* : \mathrm{SH} \to \mathrm{SH}_k$ from the classical stable homotopy category to the motivic stable homotopy category over an algebraically closed field of characteristic zero is a full and faithful embedding. Inspired by his result, in [**7**] we introduced and studied functors $c^*_{L/k} : \mathrm{SH}^G \to \mathrm{SH}_k$, where L/k is a Galois extension with Galois group G. We showed that if k is real closed and $L = k[i]$, then *after completing* at a prime p and at η, if $p \neq 2$, the functor $c^*_{L/k}$ is full and faithful. The need for the completion arose from our lack of knowledge about certain homotopy groups of the motivic sphere over $\mathbb{R}$. In the meantime advances have been made. Ananyevskiy-Levine-Panin [**1**] established a motivic version of Serre's finiteness theorem, which in particular implies that $c^*_{L/k}$ is full and faithful after η-completion. The purpose of this paper is to use Bachmann's recent results [**3**], about a localization of $\pi^{\mathbb{R}}_\star(\mathbb{S}_{\mathbb{R}})$, to remove the η-completions in the main theorem of [**7**].

2010 *Mathematics Subject Classification.* Primary: 14F42, 55P91 Secondary: 11E81, 19E15.
Key words and phrases. Equivariant and motivic stable homotopy theory, equivariant Betti realization.
The second author was partially supported by National Science Foundation Award DMS-1406327.

THEOREM 1.1. *Let k be a real closed field and $L = k[i]$ be its algebraic closure. Then the functor*

$$c^*_{L/k} : \mathrm{SH}^{C_2} \to \mathrm{SH}_k$$

is a full and faithful embedding.

One of the primary tools in the proof of Theorem 1.1 is the C_2-equivariant Betti realization functor $\mathrm{Re}_B^{C_2} : \mathrm{SH}_{\mathbb{R}} \to \mathrm{SH}^{C_2}$ which extends the functor taking a smooth $\mathbb{R}$-scheme to its $\mathbb{C}$-points with complex conjugation action. In [**6**], Dugger-Isaksen study the effect of this functor on the bigraded homotopy groups of the real motivic sphere spectrum, proving that $\mathrm{Re}_B^{C_2} : \pi^{\mathbb{R}}_{m+n\alpha}(\mathbb{S}_{\mathbb{R}})^{\wedge}_2 \to \pi^{C_2}_{m+n\sigma}(\mathbb{S}_{C_2})^{\wedge}_2$ is an isomorphism when $m \geq 2n-5$. We use odd-primary Adams spectral sequences and Bachmann's results to produce a range of bigradings in which the integral version of this map is an isomorphism. In particular, we prove the following.

THEOREM 1.2 (Theorem 3.12). *The map $\pi^{\mathbb{R}}_{m+n\alpha}(\mathbb{S}_{\mathbb{R}}) \to \pi^{C_2}_{m+n\sigma}(\mathbb{S}_{C_2})$ is*

(i) *an injection if $m = 2n - 6$ and $n \geq 0$, and*
(ii) *an isomorphism if $m \geq 2n - 5$ and $n \geq 0$.*

Outline. In Section 2 we recall Bachmann's theorem and deduce some consequences for the C_2-equivariant Betti realization functor and Morel's $\pm$-splitting of the 2-periodic stable motivic homotopy category. In Section 3 we adapt the methods of [**6**] to odd-primary Adams spectral sequences. Via an arithmetic fracture square and the results of Section 2, we deduce Theorem 3.12. Finally, in Section 4 we recall how to bootstrap Theorem 3.12 into a proof of Theorem 1.1.

Notation. We use the following notation throughout the paper.

- k is a field and L/k is a finite Galois extension of fields with Galois group G.
- SH_k is Voevodsky's stable motivic homotopy category [**13**]; hom sets in SH_k are denoted $[\ ,\ ]_k$.
- SH^G is the G-equivariant stable homotopy category in the sense of [**10**]; hom sets in SH^G are denoted $[\ ,\ ]_G$.
- $c^*_{L/k} : \mathrm{SH}^G \to \mathrm{SH}_k$ is the functor induced by the classical Galois correspondence $G/H \mapsto \mathrm{Spec}(L^H)$, constructed in [**7**, Section 4.3].[1]
- $\mathbb{S}_k$ is the sphere spectrum in SH_k and $\mathbb{S}_G$ is the sphere spectrum in SH^G.
- For integers m, n, $S^{m+n\alpha} = (S^1)^{\wedge m} \wedge (\mathbb{A}^1 \smallsetminus 0)^{\wedge n}$. If $G = C_2$, $S^{m+n\sigma} = (S^1)^{\wedge m} \wedge (S^\sigma)^{\wedge n}$ where S^σ is the one-point compactification of the real sign representation.
- For $X \in \mathrm{SH}_k$, $\pi^k_{m+n\alpha}X := [S^{m+n\alpha}, X]_k$ is the $(m + n\alpha)$-th homotopy group of X.
- For $X \in \mathrm{SH}^{C_2}$, $\pi^{C_2}_{m+n\sigma}X := [S^{m+n\sigma}, X]_{C_2}$ is the $(m + n\sigma)$-th homotopy group of X.
- For $X \in \mathrm{SH}_k$, $\pi^k_* X := \bigoplus_{m\in\mathbb{Z}} \pi_{m+0\alpha}X$ and similarly for $\pi^{C_2}_* Y$ when $Y \in \mathrm{SH}^G$.

[1] In *loc. cit.* we state that the category GsSet of G-simplicial sets is equivalent to the category $\mathrm{sPre}(\mathrm{Or}_G)$ of simplicial presheaves on the orbit category. This isn't quite true: $G\mathrm{sSet} \subseteq \mathrm{sPre}(\mathrm{Or}_G)$ is only a full subcategory. (Thanks to Tom Bachmann for drawing our attention to this inaccuracy.) This has no effect on any of the subsequent mathematics in *loc. cit.* because what is used is that the associated homotopy categories are equivalent and this is true by Elmendorf's Theorem.

- Given an embedding $\phi : k \hookrightarrow \mathbb{R}$, $\mathrm{Re}_{B,\phi}^{C_2} = \mathrm{Re}_B^{C_2} : \mathrm{SH}_k \to \mathrm{SH}^{C_2}$ is the C_2-equivariant Betti realization which extends the functor taking a smooth k-scheme X to $X(\mathbb{C})$ with the conjugation action [**7**, §4.4].
- Depending on context, η is either the motivic Hopf map arising from $\mathbb{A}^2 \smallsetminus 0 \to \mathbb{P}^1$ or the C_2-equivariant Hopf map.
- $(\ )_2^\wedge$ is the 2-completion functor and $(\ )_\eta^\wedge$ is the η-completion functor. If $a = 2$ or η we have $X_a^\wedge = \operatorname{holim} X/a^n$.
- We write $X[a^{-1}]$ or $X[1/a]$ for the homotopy colimit of $X \xrightarrow{a} X \xrightarrow{a} \cdots$. If $X \simeq X[a^{-1}]$ we say that X is *a-periodic.*

2. Preliminaries

The main new input we use in this paper is a recent theorem of Bachmann [**3**, Theorem 31]. His theorem compares a localization of $\mathrm{SH}_{\mathbb{R}}$ with the classical stable homotopy category. The most convenient form of his result for us is the following recasting.

THEOREM 2.1 (Bachmann). *Betti realization induces an equivalence of triangulated categories*

$$\mathrm{Re}_B^{C_2} : \mathrm{SH}_{\mathbb{R}}[1/2, \eta^{-1}] \xrightarrow{\simeq} \mathrm{SH}^{C_2}[1/2, \eta^{-1}].$$

PROOF. Consider the functor $\mathrm{Re}_{\mathbb{R}} : \mathrm{SH}_{\mathbb{R}} \to \mathrm{SH}$ which extends the functor $\mathrm{Sm}_{\mathbb{R}} \to \mathrm{Top}$, sending X to $X(\mathbb{R})$. Consider as well the composite $\Phi^{C_2} \circ \mathrm{Re}_B^{C_2}$, where $\Phi^{C_2} : \mathrm{SH}^{C_2} \to \mathrm{SH}$ is the geometric fixed points functor. Both $\mathrm{Re}_{\mathbb{R}}$ and $\Phi^{C_2} \circ \mathrm{Re}_B^{C_2}$ preserve homotopy colimits and if $X \in \mathrm{Sm}_{\mathbb{R}}$ and $n \in \mathbb{Z}$, they both send $\Sigma_{\mathbb{P}^1}^n \Sigma_{\mathbb{P}^1}^\infty X_+$ to the spectrum $\Sigma^n \Sigma^\infty X(\mathbb{R})_+$. It follows that $\mathrm{Re}_{\mathbb{R}} = \Phi^{C_2} \circ \mathrm{Re}_B^{C_2}$ and so we have the commutative triangle of functors

$$\begin{array}{ccc} \mathrm{SH}_{\mathbb{R}}[1/2, \eta^{-1}] & \xrightarrow{\mathrm{Re}_B^{C_2}} & \mathrm{SH}^{C_2}[1/2, \eta^{-1}] \\ & \searrow^{\mathrm{Re}_{\mathbb{R}}} & \downarrow \Phi^{C_2} \\ & & \mathrm{SH}[1/2]. \end{array}$$

Write $\rho : S^0 \to S^\sigma$ for the standard inclusion. Since $\eta^2 \rho = -2\eta$, we have the equivalence of categories $\mathrm{SH}^{C_2}[1/2, \eta^{-1}] \simeq \mathrm{SH}^{C_2}[1/2, \rho^{-1}]$. Note that Φ^{C_2} induces an equivalence $\mathrm{SH}^{C_2}[\rho^{-1}] \to \mathrm{SH}$ by [**10**, Corollary 9.6] (for a modern proof see, *e.g.*, [**11**, Theorem 6.11]), and thus it follows that Φ^{C_2} is an equivalence $\mathrm{SH}^{C_2}[1/2, \eta^{-1}] \simeq \mathrm{SH}[1/2]$. A specialization of Bachmann's theorem [**3**, Theorem 31] says that $\mathrm{Re}_{\mathbb{R}}$ in the above diagram is an equivalence. We conclude that $\mathrm{Re}_B^{C_2}$ is an equivalence as well.

□

Recall Morel's $\pm$-operations in motivic homotopy theory (see [**4**, Section 16.2]). Let ε denote the stable map induced by the twist isomorphism $S^\alpha \wedge S^\alpha \simeq S^\alpha \wedge S^\alpha$. In the C_2-equivariant setting, let ε denote the twist $S^\sigma \wedge S^\sigma \simeq S^\sigma \wedge S^\sigma$, and note that $\mathrm{Re}_B^{C_2}(\varepsilon) = \varepsilon$. In either the motivic or equivariant setting, invert 2 and note that $e_+ := (\varepsilon - 1)/2$ and $e_- := (\varepsilon + 1)/2$ are orthogonal idempotents. Let the operation $(\ )^+$ denote inversion of e_+ and let $(\ )^-$ denote inversion of e_-, *i.e.*, $(\ )^\pm$ is the cofiber of the operation $e_\pm$. For any 2-periodic motivic or C_2-equivariant spectrum X there is a natural splitting $X \simeq X^+ \vee X^-$.

LEMMA 2.2. *If X is a 2-periodic motivic or C_2-equivariant spectrum, then $X^+ \simeq X^\wedge_\eta$ and $X^- \simeq X[\eta^{-1}]$.*

PROOF. We prove the motivic version of this statement, which easily adapts to the C_2-equivariant setting. Let X be a spectrum such that $X \simeq X[1/2]$. We have $\varepsilon = -1 - \rho\eta$ whence $e_- = \frac{-1}{2}\rho\eta$. Thus inverting e_- inverts both ρ and η. Since $(2+\rho\eta)\eta = 0$, this is the same as inverting 2 and η, whence $X^- \simeq X[\eta^{-1}]$. Now apply η-completion to the splitting $X \simeq X^+ \vee X^-$ to get $X^\wedge_\eta \simeq (X^+)^\wedge_\eta \vee (X^-)^\wedge_\eta$. Since $X^- \simeq X[\eta^{-1}]$, the second summand is trivial. Since $e_+\eta = 0$, X^+ is η-complete, *i.e.*, $(X^+)^\wedge_\eta \simeq X^+$. We conclude that $X^\wedge_\eta \simeq X^+$, as desired. □

As an interesting corollary (which we will not use in the remainder of this paper) we note the following.

PROPOSITION 2.3. *The natural map $\mathrm{Re}_B^{C_2}((\mathbb{S}_\mathbb{R})^\wedge_\eta) \to (\mathbb{S}_{C_2})^\wedge_\eta$ is an equivalence.*

PROOF. Let $\mathbb{S} = \mathbb{S}_\mathbb{R}$ and η-complete the 2-primary fracture square for $\mathbb{S}$ in order to produce the bicartesian square

$$\begin{array}{ccc} \mathbb{S}^\wedge_\eta & \longrightarrow & \mathbb{S}[1/2]^\wedge_\eta \\ \downarrow & & \downarrow \\ \mathbb{S}^\wedge_{2,\eta} & \longrightarrow & (\mathbb{S}^\wedge_2[1/2])^\wedge_\eta. \end{array}$$

Applying $\mathrm{Re}_B^{C_2}$ results in a homotopy pullback square which maps to the corresponding fracture square for $(\mathbb{S}_{C_2})^\wedge_\eta$. First consider the map $\mathrm{Re}_B^{C_2}(\mathbb{S}^\wedge_{2,\eta}) \to (\mathbb{S}_{C_2})^\wedge_{2,\eta}$ between lower left corners. We may check that this is an equivalence by computing its effect on Mackey functor homotopy groups, which in turn may be done by comparing the motivic and C_2-equivariant Adams spectral sequences as in [**7**, Proposition 2.11]. Note that we also conclude that $\mathrm{Re}_B^{C_2}(\mathbb{S}^\wedge_2) \to (\mathbb{S}_{C_2})^\wedge_2$ is an equivalence by [**8**, Theorem 1] and [**7**, Theorem 2.10]. The map between the upper right vertices are equivalences by Lemma 2.2. The equivalence between bottom right corners follows from Lemma 2.2 and that $\mathrm{Re}_B^{C_2}(\mathbb{S}^\wedge_2) \simeq (\mathbb{S}_{C_2})^\wedge_2$. Since the maps on these three corners are equivalences, we conclude that the map between homotopy pullbacks $\mathrm{Re}_B^{C_2}((\mathbb{S}_\mathbb{R})^\wedge_\eta \to (\mathbb{S}_{C_2})^\wedge_\eta$ is an equivalence. □

3. Comparing stable stems

In this section we establish a range of bidegrees in which the map on stable stems $\pi^\mathbb{R}_{m+n\alpha}(\mathbb{S}_\mathbb{R}) \to \pi^{C_2}_{m+n\sigma}(\mathbb{S}_{C_2})$ induced by equivariant Betti realization is an isomorphism.

Recall that Dugger-Isaksen establish a range in which 2-complete stems are isomorphic.

THEOREM 3.1 ([**6**, Theorem 4.1]). *The map $\pi^\mathbb{R}_{m+n\alpha}((\mathbb{S}_\mathbb{R})^\wedge_2) \to \pi^{C_2}_{m+n\alpha}((\mathbb{S}_{C_2})^\wedge_2)$ is an isomorphism if $m \geq 2n-5$ and an injection if $m = 2n-6$.*

REMARK 3.2. There are isomorphisms $\pi_{m+n\alpha}((\mathbb{S}_\mathbb{R})^\wedge_2) \cong \pi_{m+n\alpha}((\mathbb{S}_\mathbb{R})^\wedge_{2,\eta})$ for all m, n by [**8**, Theorem 1]. Similarly there are isomorphisms $\pi^{C_2}_{m+n\sigma}((\mathbb{S}_{C_2})^\wedge_2) \cong \pi^{C_2}_{m+n\sigma}((\mathbb{S}_{C_2})^\wedge_{2,\eta})$ for all m, n by [**7**, Theorem 2.10]. Dugger-Isaksen's result can thus equivalently be stated as a comparison between $(2,\eta)$-complete stable stems.

Recall the discussion of motivic and C_2-equivariant cobar complexes from [**6**, §3], noting that all of these constructions may be made with $H\mathbb{F}_p$, p odd, in place of $H\mathbb{F}_2$. The significance of the p-primary cobar complex is that it forms the E_1-page of the p-primary Adams spectral sequence (in the motivic or equivariant context). In order to concisely express the properties of these spectral sequences, let $\mathbb{S}$ be $\mathbb{S}_k$ (k a field) or $\mathbb{S}_{C_2}$, let $\mathbb{M}_p$ denote the homology of a point with coefficients in $\mathbb{F}_p$ or $\underline{\mathbb{F}}_p$, let $\mathcal{A}_p$ denote the p-primary motivic or C_2-equivariant dual Steenrod algebra, let C_p^* denote the p-primary motivic or C_2-equivariant cobar complex, let γ be α or σ, depending on context, and let $\mathrm{Ext}_{\mathcal{A}_p}^{*,*+*\gamma}(\mathbb{M}_p, \mathbb{M}_p)$ denote the homology of C_p^* (which is also Ext in the category of $\mathcal{A}_p$-comodules).

THEOREM 3.3 ([**8**, Theorem 1] and [**7**, Theorem 2.10]). *The (motivic or C_2-equivariant) p-primary Adams spectral sequence has E_1-page C_p^* and E_2-page $\mathrm{Ext}_{\mathcal{A}_p}^{*,*+*\gamma}(\mathbb{M}_p, \mathbb{M}_p)$; it is strongly convergent with*

$$E_2^{s,m+n\gamma} \implies \pi_{m-s+n\gamma}\mathbb{S}_{p,\eta}^{\wedge}.$$

When $p = 2$, we have $\mathbb{S}_{2,\eta}^{\wedge} \simeq \mathbb{S}_2^{\wedge}$ as long as $\mathrm{cd}_2(k[i]) < \infty$ (in the motivic case).

The Dugger-Isaksen result is obtained by comparing cobar complexes. We first extend this method to odd p to obtain an isomorphism range on (p, η)-complete stems. We begin by recalling some facts about the motivic and equivariant Steenrod algebras at odd primes. Write $\mathcal{A}_{\mathbb{F}_p}$ for the classical mod-p dual Steenrod algebra. Recall that

$$\mathcal{A}_{\mathbb{F}_p} = \mathrm{Sym}_{\mathbb{F}_p}(\tau_0, \tau_1, \ldots, \xi_1, \xi_2, \ldots)$$

is a free graded commutative algebra, where $|\tau_i| = p^i$ and $|\xi_i| = p^i - 1$.

Recall that $\mathbb{M}_p^{\mathbb{R}} := \bigoplus_{m,n\in\mathbb{Z}} \pi_{m+n\alpha}^{\mathbb{R}} H\mathbb{F}_p = \mathbb{F}_p[\theta]$ where θ has degree $2-2\alpha$. This follows from the affirmative resolution of the Bloch-Kato conjecture [**15**, Theorem 6.1] together with [**12**, Theorem 7.4]. In the equivariant case we have $\mathbb{M}_p^{C_2} := \bigoplus_{m,n\in\mathbb{Z}} \pi_{m+n\sigma}^{C_2} H\underline{\mathbb{F}}_p = \mathbb{F}_p[\theta, \theta^{-1}]$, see *e.g.* [**5**, Theorem 2.8]. The dual motivic Steenrod algebra over $\mathbb{R}$ is equal to

$$\mathcal{A}_{\mathbb{R}} \cong \mathbb{M}_p^{\mathbb{R}} \otimes_{\mathbb{F}_p} \mathcal{A}_{\mathbb{F}_p} = \mathbb{M}_p^{\mathbb{R}}[\tau_0, \tau_1, \ldots, \xi_1, \xi_2, \cdots]/(\tau_0^2, \tau_1^2, \ldots),$$

where the elements of $\mathcal{A}_{\mathbb{F}_p}$ are considered to be bigraded by assigning weights so that $|\tau_i| = p^i + (p^i - 1)\alpha$ and $|\xi_i| = (p^i - 1) + (p^i - 1)\alpha$, see [**14**, Remark 12.12].

Similarly, the dual C_2-equivariant Steenrod algebra is equal to

$$\mathcal{A}_{C_2} \cong \mathbb{M}_p^{C_2} \otimes_{\mathbb{F}_p} \mathcal{A}_{\mathbb{F}_p} = \mathbb{M}_p^{C_2}[\tau_0, \tau_1, \ldots, \xi_1, \xi_2, \cdots]/(\tau_0^2, \tau_1^2, \ldots),$$

where in this case elements of $\mathcal{A}_{\mathbb{F}_p}$ are considered to be bigraded by assigning weights so that $|\tau_i| = p^i + (p^i - 1)\sigma$ and $|\xi_i| = (p^i - 1) + (p^i - 1)\sigma$.

Equivariant Betti realization $\mathrm{Re}_B^{C_2}$ induces maps $\mathbb{M}_p^{\mathbb{R}} \to \mathbb{M}_p^{C_2}$ and $\mathcal{A}_{\mathbb{R}} \to \mathcal{A}_{C_2}$ which have the obvious effects on the above named elements, *i.e.*, $\theta \mapsto \theta$, $\tau_i \mapsto \tau_i$ and $\xi_i \mapsto \xi_i$.

Write $\mathbb{M}_p$ for either $\mathbb{M}_p^{\mathbb{R}}$ or $\mathbb{M}_p^{C_2}$. In both cases, the dual Steenrod algebra is free over $\mathbb{M}_p$, and a basis is given by monomials $\tau_0^{\epsilon_0}\tau_1^{\epsilon_1} \cdots \xi_1^{n_1}\xi_2^{n_2} \cdots$, where $\epsilon_i \in \{0, 1\}$ and n_i is a nonnegative integer. We write $\tau^\epsilon\xi^n$ for such a monomial.

LEMMA 3.4. *Suppose that $|\tau^\epsilon\xi^n| = k + \ell\alpha$. Then $k \leq \dfrac{p}{p-1}\ell + 1$.*

PROOF. The bidegree of ξ_i satisfies $k < \frac{p}{p-1}\ell$ and the bidegree of τ_i satisfies $k \leq \frac{p}{p-1}\ell$ provided $i \geq 1$. It follows that if $\epsilon_0 = 0$ then the bidegree of $\tau^\epsilon \xi^n$ satisfies $k \leq \frac{p}{p-1}\ell$. If $\epsilon_0 = 1$ then write $\tau^\epsilon \xi^n = \tau_0 \tau^{\epsilon'} \xi^n$, where $\epsilon_0' = 0$. This element thus satisfies the inequality $k \leq \frac{p}{p-1}\ell + 1$. □

LEMMA 3.5. *The map $C^f_{\mathbb{R}} \to C^f_{C_2}$ is*

(i) *an injection in all degrees, and*

(ii) *an isomorphism if $k - f \geq \left\lfloor \frac{p}{p-1}\ell - \frac{4p-2}{p-1} \right\rfloor + 1$.*

PROOF. We have that $C^f_{C_2} \cong \mathbb{M}^{C_2}_p \otimes_{\mathbb{M}^{\mathbb{R}}_p} C^f_{\mathbb{R}}$, that $C^f_{\mathbb{R}}$ is free over $\mathbb{M}^{\mathbb{R}}_p$, and that the map $\mathbb{M}^p_{\mathbb{R}} \to \mathbb{M}^p_{C_2}$ is injective. It follows that $C^f_{\mathbb{R}} \to C^f_{C_2}$ is injective.

Let $Z = [z_1|z_2|\cdots|z_f]$ be a cobar element of Adams filtration f, where each z_i is of the form $\tau^\epsilon \xi^n$, of bidegree $k_i + \ell_i \alpha$. By Lemma 3.4 we have that $k_i \leq \frac{p}{p-1}\ell_i + 1$. Summing over i, we find that if $|Z| = k + \ell\alpha$, then $k \leq \frac{p}{p-1}\ell + f$. The cokernel of $C^f_{\mathbb{R}} \to C^f_{C_2}$ is generated by elements of the form $\theta^{-n} Z$ for $n \geq 1$. The elements θ^{-n} lie above the line of slope[2] $(p-1)/p$ passing through θ^{-1} and so we find that these satisfy the inequality $k \leq \frac{p}{p-1}\ell - \frac{4p-2}{p-1}$. It follows that the bidegree of $\theta^{-n} Z$ satisfies the inequality $k \leq \frac{p}{p-1}\ell + f - \frac{4p-2}{p-1}$. Thus the cokernel is zero in bidegrees satisfying $k > \frac{p}{p-1}\ell + f - \frac{4p-2}{p-1}$.

□

Recall the form of the motivic and C_2-equivariant Adams spectral sequences from Theorem 3.3. Equivariant Betti realization induces a map between these spectral sequences which we can now analyze.

PROPOSITION 3.6. *The map $\mathrm{Ext}^{(f,k+\ell\alpha)}_{\mathcal{A}_{\mathbb{R}}}(\mathbb{M}^{\mathbb{R}}_p, \mathbb{M}^{\mathbb{R}}_p) \to \mathrm{Ext}^{(f,k+\ell\alpha)}_{\mathcal{A}_{C_2}}(\mathbb{M}^{C_2}_p, \mathbb{M}^{C_2}_p)$ is an injection if $k-f = \left\lfloor \frac{p}{p-1}\ell - \frac{4p-2}{p-1} \right\rfloor$ and an isomorphism if $k-f \geq \left\lfloor \frac{p}{p-1}\ell - \frac{4p-2}{p-1} \right\rfloor + 1$.*

PROOF. This follows from [**6**, Lemma 3.4] and Lemma 3.5. □

LEMMA 3.7. *$\mathrm{Ext}^{(f,k+\ell\alpha)}_{\mathcal{A}_{\mathbb{R}}}(\mathbb{M}^{\mathbb{R}}_p, \mathbb{M}^{\mathbb{R}}_p)$ is a finite-dimensional $\mathbb{F}_p$-vector space for all f, k, ℓ.*

PROOF. Writing $C_{\mathbb{F}_p}$ for the classical cobar complex, we have $C_{\mathbb{R}} \cong \mathbb{M}^{\mathbb{R}}_p \otimes_{\mathbb{F}_p} C_{\mathbb{F}_p}$. This makes it clear that $C^{(f,k+\ell\alpha)}_{\mathbb{R}}$ is finite dimensional, and thus $\mathrm{Ext}^{(f,k+\ell\alpha)}_{\mathcal{A}_{\mathbb{R}}} \cong H^{(f,k+\ell\alpha)}(C_{\mathbb{R}})$ is finite dimensional as well. □

REMARK 3.8. The universal coefficient theorem further implies that

$$\mathrm{Ext}^{*,*+*\alpha}_{\mathcal{A}_{\mathbb{R}}}(\mathbb{M}^{\mathbb{R}}_p, \mathbb{M}^{\mathbb{R}}_p) \cong \mathrm{Ext}^{*,*}_{\mathcal{A}_{\mathbb{F}_p}}(\mathbb{F}_p, \mathbb{F}_p) \otimes_{\mathbb{F}_p} \mathbb{M}^{\mathbb{R}}_p$$

up to a grading shift.

THEOREM 3.9. *The map $\pi^{\mathbb{R}}_{m+n\alpha}((\mathbb{S}_{\mathbb{R}})^\wedge_{(p,\eta)}) \to \pi^{C_2}_{m+n\sigma}((\mathbb{S}_{C_2})^\wedge_{(p,\eta)})$ is*

(i) *an injection if $m = \left\lfloor \frac{p}{p-1}n - \frac{4p-2}{p-1} \right\rfloor$, and*

(ii) *an isomorphism if $m \geq \left\lfloor \frac{p}{p-1}n - \frac{4p-2}{p-1} \right\rfloor + 1$.*

[2] We use the standard convention in which the k-axis is horizontal and the $\ell\alpha$-axis is vertical.

PROOF. This follows from Proposition 3.6 using the same argument as in [**6**, Theorem 4.1]. □

REMARK 3.10. In [**6**], Dugger-Isaksen establish a second isomorphism range on 2-complete stems. Namely, $\pi_{m+n\alpha}(\mathbb{S}_{\mathbb{R}})^{\wedge}_2 \to \pi_{m+n\sigma}(\mathbb{S}_{C_2})^{\wedge}_2$ is an isomorphism if $m+n \le -1$. A version of this range appears to hold on (p,η)-complete stems, with the difference that the map might be only surjective when $m+n=-1$. However, this second isomorphism range doesn't extend to integral stems and we do not pursue it further here.

We now turn our attention to p-complete spheres.

PROPOSITION 3.11. *The map* $\pi^{\mathbb{R}}_{m+n\alpha}((\mathbb{S}_{\mathbb{R}})^{\wedge}_p) \to \pi^{C_2}_{m+n\sigma}((\mathbb{S}_{C_2})^{\wedge}_p)$ *is*

(1) *an injection if* $m = \left\lfloor \frac{p}{p-1}n - \frac{4p-2}{p-1} \right\rfloor$, *and*

(2) *an isomorphism if* $m \ge \left\lfloor \frac{p}{p-1}n - \frac{4p-2}{p-1} \right\rfloor + 1$.

PROOF. Write $\mathbb{S}$ for either of $\mathbb{S}_{\mathbb{R}}$ or $\mathbb{S}_{C_2}$. If $p=2$, the statement of the proposition is Dugger-Isaksen's Theorem 3.1, so we can assume p is odd. In this case 2 is invertible in $\mathbb{S}/p^r$ and in $\mathbb{S}^{\wedge}_p$. By Lemma 2.2 we have that $(\mathbb{S}^{\wedge}_p)^+ \simeq \mathbb{S}^{\wedge}_{(p,\eta)}$ and $(\mathbb{S}^{\wedge}_p)^- \simeq \mathbb{S}^{\wedge}_p[\eta^{-1}]$ and so we have that

$$\mathbb{S}^{\wedge}_p \simeq \mathbb{S}^{\wedge}_{(p,\eta)} \vee \mathbb{S}^{\wedge}_p[\eta^{-1}].$$

Note that we have an isomorphism $(\mathbb{S}^{\wedge}_p)^- \cong \lim_r((\mathbb{S}/p^r)^-)$. It follows from Theorem 2.1 that the map $\pi^{\mathbb{R}}_{*+*\alpha}((\mathbb{S}_{\mathbb{R}})^{\wedge}_p[\eta^{-1}]) \to \pi^{C_2}_{*+*\sigma}((\mathbb{S}_{C_2})^{\wedge}_p[\eta^{-1}])$ is an isomorphism. The result thus follows from Theorem 3.9 and the direct sum decomposition of $\mathbb{S}^{\wedge}_p$ above. □

THEOREM 3.12. *The map* $\pi^{\mathbb{R}}_{m+n\alpha}(\mathbb{S}_{\mathbb{R}}) \to \pi^{C_2}_{m+n\sigma}(\mathbb{S}_{C_2})$ *is*

(i) *an injection if* $m = 2n-6$ *and* $n \ge 0$, *and*

(ii) *an isomorphism if* $m \ge 2n-5$ *and* $n \ge 0$.

PROOF. Consider the comparison of long exact sequences of homotopy groups induced by cofiber sequences

$$\mathbb{S} \to \left(\prod_p \mathbb{S}^{\wedge}_p\right) \vee \mathbb{S}_{\mathbb{Q}} \to \left(\prod_p \mathbb{S}^{\wedge}_p\right)_{\mathbb{Q}}$$

obtained from the arithmetic fracture squares for $\mathbb{S}_{\mathbb{R}}$ and $\mathbb{S}_{C_2}$. It suffices to show that the comparison map $\pi^{\mathbb{R}}_{m+n\alpha}(-) \to \pi^{C_2}_{m+n\sigma}(-)$ at the middle and the righthand terms are injections if $m = 2n-6$ and $n \ge 0$ and an isomorphism if $m \ge 2n-5$ and $n \ge 0$.

The reader may check that when $n > 0$ the inequalities for the injectivity and isomorphism ranges of p-complete stable stems from Proposition 3.11 for odd p are dominated by Dugger-Isaksen's inequalities in Theorem 3.1, for the 2-complete stable stems. It follows that $\pi^{\mathbb{R}}_{m+n\alpha}(\prod_p (\mathbb{S}_{\mathbb{R}})^{\wedge}_p) \to \pi^{C_2}_{m+n\alpha}(\prod_p (\mathbb{S}_{C_2})^{\wedge}_p)$ is an injection if $m = 2n-6$, $n > 0$, and an isomorphism if $m \ge 2n-5$, $n > 0$. In fact, these ranges extend to $n = 0$ as well. Indeed, injectivity on $\pi^{\mathbb{R}}_{-6}(\prod_p (\mathbb{S}_{\mathbb{R}})^{\wedge}_p)$ is trivial since the group in question is 0 by connectivity of the motivic sphere spectrum. Similarly, $\pi^{\mathbb{R}}_{-5}(\prod_p (\mathbb{S}_{\mathbb{R}})^{\wedge}_p) = 0$, and $\pi^{C_2}_{-5}(\prod_p (\mathbb{S}_{C_2})^{\wedge}_p) = 0$ as well by [**2**, Proposition 7.0]. This is the only value of m, n for which the 2-complete isomorphism range does not dominate, so the isomorphism range extends.

Since we have that the map $\pi^{\mathbb{R}}_{m+n\alpha}(\prod_p(\mathbb{S}_{\mathbb{R}})^{\wedge}_p)_{\mathbb{Q}} \to \pi^{C_2}_{m+n\alpha}(\prod_p(\mathbb{S}_{C_2})^{\wedge}_p)_{\mathbb{Q}}$ is a filtered colimit of these maps, it too is an injection if $m = 2n-6$, $n \geq 0$ and an isomorphism if $m \geq 2n-5$, $n \geq 0$.

By Theorem 2.1 and Lemma 2.2, $\pi^{\mathbb{R}}_{*+*\alpha}(\mathbb{S}_{\mathbb{R}})^-_{\mathbb{Q}} \to \pi^{C_2}_{*+*\alpha}(\mathbb{S}_{C_2})^-_{\mathbb{Q}}$ is an isomorphism. By [**4**, Theorems 11 and 16.2.13], $(\mathbb{S}_{\mathbb{R}})^+_{\mathbb{Q}} \simeq \mathrm{H}\mathbb{Q}$, where $\mathrm{H}\mathbb{Q}$ is the rationalized motivic cohomology spectrum. Thus, we have that $\pi^{\mathbb{R}}_{i+j\alpha}(\mathbb{S}_{\mathbb{R}})^+_{\mathbb{Q}} = 0$ whenever $j > 0$. When $j = 0$, we have that $\pi^{\mathbb{R}}_i(\mathbb{S}_{\mathbb{R}})^+_{\mathbb{Q}} = 0$ if $i \neq 0$ and $\pi^{\mathbb{R}}_0(\mathbb{S}_{\mathbb{R}})^+_{\mathbb{Q}} = \mathbb{Q}$. We have

$$\pi^{C_2}_{i+j\alpha}(\mathbb{S}_{C_2})^+_{\mathbb{Q}} = \begin{cases} \mathbb{Q} & j \text{ is even and } i+j=0 \\ 0 & \text{else.} \end{cases}$$

The region $j \geq 0$, $i \geq 2j-5$ is entirely contained in this vanishing region. The map $\pi^{\mathbb{R}}_0(\mathbb{S}_{\mathbb{R}})^+_{\mathbb{Q}} \to \pi^{C_2}_0(\mathbb{S}_{C_2})^+_{\mathbb{Q}}$ is an ismorphism. We conclude that $\pi^{\mathbb{R}}_{m+n\alpha}(\mathbb{S}_{\mathbb{R}})^+_{\mathbb{Q}} \to \pi^{C_2}_{m+n\sigma}(\mathbb{S}_{C_2})^+_{\mathbb{Q}}$ is an injection if $m = 2n-6$ and $n \geq 0$, and an isomorphism if $m \geq 2n-5$ and $n \geq 0$. □

4. Proof of Theorem 1.1

We finish by explaining how the comparison of stable stems in the previous section implies the embedding theorem.

PROPOSITION 4.1. *If*

(i) $\mathrm{Re}^{C_2}_B : [S^n, \mathbb{S}_{\mathbb{R}}]_{\mathbb{R}} \xrightarrow{\cong} [S^n, \mathbb{S}_{C_2}]_{C_2}$, *and*

(ii) $\mathrm{Re}^{C_2}_B : [\mathrm{Spec}(\mathbb{C})_+ \wedge S^n, \mathbb{S}_{\mathbb{R}}]_{\mathbb{R}} \xrightarrow{\cong} [C_{2+} \wedge S^n, \mathbb{S}_{C_2}]_{C_2}$

are isomorphisms for all $n \in \mathbb{Z}$, then Theorem 1.1 *is true for any real closed field k.*

PROOF. Let k be a real closed field and $L = k[i]$. To prove Theorem 1.1, it suffices to prove that

(a) $c^*_{L/k} : [S^n, \mathbb{S}_{C_2}]_{C_2} \xrightarrow{\cong} [S^n, \mathbb{S}_k]_k$, and

(b) $c^*_{L/k} : [C_{2+} \wedge S^n, \mathbb{S}_{C_2}]_{C_2} \xrightarrow{\cong} [\mathrm{Spec}(L)_+ \wedge S^n, \mathbb{S}_k]_k$

are isomorphisms for all $n \in \mathbb{Z}$, by the same argument as in the beginning of the proof of [**7**, Theorem 2.21].

To prove that the maps in (a), (b) are isomorphisms, we can assume that $k = \mathbb{R}$ and $L = \mathbb{C}$, by the same argument as in [**7**, Proposition 2.20]. We now consider the C_2-equivariant Betti realization functor $\mathrm{Re}^{C_2}_B : \mathrm{SH}_{\mathbb{R}} \to \mathrm{SH}^{C_2}$. Since $\mathrm{Re}^{C_2}_B \circ c^*_{\mathbb{C}/\mathbb{R}} = \mathrm{id}$, it follows that (a) and (b) are isomorphisms. □

COROLLARY 4.2 (Theorem 1.1). *Let k be a real closed field and $L = k[i]$ be its algebraic closure. Then the functor*

$$c^*_{L/k} : \mathrm{SH}^{C_2} \to \mathrm{SH}_k$$

is a full and faithful embedding.

PROOF. If $i < 0$ then $\pi^{\mathbb{R}}_i(\mathbb{S}_{\mathbb{R}}) = \pi^{C_2}_i(\mathbb{S}_{C_2}) = 0$ and so the map in 4.1(i) is an isomorphism for $i < 0$. It is an isomorphism for $i \geq 0$ by setting $n = 0$ in Theorem 3.12. The map in 4.1(ii) is identical to the map $[S^n, \mathbb{S}_{\mathbb{C}}]_{\mathbb{C}} \to [S^n, \mathbb{S}]$ induced by complex Betti realization. This is an isomorphism by Levine's theorem [**9**]. □

Acknowledgments. The authors thank the anonymous referee for a careful reading and several improvements to the exposition.

References

[1] A. Ananyevskiy, M. Levine, and I. Panin, *Witt sheaves and the η-inverted sphere spectrum*, J. Topol. **10** (2017), no. 2, 370–385, DOI 10.1112/topo.12015. MR3653315

[2] S. Araki and K. Iriye, *Equivariant stable homotopy groups of spheres with involutions. I*, Osaka J. Math. **19** (1982), no. 1, 1–55. MR656233

[3] T. Bachmann. *Motivic and real étale stable homotopy theory.* ArXiv e-prints, 1608.08855.

[4] D.-C. Cisinski and F. Déglise. *Triangulated categories of mixed motives.* ArXiv e-prints, 0912.2110v3.

[5] D. Dugger, *An Atiyah-Hirzebruch spectral sequence for KR-theory*, K-Theory **35** (2005), no. 3-4, 213–256 (2006), DOI 10.1007/s10977-005-1552-9. MR2240234

[6] D. Dugger and D. C. Isaksen, *$\mathbb{Z}/2$-equivariant and $\mathbb{R}$-motivic stable stems*, Proc. Amer. Math. Soc. **145** (2017), no. 8, 3617–3627, DOI 10.1090/proc/13505. MR3652813

[7] J. Heller and K. Ormsby, *Galois equivariance and stable motivic homotopy theory*, Trans. Amer. Math. Soc. **368** (2016), no. 11, 8047–8077, DOI 10.1090/tran6647. MR3546793

[8] P. Hu, I. Kriz, and K. Ormsby, *Convergence of the motivic Adams spectral sequence*, J. K-Theory **7** (2011), no. 3, 573–596, DOI 10.1017/is011003012jkt150. MR2811716

[9] M. Levine, *A comparison of motivic and classical stable homotopy theories*, J. Topol. **7** (2014), no. 2, 327–362, DOI 10.1112/jtopol/jtt031. MR3217623

[10] L. G. Lewis Jr., J. P. May, M. Steinberger, and J. E. McClure, *Equivariant stable homotopy theory*, Lecture Notes in Mathematics, vol. 1213, Springer-Verlag, Berlin, 1986. With contributions by J. E. McClure. MR866482

[11] A. Mathew, N. Naumann, and J. Noel, *Nilpotence and descent in equivariant stable homotopy theory*, Adv. Math. **305** (2017), 994–1084, DOI 10.1016/j.aim.2016.09.027. MR3570153

[12] A. Suslin and V. Voevodsky, *Bloch-Kato conjecture and motivic cohomology with finite coefficients*, The arithmetic and geometry of algebraic cycles (Banff, AB, 1998), NATO Sci. Ser. C Math. Phys. Sci., vol. 548, Kluwer Acad. Publ., Dordrecht, 2000, pp. 117–189. MR1744945

[13] V. Voevodsky, *$\mathbf{A}^1$-homotopy theory*, Proceedings of the International Congress of Mathematicians, Vol. I (Berlin, 1998), Doc. Math. **Extra Vol. I** (1998), 579–604. MR1648048

[14] V. Voevodsky, *Reduced power operations in motivic cohomology*, Publ. Math. Inst. Hautes Études Sci. **98** (2003), 1–57, DOI 10.1007/s10240-003-0009-z. MR2031198

[15] V. Voevodsky, *On motivic cohomology with $\mathbf{Z}/l$-coefficients*, Ann. of Math. (2) **174** (2011), no. 1, 401–438, DOI 10.4007/annals.2011.174.1.11. MR2811603

University of Illinois Urbana-Champaign
Email address: jbheller@illinois.edu

Reed College
Email address: ormsbyk@reed.edu

Contemporary Mathematics
Volume **707**, 2018
http://dx.doi.org/10.1090/conm/707/14251

An Étale realization which does NOT exist

Jesse Leo Kass and Kirsten Wickelgren

Abstract. For a global field, local field, or finite field k with infinite Galois group, we show that there can not exist a functor from the Morel–Voevodsky $\mathbf{A}^1$-homotopy category of schemes over k to a genuine Galois equivariant homotopy category satisfying a list of hypotheses one might expect from a genuine equivariant category and an étale realization functor. For example, these hypotheses are satisfied by genuine $\mathbb{Z}/2$-spaces and the $\mathbf{R}$-realization functor constructed by Morel–Voevodsky. This result does not contradict the existence of étale realization functors to (pro-)spaces, (pro-)spectra or complexes of modules with actions of the absolute Galois group when the endomorphisms of the unit is not enriched in a certain sense. It does restrict enrichments to representation rings of Galois groups.

1. Introduction

Grothendieck envisioned that cohomological invariants of algebraic varieties are controlled by certain motives, with transcendental invariants, such as the Galois action on étale cohomology, being recovered by realization functors. Work of Morel–Voevodsky on $\mathbf{A}^1$-homotopy theory provides candidate categories of motives, and realization functors have been constructed from the Morel–Voevodsky $\mathbf{A}^1$-homotopy category and its stabilization [**MV99**] [**Isa04**] [**Qui07**] [**Ayo14a**], [**CD16**, Section 7.2].

For example, Morel–Voevodsky construct an $\mathbf{R}$-realization functor [**MV99**, 3.3]. On the level of schemes, this functor takes an $\mathbf{R}$-scheme X to the topological space (or simplicial set) of its complex points $X(\mathbf{C})$ together with the action of the Galois group $\mathrm{Gal}(\mathbf{C}/\mathbf{R})$ on $X(\mathbf{C})$. The association $X \mapsto X(\mathbf{C})$ gives rise to functors from the $\mathbf{A}^1$-homotopy categories of spaces and of $\mathbf{P}^1$-spectra. The target category of this functor may be taken to be the genuine $\mathrm{Gal}(\mathbf{C}/\mathbf{R})$-homotopy category in the former case and the genuine $\mathrm{Gal}(\mathbf{C}/\mathbf{R})$-equivariant stable homotopy category in

2010 *Mathematics Subject Classification.* Primary 14F42; Secondary 55P91,14F05.

Jesse Leo Kass was partially sponsored by the Simons Foundation under Grant Number 429929, and the National Security Agency under Grant Number H98230-15-1-0264. The United States Government is authorized to reproduce and distribute reprints notwithstanding any copyright notation herein. This manuscript is submitted for publication with the understanding that the United States Government is authorized to reproduce and distribute reprints.

Kirsten Wickelgren was partially supported by National Science Foundation Awards DMS-1406380 and DMS-1552730 and an AIM 5-year fellowship. She wishes to thank the Institut Mittag-Leffler for a very pleasant stay at the special program "Algebro-geometric and homotopical methods" while working on this project.

the latter. The adjective genuine (or *fine* in the terminology of [**MV99**, 3.3]) refers to the homotopy theory in which a weak equivalence of $\mathrm{Gal}(\mathbf{C}/\mathbf{R})$-spaces $X \to Y$ is not only an equivariant map which is a weak equivalence in the non-equivariant sense, but also satisfies the requirement that the map on fixed points $X^H \to Y^H$ is an equivalence for all subgroups of the group acting on the spaces, in this case $\mathrm{Gal}(\mathbf{C}/\mathbf{R})$ [**LMSM86**]. In the stable case, more notation would be required to define genuine $\mathrm{Gal}(\mathbf{C}/\mathbf{R})$-equivariant spectra, but one consequence of the more stringent notion of weak equivalence is that the group of endomorphisms of the sphere spectrum encodes interesting information about $\mathrm{Gal}(\mathbf{C}/\mathbf{R})$-sets, in contrast to $\mathbf{Z}$, which is isomorphic to the endomorphisms of the sphere in ordinary stable homotopy. It follows that the Euler characteristic in genuine equivariant homotopy theory can be enriched from an element of $\mathbf{Z}$ to an element of the Burnside ring $A(\mathrm{Gal}(\mathbf{C}/\mathbf{R}))$ of formal differences of finite $\mathrm{Gal}(\mathbf{C}/\mathbf{R})$-sets.

One might hope to generalize this construction to fields k with infinite Galois groups. Namely, let $G = \mathrm{Gal}_k$ denote the absolute Galois group of k. One could hope to construct a category of genuine G-spectra or G-pro-spectra receiving an étale realization functor from $\mathbf{P}^1$-spectra over k. Indeed, for profinite groups, homotopy theories of genuine G-(pro)-spectra are constructed in [**Fau08**] and [**Bar17**], and [**Qui13**] contains a construction of a homotopy theory of profinite G-equivariant spectra. Étale realization functors have also been constructed over k [**Isa04**] [**Qui07**] [**Ayo14a**], but not so that the target is a genuine G-equivariant homotopy theory.

We show here that there is a reason for this lacuna. Namely, for k a global field, local field, or finite field with infinite Galois group, there does not exist a symmetric monoidal étale realization functor from $\mathbf{P}^1$-spectra over k to a genuine G-equivariant homotopy theory under a set of hypothesis on what the terms *étale realization* and *genuine* are expected to imply. The hypotheses are best explained by the proof, and the proof is based on a result of Hoyois [**Hoy14**, Theorem 1.5]. Namely, Hoyois shows that the $\mathbf{A}^1$-Euler characteristic of a smooth proper variety with trivial cotangent sheaf is 0, which implies, in particular, that the $\mathbf{A}^1$-Euler characteristic of an elliptic curve E is 0. The outline of the proof is as follows. Suppose there is a symmetric monoidal étale realization functor $L\,\mathrm{Et}$. Then the Euler characteristic of $L\,\mathrm{Et}(E)$ is well-defined and equal to 0. On the other hand, in genuine G-spectra, one could expect that the Euler characteristic is related to the alternating sum of the cohomology groups (with coefficients in some ring or field) considered with their G-actions. For example, this is the case for G a finite group (Proposition 2.3). The theory of weights for Galois representations shows that the alternating sum of the étale cohomology groups for E is non-zero in a representation ring. Since one could expect the cohomology groups of the étale realization to be the étale cohomology groups, at least for finite coefficients (as in [**Fri82**, Proposition 5.9]), this is a contradiction.

A precise formulation is as follows. First, we introduce a definition. Let k be a field and let R be a ring. Let $\mathrm{Funct}(\mathrm{Gal}_k, R)$ denote the ring of functions $\mathrm{Gal}_k \to R$ with point-wise addition and multiplication.

DEFINITION 1.1. By a *representation ring* of Gal_k with coefficients in R, we mean a ring $\mathrm{Rep}(\mathrm{Gal}_k, R)$ such that:

(1) The isomorphism class of a finitely generated R-module A with a continuous Gal_k-action given by $A = H^i_{\text{ét}}(X_{\overline{k}}, R)$, where X is a smooth, proper variety over k, determines an element $[A]$ of $\mathrm{Rep}(\mathrm{Gal}_k, R)$.

(2) The trace map which takes A to the function $\mathrm{Gal}_k \to R$ defined by taking g in G to the trace of g acting on A,

$$g \mapsto \mathrm{Tr}\, g|A$$

extends to a homomorphism $\mathrm{Rep}(\mathrm{Gal}_k, R) \to \mathrm{Funct}(\mathrm{Gal}_k, R)$.

THEOREM 1.2. *Let k be a global field, local field with infinite Galois group, or finite field. Let p be a prime different from the characteristic of k. It is impossible to simultaneously construct all of the following*

(1) *A symmetric monoidal category $\mathcal{H}_{\mathrm{Gal}_k}(\mathbf{Spt})$, enriched over abelian groups. Let $A(\mathrm{Gal}_k) = \mathrm{End}_{\mathcal{H}_{\mathrm{Gal}_k}(\mathbf{Spt})}(\mathbf{S})$ denote the endomorphisms of the symmetric monoidal unit $\mathbf{S}$.*

Let $\mathbf{S}$ denote the symmetric monoidal unit, and define $A(\mathrm{Gal}_k)$ to be the ring

$$A(\mathrm{Gal}_k) = \mathrm{End}_{\mathcal{H}_{\mathrm{Gal}_k}(\mathbf{Spt})}(\mathbf{S}).$$

(2) *For all n, a representation ring $\mathrm{Rep}(\mathrm{Gal}_k, \mathbb{Z}/p^n)$ with coefficients in $\mathbb{Z}/p^n$.*

(3) *A symmetric monoidal additive functor $\mathrm{L\,Et} : \mathcal{H}_{\mathbf{A}^1}(\mathbf{Spt}^{\mathbf{P}^1}(k)) \to \mathcal{H}_{\mathrm{Gal}_k}(\mathbf{Spt})$, and ring homomorphisms $\theta_n : A(G) \to \mathrm{Rep}(\mathrm{Gal}_k, \mathbb{Z}/p^n)$ such that*

$$\theta_n \chi(\mathrm{L\,Et}\, \Sigma^{\infty}_{\mathbf{P}^1} X_+) = \sum_{i=-\infty}^{\infty} (-1)^i [H^i_{ét}(X_{\overline{k}}, \mathbb{Z}/p^n)]$$

for smooth proper schemes X over k.

We are thinking of $\mathcal{H}_{\mathrm{Gal}_k}(\mathbf{Spt})$ as a genuine Gal_k-equivariant stable homotopy category, $A(\mathrm{Gal}_k)$ as a Burnside ring of Gal_k, $\mathrm{Rep}(\mathrm{Gal}_k, \mathbb{Z}/p^n)$ as a representation ring of $\mathbb{Z}/(p^n)$-modules with Gal_k-action, and $\mathrm{L\,Et}$ as an étale realization functor. Since there are interesting contexts in which one may require a continuous Galois representation to satisfy additional conditions (e.g., be de Rham, crystalline, semi-stable, potentially semi-stable, etc.), we do not require that $\mathrm{Rep}(\mathrm{Gal}_k, \mathbb{Z}/p^n)$ be the Grothendieck ring of continuous Galois representations, although this is certainly a possibility for (2). Theorem 1.2 is proven as Theorem 3.1 below and there is another variant given in Theorem 3.3, where the representation rings involved only have coefficients in fields.

In Section 2, we show that the constructions of Theorem 1.2 do exist when $k = \mathbf{R}$, the realization functor (3) being that of Morel–Voevodsky [**MV99**, 3.3], and (1) and (2) being the usual genuine $\mathbb{Z}/2$-equivariant stable homotopy category and $\mathbb{Z}/2$-representation rings, respectively. Section 3 contains the main result. As described above, the basic idea of the proof is that the $\mathbf{A}^1$-Euler characteristic of an elliptic curve and the alternating sum of its étale cohomology groups are incompatible as Galois representations.

2. Enriched Euler characteristics and genuine $\mathrm{Gal}_{\mathbb{R}}$-realization

Let G be a finite group. Let $A(G)$ denote the Burnside ring, defined as the group completion (or Grothendieck group) of the semi-ring of isomorphism classes of finite G-sets with addition and multiplication given by disjoint union and product respectively.

The Burnside ring is the endomorphisms of the monoidal unit in the Burnside category $\mathrm{Burn}(G)$, whose definition we now recall. The objects of $\mathrm{Burn}(G)$ are finite G-sets. Given finite G-sets T and S, consider the set of equivalence classes of diagrams of G-sets

$$S \leftarrow U \to T$$

under the equivalence relation $[S \leftarrow U \to T] \sim [S \leftarrow U' \to T]$ when there is a G-equivariant bijection $\alpha : U \to U'$ such that the diagram

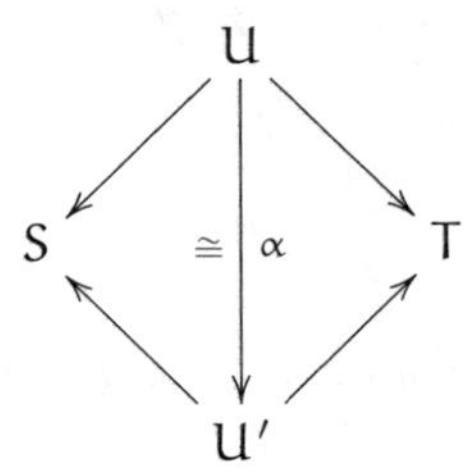

commutes. The morphisms $\mathrm{Burn}(G)(S,T)$ from S to T are the elements of the group completion of this set of equivalence classes under the operation

$$[S \leftarrow U \to T] + [S \leftarrow U' \to T] = [S \leftarrow U \textstyle\coprod U' \to T].$$

Composition of morphisms $S \leftarrow U \to T$ and $T \leftarrow U' \to Q$ is given by pullback

(2.1)

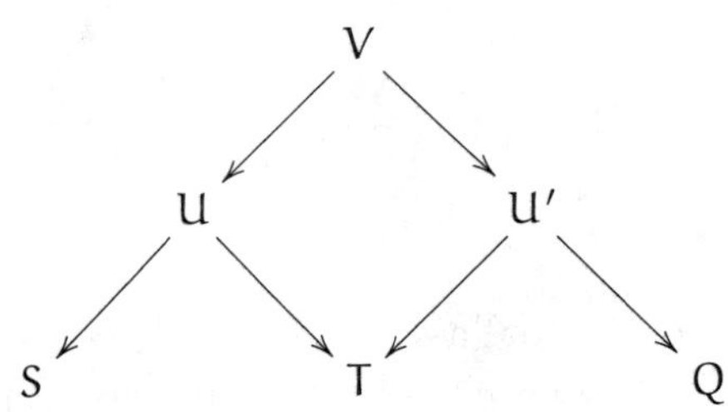

and the composition

$$\mathrm{Burn}(G)(S,T) \times \mathrm{Burn}(G)(T,Q) \to \mathrm{Burn}(G)(S,Q)$$

is defined by (2.1) and bilinearity. The symmetric monoidal structure takes G-sets S and T to their product $S \times T$, so $A(G) = \mathrm{Burn}(G)(*,*)$, where $*$ denotes the one-point G-set.

An object X of a symmetric monoidal category with unit $\mathbf{S}$ and monoidal product $\wedge$ is said to be fully dualizable if there is a dual object DX together with coevaluation and evaluation morphisms

$$\eta : \mathbf{S} \to X \wedge DX, \quad \epsilon : DX \wedge X \to \mathbf{S}$$

such that the compositions

$$X \overset{\eta \wedge 1_X}{\to} X \wedge DX \wedge X \overset{1_X \wedge \epsilon}{\to} X,$$

$$DX \overset{1_{DX} \wedge \eta}{\to} DX \wedge X \wedge DX \overset{\epsilon \wedge 1_{DX}}{\to} DX$$

are the identities. For example, every element of Burn(G) is fully dualizable with $DX = X$.

Given an endomorphism f of a fully dualizable object X, the trace $\mathrm{Tr}(f)$ is the endomorphism of $\mathbf{S}$ determined by the composition

$$\mathbf{S} \xrightarrow{\eta} X \wedge DX \xrightarrow{f \wedge 1_{DX}} X \wedge DX \xrightarrow{\tau} DX \wedge X \xrightarrow{\epsilon} \mathbf{S},$$

where τ is the map which switches the order of the two factors. The Euler characteristic $\chi(X)$ is the trace of the identity $\chi(X) = \mathrm{Tr}(1_X)$. The following is a straightforward consequence of the definitions and is not original, but it is included for convenience and clarity.

LEMMA 2.1. *Let T and U be finite G-sets, and let $T \xleftarrow{i} U \xrightarrow{j} T$ be a diagram of finite G-sets. For elements t and t' in T, let $U_{t,t'}$ be the set $U_{t,t'} = i^{-1}(t) \cap j^{-1}(t')$. Then there are the following equalities in the Burnside ring $A(G)$.*

(1) $\mathrm{Tr}[T \leftarrow U \rightarrow T] = \coprod_{t \in T} U_{t,t}$.
(2) $\chi(T) = T$.

PROOF. We first show (1). Let $\Delta : T \rightarrow T \times X$ denote the diagonal. Then η and ϵ are the morphisms $* \leftarrow T \xrightarrow{\Delta} T \times T$ and $T \times T \xleftarrow{\Delta} T \rightarrow *$ respectively. The composition of $[T \leftarrow U \rightarrow T] \times 1_T$ and τ is $T \times T \xleftarrow{i \times 1_T} U \times T \xrightarrow{1_T \times j} T \times T$. The pullback diagram

$$\begin{array}{ccc} U & \xrightarrow{1_U \times i} & U \times T \\ \downarrow{\scriptstyle i} & & \downarrow{\scriptstyle i \times 1_T} \\ T & \xrightarrow[\Delta]{} & T \times T \end{array}$$

shows that the composition $\eta \circ ([T \leftarrow U \rightarrow T] \times 1_T) \circ \tau$ is $* \leftarrow U \xrightarrow{i \times j} T \times T$. The composition of this last map with η shows $\mathrm{Tr}[T \leftarrow U \rightarrow T] = \coprod_{t \in T} U_{t,t}$. (2) follows from (1). □

Let $\mathcal{H}_G(\mathbf{Spt})$ denote the homotopy category of genuine G-spectra [**May96**, XII 5], $\mathbf{S}$ denote the sphere spectrum, and $\mathcal{H}_G(\mathbf{Spt})^d$ denote the full subcategory of fully dualizable objects.

It is a result of Segal [**Seg**] and tom Dieck [**tD79**] that there is a ring isomorphism

$$A(G) \cong \mathrm{End}_{\mathcal{H}_G(\mathbf{Spt})}(\mathbf{S}), \tag{2.2}$$

so for X in $\mathcal{H}_G(\mathbf{Spt})^d$ we consider $\chi(X)$ to be an element of the Burnside ring $A(G)$. The ring homomorphism (2.2) is induced from the fully faithful, symmetric, monoidal functor $\Sigma^\infty_+ : \mathrm{Burn}(G) \rightarrow \mathcal{H}_G(\mathbf{Spt})$ taking a finite G-set T to the suspension spectrum of T_+ [**May96**, Corollary 3.2 XIX].

Let R be a ring. Let $\mathrm{Rep}(G, R)$ denote the ring with generators given by isomorphism classes of finitely generated R-modules with G-action subject to the relation that $B = A \oplus C$ when there is a short exact sequence

$$A \rightarrow B \rightarrow C.$$

Addition and multiplication in $\mathrm{Rep}(G, R)$ are induced from $\oplus$ and $\otimes$ respectively. For X in $\mathcal{H}_G(\mathbf{Spt})$, the cohomology groups $H^i(X, R)$ determine elements of $\mathrm{Rep}(G, R)$.

DEFINITION 2.2. Let $\theta : A(G) \to \mathrm{Rep}(G, R)$ be the map determined by taking a finite G-set T to the permutation representation on $\oplus_T R$.

PROPOSITION 2.3. (1) *Let* R *be any ring. For all finite* G*-CW complexes* X *we have that*

$$\theta_X(X) = \sum_{i=-\infty}^{\infty} (-1)^i H^i(X, R).$$

(2) *Let* R *be a field of characteristic not dividing the order of* G. *Then for all* X *in* $\mathcal{H}_G(\mathbf{Spt})^d$

$$\theta_X(X) = \sum_{i=-\infty}^{\infty} (-1)^i H^i(X, R),$$

and in particular the right hand side is a well-defined element of $\mathrm{Rep}(G, R)$ *in the sense that only finitely many of the summands are non-zero and the non-zero summands are finitely generated* R*-modules.*

The proof of Proposition 2.3 (1) is a straightforward induction on the cells of X. To prove (2), we will have use of the following lemmas.

LEMMA 2.4. *Let* G *be a finite cyclic group. Let* R *be a field of characteristic not dividing the order of* G. *Let* T *be a finite* G*-set, and* $V \subseteq \theta T$ *be a* G*-fixed subspace such that the corresponding representation* $\rho : G \to \mathrm{GL}\, V$ *satisfies the property that* $\mathrm{Tr}(\rho(g))$ *is the image of an integer under the canonical map* $\mathbb{Z} \to R$ *for all* g *in* G. *Then there are* G*-sets* E *and* I *such that* $V \oplus \theta E \cong \theta I$.

PROOF. If G has order 1, then the lemma is true. Let G have order $n > 1$ and assume inductively that the lemma holds for cyclic groups of smaller orders. Because matrices which are similar over the algebraic closure of R are similar over R, we may assume R is algebraically closed. Thus θT and V decompose into simultaneous 1-dimensional eigenspaces whose eigenvalues are nth roots of unity. Let d denote the largest order of an eigenvalue of V, and note that d divides n. If $d = 1$, then $V \cong \theta I$ where I is a set with trivial G-action of cardinality equal to the dimension of V, so the lemma is true.

We now induct on d. Let g denote a generator of G. The condition that $\mathrm{Tr}(\rho(g))$ is an integer implies that the eigenspaces of g in V associated to the primitive dth roots of 1 are all the same dimension. Call this dimension a. Let $d = \prod_{j=1}^{l} p_j^{e_j}$ be the prime factorization of d. Let E_j denote the G-set consisting of a single orbit of cardinality d/p_j, and let $E' = \coprod_{i=1}^{a} \coprod_{j=1}^{l} E_j$. Then for each dth root of unity, we may choose an a dimensional eigenspace of $V \oplus \theta E'$. Let V' denote the direct sum of these chosen eigenspaces and let $V'' \subset V \oplus \theta E'$ denote a complementary G-representation to V'. By construction $V' \cong \theta(\coprod_{j=1}^{a} I')$, where I' is the G-set with a single orbit of cardinality d. In particular, this representation satisfies the property that the trace of every element of G is an integer. It follows that the same property holds for V''. By induction on d, there are G-sets E'' and I'' such that $V'' \oplus \theta E'' \cong \theta I''$. Thus $V \oplus \theta E' \oplus \theta E'' \cong \theta(\coprod_{j=1}^{a} I') \oplus \theta I''$. We may therefore let $E = E' \coprod E''$ and $I = (\coprod_{j=1}^{a} I') \coprod I''$, and the lemma is true. □

LEMMA 2.5. *Let* G *be a finite cyclic group. Assume* R *has a prime ideal of residue characteristic not dividing the order of* G. *If* T *and* T' *are two finite* G*-sets such that* $\theta T \cong \theta T'$, *then* $T \cong T'$.

REMARK 2.6. Lemma 2.5 becomes false when G is $\mathbb{Z}/2 \times \mathbb{Z}/2$ [**Rob85**, §7 Example 2].

PROOF. By assumption, there is a ring map $R \to K$ where K is an algebraically closed field of characteristic not dividing the order of G. We may replace R by K, allowing us to decompose $\theta T \cong \theta T'$ into simultaneous 1-dimensional eigenspaces whose eigenvalues are nth roots of unity. Let d denote the largest order of such an eigenvalue. If $d = 1$, then both T and T' have trivial G-actions and $|T| = |T'| = \dim \theta T'$, so the lemma is true. Assume by induction that the lemma holds for all smaller values of d. Let g be a generator of G and ζ be a primitive dth root of unity. The number of d-cycles in θT is the dimension of the ζ-eigenspace of T. Since the same holds for T', it follows that T and T' contain the same number of d-cycles. Furthermore, removing the d-cycles from both T and T' results in finite G-sets with isomorphic associated permutation representations for which the lemma holds inductively. Thus $T \cong T'$. □

LEMMA 2.7. *Let* G *be a finite group and let* $f : T \to T$ *be an idempotent in the Burnside category of* G. *Let* R *be a field of characteristic not dividing the order of* G. *Then* $\theta \operatorname{Tr}(f) = \operatorname{Image} H^0(f)$ *in* $\operatorname{Rep}(G, R)$.

We establish a few preliminaries before proving Lemma 2.7.

REMARK 2.8. We may directly calculate the trace functions $\chi_{\theta \operatorname{Tr}(f)}$ and $\chi_{\operatorname{Image} H^0(f)}$ of $\theta \operatorname{Tr}(f)$ and $\operatorname{Image} H^0(f)$. Since f is idempotent,

$$\chi_{\operatorname{Image} H^0(f)}(g) = \operatorname{Tr}(H^0(f) | \operatorname{Image} H^0(f)) = \operatorname{Tr}(H^0(g)H^0(f)|H^0(T)).$$

Let $f = [T \xleftarrow{i} U \xrightarrow{j} T] - [T \xleftarrow{i} V \xrightarrow{j} T]$. Let δ_t be the function $\delta_t : T \to R$ such that $\delta_t(t) = 1$ and $\delta_t(t') = 0$ for $t' \neq t$. In the below, i_* denotes the push forward on H^0 associated to i, and g^*, j^* are the corresponding pullbacks. We compute

$$\begin{aligned}\chi_{\operatorname{Image} H^0(f)}(g) = \operatorname{Tr}(H^0(g)H^0(f)|H^0(T)) &= \sum_{t \in T} (g^* i_* j^* \delta_t)(t) = \sum_{t \in T} (i_* g^* j^* \delta_t)(t) \\ &= \sum_{t \in T} \Big(\sum_{u \in U : i(u) = t} (g^* j^* \delta_t)(u) - \sum_{v \in V : i(v) = t} (g^* j^* \delta_t)(v) \Big) \\ &= \sum_{t \in T} \Big(\sum_{u \in U : i(u) = t} \delta_t(gju) - \sum_{v \in V : i(v) = t} \delta_t(gjv) \Big) \\ &= |\{u \in U : i(u) = gj(u)\}| - |\{v \in V : i(v) = gj(v)\}|.\end{aligned}$$

It follows from Lemma 2.1 that

$$\chi_{\theta \operatorname{Tr}(f)}(g) = |\{u \in U : gu = u, i(u) = j(u)\}| - |\{v \in V : gv = v, i(v) = j(v)\}|.$$

It is not clear from these calculations that $\chi_{\operatorname{Image} H^0(f)} = \chi_{\theta \operatorname{Tr}(f)}$. We will show in Lemma 2.7 that they are.

It is clear that $\chi_{\operatorname{Image} H^0(f)}$ takes values in the integers, and it follows from

$$\chi_{\operatorname{Image} H^0(f)} + \chi_{\operatorname{Ker} H^0(f)} = \chi_{H^0(T)},$$

that $\chi_{\operatorname{Ker} H^0(f)}$ also takes values in the integers.

To prove Lemma 2.7, we will make use of the following variation of the Burnside category.

DEFINITION 2.9. Define the category $\mathrm{Burn}(G,R)$ to have objects G-sets, and for G-sets S and T, the morphisms $\mathrm{Burn}(G,R)(S,T)$ are R-linear combinations of equivalence classes of diagrams of G-sets $S \leftarrow U \rightarrow T$. Composition is defined to be R-bilinear and induced from (2.1) as in the definition of the Burnside category.

$\mathrm{Burn}(G,R)(S,T)$ is a symmetric, monoidal category with the monoidal structure defined by the product of G-sets, and $\mathrm{Tr}[T \leftarrow U \rightarrow T] = \coprod_{t\in T} U_{t,t}$ by the proof of Lemma 2.1.

LEMMA 2.10. *Let* A *and* B *be morphisms in* $\mathrm{Burn}(G,R)(T,T)$. *Then* $\mathrm{Tr}(A\circ B) = \mathrm{Tr}(B \circ A)$.

PROOF. Since composition of morphisms is R-bilinear, and Tr is R-linear, it suffices to prove the claim where A and B are determined by diagrams of G-sets $T \leftarrow A \rightarrow T$ and $T \leftarrow B \rightarrow T$, respectively. Then:

$$\begin{aligned}
\mathrm{Tr}(A \circ B) &\cong \coprod_{t\in T}(A \circ B)_{tt} \\
&\cong \coprod_{t\in T}\coprod_{t'\in T} A_{t,t'} \times B_{t',t} \\
&\cong \coprod_{t,t'\in T} A_{t,t'} \times B_{t',t} \\
&\cong \coprod_{t,t'\in T} B_{t',t} \times A_{t,t'} \\
&\cong \coprod_{t'\in T}\coprod_{t\in T} B_{t',t} \times A_{t,t'} \\
&\cong \coprod_{t'\in T} (A \circ B)_{t',t'} \cong \mathrm{Tr}(B \circ A).
\end{aligned}$$

□

PROOF. (of Lemma 2.7) By Maschke's theorem [**EGH$^+$11**, Theorem 3.1, Theorem 3.5], it suffices to show that the trace functions $\chi_{\theta\,\mathrm{Tr}(f)}$ and $\chi_{\mathrm{Image}\,H^0(f)}$ of $\theta\,\mathrm{Tr}(f)$ and $\mathrm{Image}\,H^0(f)$ are the same. To do this, we may show for each g in G that $\chi_{\theta\,\mathrm{Tr}(f)}(g) = \chi_{\mathrm{Image}\,H^0(f)}(g)$, thus reducing to the case where G is a finite cyclic group.

By Remark 2.7, $\chi_{\mathrm{Image}\,H^0(f)}$ takes values in the integers, so we may apply Lemma 2.4 to $\mathrm{Image}\,H^0(f) \subseteq \theta T$. Thus there are G-sets E and I such that $\mathrm{Image}\,H^0(f) \oplus \theta E \cong \theta I$. Let $f' : T\coprod E \rightarrow T\coprod E$ be the idempotent in the Burnside category of G defined $f' = f\coprod 1_E$. Applying Remark 2.7 and Lemma 2.4 to $\mathrm{Ker}\,H^0(f') \subseteq \theta(T\coprod)$, there are G-sets E' and I' such that $\mathrm{Ker}\,H^0(f') \oplus \theta E' \cong \theta I'$. Let $f'' : T\coprod E \coprod E' \rightarrow T\coprod E\coprod E'$ be the idempotent in the Burnside category of G defined $f'' = f'\coprod 0_{E'}$.

By Lemma 2.1, $\mathrm{Tr}\,f'' \cong E\coprod \mathrm{Tr}\,f$, and $\mathrm{Image}\,H^0(f'') \cong \mathrm{Image}\,H^0(f)\oplus\theta E$. Therefore it suffices to prove the lemma with f replaced by f''.

Since f'' is an idempotent,

$$\theta(T\coprod E\coprod E' \rightarrow T) \cong \mathrm{Ker}\,H^0(f'') \oplus \mathrm{Image}\,H^0(f'').$$

By construction, $\operatorname{Image} H^0(f'') \cong \operatorname{Image} H^0(f') \cong \operatorname{Image} H^0(f) \oplus \theta E \cong I$, and $\operatorname{Ker} H^0(f'') \cong \operatorname{Ker} H^0(f') \oplus \theta E' \cong \theta I'$. Replacing T by $T \coprod E \coprod E'$ and f by f'', we may therefore assume that there exist G-sets I and I' such that $\operatorname{Image} H^0(f) \cong \theta I$ and $\operatorname{Ker} H^0(f) \cong \theta I'$.

Since f is an idempotent, there is an isomorphism $\theta T \cong \operatorname{Image} H^0(f) \oplus \operatorname{Ker} H^0(f)$, whence an isomorphism $\eta : \theta T \to \theta(I \coprod I')$. Define $T' = I \coprod I'$. For $t \in T$ and $t' \in T'$, let $\eta_{t',t}$ denote the corresponding entry of the matrix of η with respect to the bases T and T', and similarly define $(\eta^{-1})_{t,t'}$ to be the matrix entry of the inverse of η. Note that $\eta_{t',t}(t',t)$ is an element of $\theta(T' \times T)$. The subset $\coprod_{t\in T, t'\in T'} \eta_{t',t}(t',t)$ of $\theta(T' \times T)$ is invariant under the action of G because η is a G-isomorphism. It follows that we may view $\coprod_{t\in T, t'\in T'} \eta_{t',t}(t',t)$ as an R-linear combination of G-subsets of $T' \times T$. The analogous assertion holds for η'.

The R-linear combination of G-invariant subsets of $T' \times T$ associated to

$$\coprod_{t\in T, t'\in T'} \eta_{t',t}(t',t)$$

determines a morphism in $\operatorname{Burn}(G,R)(T,T')$, which we will denote by $\operatorname{Burn}(\eta)$. Similarly define $\operatorname{Burn}(\eta^{-1})$ in $\operatorname{Burn}(G,R)(T',T)$. By construction, $\operatorname{Burn}(\eta) \circ f \circ \operatorname{Burn}(\eta^{-1}) = I$ where the two maps to T' are both the canonical inclusion $I \to T'$.

We claim that Lemma 2.10 implies that

$$\operatorname{Tr}(\operatorname{Burn}(\eta) \circ f \circ \operatorname{Burn}(\eta^{-1})) = \operatorname{Tr}(f).$$

To see that we may apply Lemma 2.10 in this case, note that by Lemma 2.5, we may fix an isomorphism of G-sets $T \cong T'$, and consider $\operatorname{Burn}(\eta)$ and $\operatorname{Burn}(\eta^{-1})$ to both be elements of $\operatorname{Burn}(G,R)(T,T)$ allowing us to conclude $\operatorname{Tr}(\operatorname{Burn}(\eta) \circ f \circ \operatorname{Burn}(\eta^{-1})) = \operatorname{Tr}(\operatorname{Burn}(\eta^{-1}) \circ \operatorname{Burn}(\eta) \circ f)$. Thus $\operatorname{Tr}(f) = I$. Since $\theta I \cong \operatorname{Image} H^0(f)$, we have shown the lemma. □

The following corollary of Lemma 2.7 is not needed for the rest of the article, but is included as a curiosity. To put this in context, we remark that there are many interesting idempotents of the Burnside ring of a finite group G. The idempotents of $A(G) \otimes \mathbb{Q}$ were computed by [**Sol67**], [**Yos83**], and [**Glu81**]. The integral idempotents, i.e., the idempotents of $A(G)$, were computed by Dress [**Dre69**], see [**Bou00**, Theorem 3.3.7, Corollary 3.3.9].

COROLLARY 2.11. *Let G be a finite group and let R be a field of characteristic not dividing the order of G. Suppose f is an idempotent in the Burnside ring* $A(G)$. *Then* θf *is either* 0 *or the trivial representation in* $\operatorname{Rep}(G,R)$.

PROOF. Since $\operatorname{Image} H^0(f)$ is a submodule of $H^0(*)$, we have that $\operatorname{Image} H^0(f)$ is either 0 or R with the trivial action. The corollary then follows from Lemma 2.7. □

We wish to thank Serge Bouc and Alexander Duncan for useful correspondence concerning Corollary 2.11. In particular, they explicitly computed an interesting (integral) idempotent of the Burnside ring of the alternating group $G = A_5$ such that the associated element of $\operatorname{Rep}(G,\mathbb{Q})$ is 0.

PROOF. (of Proposition 2.3) We first make an observation useful for proving both (1) and (2): For a cofiber sequence $X \to Y \to Z$ in $\mathcal{H}_G(\mathbf{Spt})^d$, there is an equality $\chi(X) + \chi(Z) = \chi(Y)$ in $A(G)$ (see for example [**May96**, XVII Theorem 1.6]), and therefore an equality $\theta\chi(X) + \theta\chi(Z) = \theta\chi(Y)$.

Since a short exact sequence $A \to B \to C$ of G-modules induces the relation $A + C = B$ in $\mathrm{Rep}(G, R)$, it follows by induction that an exact sequence

$$0 \to M_n \to M_{n+1} \to \ldots \to M_{m-1} \to M_m \to 0$$

of G-modules induces the relation $\sum_{i=-\infty}^{\infty}(-1)^i M_i = 0$. The cofiber sequence $X \to Y \to Z$ induces the long exact sequence

$$\ldots \to H^i(Z, R) \to H^i(Y, R) \to H^i(X, R) \to H^{i+1}(Z, R) \to \ldots$$

in cohomology, whence we have that

$$\sum_{i=-\infty}^{\infty} (-1)^i H^i(X, R) + \sum_{i=-\infty}^{\infty} (-1)^i H^i(Z, R) = \sum_{i=-\infty}^{\infty} (-1)^i H^i(Y, R)$$

under the hypothesis that for two (and thus all three) of X, Y, and Z, only finitely many terms in each sum are non-zero. Therefore if the claim holds for any two of X, Y, and Z in a cofiber sequence $X \to Y \to Z$, it holds for the third.

We now prove (1): By induction on the number of cells of the finite G-CW complex X, it therefore suffices to show the claim for X a finite G-set. This then follows from Lemma 2.1 (2).

We now prove (2): Let A be a G-CW complex which is a retract in the homotopy category of a finite G-CW complex Y. Let ι be the endomorphism of Y given by the composition of the retract and the inclusion $\iota : Y \to A \to Y$. It is formal that $\mathrm{Tr}(\iota) = \chi(A)$. We likewise have that the image $\mathrm{Image}\, H^i(\iota)$ of $H^i(\iota)$ is a direct summand of $H^i(Y, R)$ isomorphic to $H^i(A, R)$. We show the claim for A by proving that for a G-equivariant map $f : Y \to Y$ which is idempotent in the homotopy category, we have

$$\theta\, \mathrm{Tr}(f) = \sum_{i=-\infty}^{\infty} (-1)^i [\mathrm{Image}\, H^i(f)], \tag{2.3}$$

and applying (2.3) for $f = \iota$. Both sides of are unchanged under equivariant homotopy, and it follows that we may assume that f is cellular [**May96**, I. Theorem 3.4]. This allows us to induct on the dimension of the top cells of Y, reducing to the case where $Y = T_+ \wedge S^n$ for a G-set T. Since replacing f by its suspension multiplies both sides of (2.3) by -1, we replace f by $f \wedge S^{-n}$, reducing to the case where $Y = T_+$ and the map f is now is the stable homotopy category. Then as remarked above f corresponds to an idempotent $f : T \to T$ in $\mathrm{Burn}(G)$ [**May96**, Corollary 3.2 XIX]. Thus $\sum_{i=-\infty}^{\infty}(-1)^i [\mathrm{Image}\, H^i(f)] = [\mathrm{Image}\, H^0(f)]$. The equality (2.3) then follows by Lemma 2.7.

Let V be an n-dimensional representation of G and let S^{-V} be the dual of the one-point compactification S^V of V. If X is a strongly dualizable G-spectra, then so is DX and $\chi(X) = \chi(DX)$. Thus $\chi(S^{-V}) = \chi(S^V)$. Since S^V is a finite G-CW complex, we have that

$$\theta\chi(S^V) = \sum_{i=-\infty}^{\infty} (-1)^i H^i(S^V, R) = R + (-1)^n R,$$

where the action of G on the first summand is trivial, and the action of G on the second summand is via the sign on the determinant, i.e. $g \in G$ acts by ± 1

depending on if g preserves or reverses the orientation of S^V. Combining with the previous, we have

$$\theta\chi(S^{-V}) = R + (-1)^n R.$$

By definition, $H^i(DX, R) = \pi_{-i}F(DX, R) = \pi_{-i}(X \wedge R) = H_{-i}(X, R)$. Thus

$$H^i(S^{-V}, R) = \begin{cases} R & \text{if } i = 0 \\ R & \text{if } i = -n \\ 0 & \text{otherwise} \end{cases},$$

where for $i = 0$, the action is trivial, and for $i = -n$, G acts on R by the sign of the determinant. Thus the claim holds for S^{-V}.

By [**FLM01**, Proposition 2.1], X in $\mathcal{H}_G(\mathbf{Spt})^d$ is equivalent to $\Sigma^{-V}\Sigma^\infty A$ where A is a finitely dominated based G-CW complex and V is a representation of G. Since A is finitely dominated, by definition we have a finite G-CW complex Y such that A is a retract of Y in the homotopy category of G-CW complexes. By the above, the claim holds for A and S^{-V}. Since the smash product of two endomorphisms of $\mathbf{S}$ induces the multiplication of $A(G)$, we have $\chi(\Sigma^{-V}\Sigma^\infty A) = \chi(S^{-V})\chi(A)$. Thus $\theta\chi(\Sigma^{-V}\Sigma^\infty A) = \theta\chi(S^{-V})\theta\chi(A)$. By the Künneth spectral sequence, we have the equality

$$\sum_{i=-\infty}^{\infty} (-1)^i H^i(\Sigma^{-V}\Sigma^\infty A, R) = (\sum_{i=-\infty}^{\infty} (-1)^i H^i(S^{-V}, R))(\sum_{i=-\infty}^{\infty} (-1)^i H^i(A, R))$$

in $\mathrm{Rep}(G, R)$. The claim for X then follows from the claim for S^{-V} and A.

□

This concludes the facts we need about genuine G-spectra. We now turn to the relationship between $\mathbf{A}^1$-homotopy theory over $\mathbf{R}$ and genuine $\mathrm{Gal}_\mathbf{R}$-spectra.

Let $\mathcal{H}_{\mathbf{A}^1}(\mathbf{sPre}(\mathbf{Sm}_\mathbf{R}))$ denote the $\mathbf{A}^1$-homotopy category of simplicial presheaves on smooth schemes over $\mathbf{R}$ in the sense of Morel-Voevodsky. Let $\mathcal{H}_{\mathrm{Gal}_\mathbf{R}}(\mathbf{sSet})$ denote the homotopy category of genuine G-spaces for $G = \mathrm{Gal}_\mathbf{R}$. There is a symmetric monoidal functor

$$\mathrm{LB} : \mathcal{H}_{\mathbf{A}^1}(\mathbf{sPre}(\mathbf{Sm}_\mathbf{R})) \to \mathcal{H}_{\mathrm{Gal}_\mathbf{R}}(\mathbf{sSet}),$$

called Betti realization, which takes a smooth scheme X over $\mathbf{R}$ to the complex points $X(\mathbf{C})$ with the $\mathrm{Gal}_\mathbf{R}$-action induced from the tautological action of $\mathrm{Gal}_\mathbf{R}$ on $\mathbf{C}$. See [**MV99**, 3, 3.3].

LB furthermore determines a functor after stabilization. Namely, let

$$\mathcal{H}_{\mathbf{A}^1}(\mathbf{Spt}^{\mathbf{P}^1}(\mathbf{R}))$$

denote the $\mathbf{A}^1$-homotopy category of $\mathbf{P}^1$-spectra over $\mathbf{R}$. As above, let $\mathcal{H}_{\mathrm{Gal}_\mathbf{R}}(\mathbf{Spt})$ denote the homotopy category of genuine G-spectra for $G = \mathrm{Gal}_\mathbf{R}$. There is a commutative diagram of Betti-realization functors [**HO16**, 4.4]

$$\begin{array}{ccc} \mathcal{H}_{\mathbf{A}^1}(\mathbf{sPre}(\mathbf{Sm}_\mathbf{R})) & \xrightarrow{\mathrm{LB}} & \mathcal{H}_{\mathrm{Gal}_\mathbf{R}}(\mathbf{sSet}) \\ \downarrow{\scriptstyle (\Sigma^\infty_{\mathbf{P}^1})_+} & & \downarrow{\scriptstyle (\Sigma^\infty_{S^\mathbf{C}})_+} \\ \mathcal{H}_{\mathbf{A}^1}(\mathbf{Spt}^{\mathbf{P}^1}(\mathbf{R})) & \xrightarrow{\mathrm{LB}} & \mathcal{H}_{\mathrm{Gal}_\mathbf{R}}(\mathbf{Spt}). \end{array}$$

For k a field, Morel has shown that the endomorphisms of the sphere,

$$\mathrm{End}_{\mathcal{H}_{\mathbf{A}^1}(\mathbf{Spt}^{\mathbf{P}^1}(k))}(\mathbf{S}),$$

are isomorphic to the Grothendieck-Witt group $\mathrm{GW}(k)$ of k [**Mor12**, Corollary 1.24] (see [**Hoy14**, footnote p 2] about the non-perfect case), defined to be the group completion of the semi-ring of symmetric bilinear forms. $\mathrm{GW}(k)$ is generated by $\langle a \rangle$ for $a \in k^*/(k^*)^2$ and has relations given by $\langle u \rangle + \langle -u \rangle = \langle 1 \rangle + \langle -1 \rangle$ and $\langle u \rangle + \langle v \rangle = \langle u+v \rangle + \langle (u+v)uv \rangle$ for $u, v \in k^*$ and $u+v \neq 0$. See [**Mor12**, Lemma 3.9] who cites [**MH73**]. The element $\langle a \rangle$ corresponds to the bilinear form $(x, y) \mapsto axy$. We will compare $\mathrm{GW}(\mathbf{R})$ and $A(\mathrm{Gal}_{\mathbf{R}})$ and for this, we need the following well-known constructions.

For a separable field extension $k \subseteq L$, there is a map $\mathrm{Tr}_{L/k} : \mathrm{GW}(L) \to \mathrm{GW}(k)$ defined by sending a symmetric bilinear form $b : V \times V \to L$ to the composition $\mathrm{Tr}_{L/k} b : V \times V \to k$ of the form b with the field trace map $\mathrm{Tr}_{L/k} : L \to k$. In $\mathrm{Tr}_{L/k} b$, the vector space V is now viewed as a vector space over k.

The map sending a bilinear form to the ordered pair of its rank and signature determines an isomorphism

$$\mathrm{GW}(\mathbf{R}) \xrightarrow[\cong]{\mathrm{rank} \times \mathrm{signature}} \mathbb{Z} \times \mathbb{Z},$$

as can be checked using the generators and relations described above and Sylvester's law of inertia.

Since LB is a functor, it determines a map

$$\mathsf{LB} : \mathrm{GW}(\mathbf{R}) \cong \mathrm{End}_{\mathcal{H}_{\mathbf{A}^1}(\mathbf{Spt}^{\mathbf{P}^1}(\mathbf{R}))}(\mathbf{S}) \to \mathrm{End}_{\mathcal{H}_{\mathrm{Gal}_{\mathbf{R}}}(\mathbf{Spt})}(\mathbf{S}) \cong A(\mathrm{Gal}_{\mathbf{R}}).$$

Since LB is symmetric monoidal, $\mathsf{LB}(\chi Y) = \chi(\mathsf{LB} Y)$ for every strongly dualizable object Y of $\mathcal{H}_{\mathbf{A}^1}(\mathbf{Spt}^{\mathbf{P}^1}(\mathbf{R}))$. It is a result of Hoyois [**Hoy14**, Theorem 1.9] that for a separable field extension $k \subseteq L$, the fully dualizable spectrum $\Sigma^\infty_{\mathbf{P}^1} \operatorname{Spec} L_+$ has Euler characteristic $\chi \operatorname{Spec} L = \mathrm{Tr}_{L/k}\langle 1 \rangle$. Thus in $\mathrm{End}_{\mathcal{H}_{\mathbf{A}^1}(\mathbf{Spt}^{\mathbf{P}^1}(\mathbf{R}))}(\mathbf{S})$, we have that $\chi(\Sigma^\infty_{\mathbf{P}^1} \operatorname{Spec} \mathbb{C}_+) = \langle 1 \rangle + \langle -1 \rangle$. Since $\mathsf{LB}(\Sigma^\infty_{\mathbf{P}^1} \operatorname{Spec} \mathbb{C}_+)$ is the element of $\mathcal{H}_{\mathrm{Gal}_{\mathbf{R}}}(\mathbf{Spt})$ corresponding to the finite $\mathrm{Gal}_{\mathbf{R}}$-set given by $\mathrm{Gal}_{\mathbf{R}}$ with left translation, it follows that the map

$$\mathrm{GW}(\mathbf{R}) \to A(\mathrm{Gal}_{\mathbf{R}})$$

is determined by

$$\begin{array}{ccc} \mathrm{GW}(\mathbf{R}) & \longrightarrow & A(\mathrm{Gal}_{\mathbf{R}}) \\ {\scriptstyle \mathrm{rank} \times \mathrm{signature}} \downarrow & & \downarrow {\scriptstyle (T \mapsto |T|) \times (T \mapsto |T^{\mathrm{Gal}_{\mathbf{R}}}|)} \\ \mathbf{Z}^2 & \xrightarrow{\ 1\ } & \mathbf{Z}^2 \end{array},$$

and in particular is an isomorphism. Here the notation $|T|$ for a finite set T denotes the cardinality of T. We remark that there is much more to say about the relationship between the Burnside ring and the Grothendieck-Witt group. See, for example, the recent work of Kyle Ormsby and Jeremiah Heller [**HO16**].

The following proposition about an elliptic curve is one consequence of the existence of a realization over $k = \mathbb{R}$. In the next section, we show that the analogous result often fails to hold when k is a more complicated field and then use this fact to show that a suitable realization functor cannot exist.

PROPOSITION 2.12. *Let E be an elliptic curve over $\mathbb{R}$. Then*

(1) *For any ring* R, *we have* $\sum_{i=0}^{2}(-1)^i H^i(E(\mathbf{C}), R) = 0$ *in* $\mathrm{Rep}(\mathrm{Gal}_{\mathbf{R}}, R)$.

(2) *For a finite ring* R, *we have* $\sum_{i=0}^{2}(-1)^i H^i_{ét}(E_{\mathbf{C}}, R) = 0$ *in* $\mathrm{Rep}(\mathrm{Gal}_{\mathbf{R}}, R)$.

PROOF. E is strongly dualizable in $\mathcal{H}_{\mathbf{A}^1}(\mathbf{sPre}(\mathbf{Sm}_{\mathbf{R}}))$, and by [**Hoy14**, Theorem 1.5], the Euler characteristic of E is 0. Since LB is a symmetric monoidal functor, it follows that LBE is strongly dualizable in $\mathcal{H}_{\mathrm{Gal}_{\mathbf{R}}}(\mathbf{Spt})$ and that $\chi LBE = 0$.
(1): By [**Ill00**], LBE is a finite $\mathrm{Gal}_{\mathbf{R}}$-CW complex, so we may apply Proposition 2.3 (1) to LBE. It follows that $\sum_{i=-\infty}^{\infty}(-1)^i H^i(LBE, R) = 0$. Since $LBE = E(\mathbb{C})$ and $H^i(E(\mathbb{C}), R) = 0$ for $i < 0$ and $i > 2$, we have $\sum_{i=0}^{2}(-1)^i H^i(E(\mathbf{C}), R) = 0$ as claimed.
(2): For R finite, we have a natural isomorphism $H^i(E(\mathbf{C}), R) \cong H^i_{ét}(E_{\mathbf{C}}, R)$ by [**Mil80**, III Theorem 3.12] [**sga73**, XI], so (2) follows from (1). □

REMARK 2.13. We can also verify Proposition (2.12) directly. Indeed, when $R = \mathbb{C}$, the representations $H^i(E(\mathbb{C}), R)$ can be described explicitly as follows. Let $\mathbb{C}$ denote the trivial representation and $\mathbb{C}(1)$ denote the sign representation in $\mathrm{Rep}(\mathrm{Gal}_{\mathbf{R}}, \mathbb{C})$. Then in $\mathrm{Rep}(\mathrm{Gal}_{\mathbf{R}}, \mathbb{C})$, we have equalities $H^0(E(\mathbf{C}), \mathbb{C}) = \mathbb{C}$, $H^2(E(\mathbf{C}), \mathbb{C}) = \mathbb{C}(1)$, $H^1(E(\mathbf{C}), \mathbb{C}) = \mathbb{C} + \mathbb{C}(1)$. Indeed, the first equality follows because $\mathrm{Gal}_{\mathbb{R}}$ acts trivially on the single connected component of $E(\mathbb{C})$; the second equality follows because the non-trivial element of $\mathrm{Gal}_{\mathbb{R}}$ reverses orientation; the third equality follows because the cup product gives an isomorphism of $\mathrm{Gal}_{\mathbb{R}}$-representations $\wedge^2 H^1(E(\mathbf{C}), \mathbb{C}) \cong H^2(E(\mathbf{C}), \mathbb{C})$, and in the representation ring, $H^1(E(\mathbf{C}), \mathbb{C})$ is a direct sum of one-dimensional representations.

3. Enriched Euler characteristics and restrictions on genuine Gal_k-realization

Now let G be a profinite group, or more specifically a Galois group. One could hope to construct a homotopy category $\mathcal{H}_G(\mathbf{Spt})$ of genuine G-spectra or G-pro-spectra, and a Burnside ring $A(G) = \mathrm{End}_{\mathcal{H}_G(\mathbf{Spt})}(\mathbf{S})$. For example, homotopy theories of G-spectra are constructed in [**Fau08**] [**Qui13**] and [**Bar17**]. We have in mind that there is some appropriate sort of G-set and corresponding Burnside category such that a suspension spectrum functor is fully faithful into $\mathcal{H}_G(\mathbf{Spt})$, and that objects in $\mathcal{H}_G(\mathbf{Spt})$ can be constructed from colimits of suspensions of these G-sets, as is the case when G is finite. The point here being that the Euler characteristic of a G-set would be itself, and spaces would be built from G-sets, giving credibility to an analogue of Proposition 2.3. For any profinite group, a category of genuine G-equivariant spectra has been constructed from a Burnside category by Barwick in [**Bar17**], and studied by Barwick, Glasman, and Shah in [**BGS16**].

Let k be a field, and let $\overline{k}$ denote an algebraic closure of k and $\mathrm{Gal}_k = \mathrm{Gal}(\overline{k}/k)$. One could furthermore hope to construct an *étale realization functor*

$$L\,\mathrm{Et} : \mathcal{H}_{\mathbf{A}^1}(\mathbf{Spt}^{\mathbf{P}^1}(k)) \to \mathcal{H}_{\mathrm{Gal}_k}(\mathbf{Spt})$$

from the stable $\mathbf{A}^1$-homotopy category of $\mathbf{P}^1$-spectra over k to genuine Gal_k-spectra or pro-spectra, appropriately completed away from the characteristic of k. For example, Quick has constructed an étale realization functor from the stable $\mathbf{A}^1$-homotopy category to a stable homotopy category of profinite spaces in [**Qui07**]. Étale realization functors have been constructed and studied by Ayoub [**Ayo14a**] and [**CD16**, Section 7.2] in a generalization of the following context. The category of Voevodsky motives is analogous to the stable $\mathbf{A}^1$-category [**Ayo14b**, Section 2]. Let Λ be the ring $\Lambda = \mathbf{Z}/\ell$ for a prime ℓ different from the characteristic of k, and assume that k is perfect. Let $D(\mathbf{Sh}(k_{ét}, \Lambda))$ denote the derived category of sheaves of Λ-modules on the small étale site of Spec k. There is an étale realization functor from Voevodsky motives to $D(\mathbf{Sh}(k_{ét}, \Lambda))$. Since a sheaf of Λ-modules on the small étale site of Spec k is a Λ-module with an action of Gal_k, the derived category $D(\mathbf{Sh}(k_{ét}, \Lambda))$ is similar to spectra (of $H\Lambda$-modules) equipped with an action of Gal_k. We wish to draw a similarity between $D(\mathbf{Sh}(k_{ét}, \Lambda))$ and a homotopy theory of spectra with a Gal_k-action and contrast $D(\mathbf{Sh}(k_{ét}, \Lambda))$ with a notion of genuine Gal_k-spectra. For example, the endomorphisms of the symmetric monoidal unit of $D(\mathbf{Sh}(k_{ét}, \Lambda))$ is Λ, in contrast to the Burnside ring. Since étale realization functors exist in powerful contexts, one could hope for an étale realization functor to genuine Galois equivariant spectra or pro-spectra.

We have in mind that applying the étale realization functor $\mathrm{L\,Et}$ to the suspension spectrum of a smooth scheme gives an appropriate suspension spectrum of the étale topological type of Artin-Mazur [**AM86**] and Friedlander [**Fri82**]. In other words, we have in mind that applying $\mathrm{L\,Et}$ is compatible with an unstable étale realization functor $\mathrm{L\,Et}$ as constructed by Isaksen [**Isa04**] in the non-equivariant context. From this, we are lead to the following two expectations.

- Since the étale topological type $\mathrm{Et}((X \times Y)_{\overline{k}})$ is equivalent to the product $\mathrm{Et}(X_{\overline{k}}) \times \mathrm{Et}(Y_{\overline{k}})$, cf. [**Del77**, Corollarie 1.11], it is reasonable to hope that $\mathrm{L\,Et}$ is symmetric monoidal.
- Since the étale cohomology of a smooth scheme X with finite coefficients R is isomorphic to the cohomology of the étale topological type Et X with coefficients in R, this would result in an isomorphism of Gal_k-representations
$$H^i_{ét}(X_{\overline{k}}, R) \cong H^i(\mathrm{L\,Et}\, X_{\overline{k}}, R).$$
This implies that if the Euler characteristic of a genuine G-spectrum or pro-spectrum is connected to the Gal_k-representations given by its cohomology groups as in Proposition 2.3, then applying a homomorphism θ from A(G) to a representation ring sends this Euler characteristic to $\sum_{i=-\infty}^{\infty} H^i_{ét}(X_{\overline{k}}, R)$.

The purpose of this paper is to show that, for many fields k, it is impossible to simultaneously satisfy these hopes, i.e., such an étale realization functor does not exist.

THEOREM 3.1. *Let* k *be a global field, local field with infinite Galois group, or finite field. Let* p *be a prime different from the characteristic of* k*. It is impossible to simultaneously construct all of the following*

(1) *A symmetric monoidal category* $\mathcal{H}_{\mathrm{Gal}_k}(\mathbf{Spt})$*, enriched over abelian groups. Let* $A(\mathrm{Gal}_k) = \mathrm{End}_{\mathcal{H}_{\mathrm{Gal}_k}(\mathbf{Spt})}(\mathbf{S})$ *denote the endomorphisms of the symmetric monoidal unit* $\mathbf{S}$.
(2) *For all* n*, a representation ring* $\mathrm{Rep}(\mathrm{Gal}_k, \mathbb{Z}/p^n)$ *with coefficients in* $\mathbb{Z}/p^n$.
(3) *A symmetric monoidal additive functor* $\mathrm{L\,Et} : \mathcal{H}_{\mathbf{A}^1}(\mathbf{Spt}^{\mathbf{P}^1}(k)) \to \mathcal{H}_{\mathrm{Gal}_k}(\mathbf{Spt})$*, and ring homomorphisms* $\theta_n : A(G) \to \mathrm{Rep}(\mathrm{Gal}_k, \mathbb{Z}/p^n)$ *such that*
$$\theta_n \chi(\mathrm{L\,Et}\, \Sigma^\infty_{\mathbf{P}^1} X_+) = \sum_{i=-\infty}^{\infty} (-1)^i [H^i_{ét}(X_{\overline{k}}, \mathbb{Z}/p^n)]$$

for smooth proper schemes X *over* k.

LEMMA 3.2. *Let* E *be an elliptic curve over a finite field* $\mathbf{F}_\ell$, *and let* p *be any prime not dividing* ℓ. *Suppose that for all* n, *we have a ring* $\mathrm{Rep}(\mathrm{Gal}_k, \mathbb{Z}/p^n)$ *as in Theorem* 3.1. *There exists a positive integer* n, *such that* $\sum_{i=-\infty}^{\infty}(-1)^i[H^i_{ét}(E_{\bar{k}}, \mathbb{Z}/p^n)]$ *is non-zero in* $\mathrm{Rep}(\mathrm{Gal}_k, \mathbf{Z}/p^n)$.

PROOF. Suppose the contrary. Then for all g in Gal_k and all n, we have the equality

$$\mathrm{Tr}\, g|H^1(E_{\bar{k}}, \mathbb{Z}/p^n) = \mathrm{Tr}\, g|H^0(E_{\bar{k}}, \mathbb{Z}/p^n) + \mathrm{Tr}\, g|H^2(E_{\bar{k}}, \mathbb{Z}/p^n)$$

in $\mathbb{Z}/p^n$. By definition, $H^i(E_{\bar{k}}, \mathbb{Q}_p) = \mathbb{Q}_p \otimes \varprojlim_n H^i(E_{\bar{k}}, \mathbb{Z}/p^n)$ [**Mil80**, V I]. Therefore

$$\mathrm{Tr}\, g|H^1(E_{\bar{k}}, \mathbb{Q}_p) = \mathrm{Tr}\, g|H^0(E_{\bar{k}}, \mathbb{Q}_p) + \mathrm{Tr}\, g|H^2(E_{\bar{k}}, \mathbb{Q}_p) \tag{3.1}$$

in $\mathbb{Q}_p$. Let F be the geometric Frobenius in Gal_k, i.e., the inverse of the arithmetic Frobenius $a \mapsto a^\ell$ for all a in $\bar{k}$. By the Weil Conjectures, $\mathrm{Tr}\, F|H^1(E_{\bar{k}}, \mathbb{Q}_p) = a_1 + a_2$, where a_i are algebraic integers with absolute value $\ell^{1/2}$ [**FK88**, Theorem IV.1.2]. Since Gal_k acts trivially on $H^0(E_{\bar{k}}, \mathbb{Q}_p)$ and there is an isomorphism $H^2(E_{\bar{k}}, \mathbb{Q}_p) \cong \mathbb{Q}_p(1)$, Equation (3.1) for $g = F$ becomes

$$a_1 + a_2 = 1 + \ell.$$

Taking the absolute value of both sides, we have

$$2\ell^{1/2} = |a_1| + |a_2| \geq |a_1 + a_2| = 1 + \ell,$$

which is impossible for any prime power ℓ. □

PROOF. (of Theorem 3.1) Suppose k is a finite field $k = \mathbf{F}_\ell$. Choose an elliptic curve over $k = \mathbf{F}_\ell$. To see that this is indeed possible, note that the Weierstraß equation $y^2 = x(x-1)(x+1)$ determines an elliptic curve when $\mathbf{F}_\ell$ has odd characteristic. When $\ell = 2^d$, the equation $y^2 + y = x^3$ produces the desired elliptic curve. By Lemma 3.2, there exists a positive integer n such that $\sum_{i=-\infty}^{\infty}(-1)^i[H^i_{ét}(E_{\bar{k}}, \mathbb{Z}/p^n)]$ is non-zero in $\mathrm{Rep}(\mathrm{Gal}_k, \mathbf{Z}/p^n)$. By hypothesis (3) of the theorem, $\theta_n\chi(\mathrm{L\,Et}\,\Sigma^\infty_{\mathbf{P}^1} E_+) = \sum_{i=-\infty}^{\infty}(-1)^i[H^i_{ét}(E_{\bar{k}}, \mathbb{Z}/p^n)]$. It follows that $\chi(\mathrm{L\,Et}\,\Sigma^\infty_{\mathbf{P}^1} E_+)$ is non-zero. Since $\mathrm{L\,Et}$ is a symmetric monoidal functor (by (3)), it follows that $\mathrm{L\,Et}\,\chi(\Sigma^\infty_{\mathbf{P}^1} E_+) = \chi(\mathrm{L\,Et}\,\Sigma^\infty_{\mathbf{P}^1} E_+)$, and therefore $\chi(\Sigma^\infty_{\mathbf{P}^1} E_+)$ is non-zero. This contradicts [**Hoy14**, Theorem 1.5].

Suppose k is a local field with infinite Galois group, i.e., k is a finite extension of $\mathbb{Q}_\ell$ or $\mathbf{F}_\ell((t))$, and let $\mathcal{O}_k$ denote the integral closure of $\mathbb{Z}_\ell$ or $\mathbf{F}_\ell[[t]]$ in k. We may choose an elliptic curve E over k with good reduction by choosing an elliptic curve over the residue field (as discussed in the previous paragraph) and then applying [**sga03**, III 7.3]. Thus we have an elliptic curve $f : \mathcal{E} \to \mathrm{Spec}\,\mathcal{O}_k$. Let $E_0 = \mathcal{E} \times_{\mathrm{Spec}\,\mathcal{O}_k} \mathrm{Spec}\,\mathbf{F}_\ell$ denote the fiber over the closed point. By smooth-proper base change, [**Mil80**, VI §4 Corollary 4.2] $R^i f_* \mathbb{Z}/p^n$ is a locally constant sheaf on $\mathrm{Spec}\,\mathcal{O}_k$ whose stalk at a geometric point $s : \mathrm{Spec}\,\Omega \to \mathrm{Spec}\,\mathcal{O}_k$ is $H^i(\mathcal{E}_s, \mathbb{Z}/p^n)$, where $\mathcal{E}_s = \mathcal{E} \times_{\mathrm{Spec}\,\mathcal{O}_k} \mathrm{Spec}\,\Omega$. It follows that the Gal_k-action on $H^i(E_{\bar{k}}, \mathbb{Z}/p^n)$ factors through the quotient map

$$\mathrm{Gal}_k \to \mathrm{Gal}(k^{un}/k) \cong \pi_1(\mathrm{Spec}\,\mathcal{O}_k) \cong \mathrm{Gal}_{\mathbf{F}_{\ell^d}},$$

where k^{un} is the maximal unramified extension of k and $\mathbf{F}_{\ell^d}$ is the residue field. It likewise follows that $H^i(E_{\overline{k}}, \mathbb{Z}/p^n)$ is isomorphic to $H^i((E_0)_{\overline{\mathbf{F}_\ell}}, \mathbb{Z}/p^n)$ as $\mathrm{Gal}_{\mathbf{F}_{\ell^d}}$-modules. This reduces the claim to the case where k is a finite field.

Lastly, suppose that k is a global field. Choose an elliptic curve E over k, and let v be a place not dividing the discriminant, so E has good reduction at v. Let k_v be the completion of k at v. The decomposition group of v in Gal_k gives an injection $\mathrm{Gal}_{k_v} \to \mathrm{Gal}_k$, and the Gal_{k_v}-action on $H^i(E_{\overline{k}}, \mathbb{Z}/p^n)$ is isomorphic to the Gal_{k_v}-action on $H^i(E_{\overline{k_v}}, \mathbb{Z}/p^n)$. By the previous paragraph, we may again reduce the claim to the case where k is a finite field. □

We give a slight variation on Theorem 3.1, where only the representation rings $\mathrm{Rep}(\mathrm{Gal}_k, \mathbb{Z}/p)$ with p prime appear.

THEOREM 3.3. *Let k be a global field, a local field with infinite Galois group, or a finite field. It is impossible to simultaneously construct all of the following*

(1) *A symmetric monoidal category $\mathcal{H}_{\mathrm{Gal}_k}(\mathbf{Spt})$, enriched over abelian groups. Let $A(\mathrm{Gal}_k) = \mathrm{End}_{\mathcal{H}_{\mathrm{Gal}_k}(\mathbf{Spt})}(\mathbf{S})$ denote the endomorphisms of the symmetric monoidal unit $\mathbf{S}$.*
(2) *For infinitely many primes p, a representation ring $\mathrm{Rep}(\mathrm{Gal}_k, \mathbb{Z}/p)$ with coefficients in $\mathbb{Z}/p$.*
(3) *A symmetric monoidal additive functor $\mathrm{L\,Et} : \mathcal{H}_{\mathbf{A}^1}(\mathbf{Spt}^{\mathbf{P}^1}(k)) \to \mathcal{H}_{\mathrm{Gal}_k}(\mathbf{Spt})$, and a homomorphism of rings $\theta : A(G) \to \mathrm{Rep}(\mathrm{Gal}_k, \mathbb{Z}/p)$ such that $\theta\chi(\mathrm{L\,Et}\,\Sigma^\infty_{\mathbf{P}^1} X_+) = \sum_{i=-\infty}^{\infty} (-1)^i H^i_{\acute{e}t}(X_{\overline{k}}, \mathbb{Z}/p)$ for smooth proper scheme X over k.*

PROOF. Replace Lemma 3.2 with Lemma 3.4 below in the proof of Theorem 3.1. □

LEMMA 3.4. *Let E be an elliptic curve over a finite field $\mathbf{F}_\ell$. Let p be any prime greater than the cardinality of $E(\mathbf{F}_\ell)$. Then $\sum_{i=-\infty}^{\infty} (-1)^i [H^i_{\acute{e}t}(E_{\overline{k}}, \mathbb{Z}/p)]$ is non-zero in $\mathrm{Rep}(\mathrm{Gal}_k, \mathbf{Z}/p)$.*

PROOF. Let $k = \mathbf{F}_\ell$ and let F be the geometric Frobenius in Gal_k, i.e., the inverse of the arithmetic Frobenius $a \mapsto a^\ell$ for all a in $\overline{k}$. By assumption (2), it suffices to show that $\sum_{i=0}^{2} (-1)^i \operatorname{Tr} F | H^i_{\acute{e}t}(E_{\overline{k}}, \mathbb{Z}/p)$ is non-zero in $\mathbb{Z}/p$. By [**Mil80**, pg. 292] the trace of F equals the trace of the Frobenius endomorphism of E. By the Lefschetz Trace Formula,

$$\sum_{i=0}^{2} (-1)^i \operatorname{Tr} F | H^i_{\acute{e}t}(E_{\overline{k}}, \mathbb{Z}/p) = |E(\mathbf{F}_\ell)|$$

in $\mathbb{Z}/p$ (see [**DPSW16**, Proposition 7.1] for further explanation on the mod p version: it is contained in [**Del77**] and can also be deduced from [**Hoy14**]). Since p is greater than $|E(\mathbf{F}_\ell)|$ by assumption and all elliptic curves contain at least one rational point, it follows that $\sum_{i=0}^{2} (-1)^i \operatorname{Tr} F | H^i_{\acute{e}t}(E_{\overline{k}}, \mathbb{Z}/p)$ is non-zero as desired. □

REMARK 3.5. By Section 2, when $k = \mathbb{R}$, there does exist a symmetric monoidal category $\mathcal{H}_{\mathrm{Gal}_k}(\mathbf{Spt})$ and a symmetric monoidal functor $\mathrm{L\,Et} : \mathcal{H}_{\mathbf{A}^1}(\mathbf{Spt}^{\mathbf{P}^1}(k)) \to \mathcal{H}_{\mathrm{Gal}_k}(\mathbf{Spt})$ satisfying the conditions of Theorems 3.1 and 3.3.

Acknowledgements

We wish to thank Joseph Ayoub, Tom Bachmann, Gunnar Carlsson, Akhil Mathew, and all the participants of the AIM workshop on Derived Equivariant Algebraic Geometry, June 13–17, 2016, organized by Andrew Blumberg, Teena Gerhardt, Michael Hill, and Kyle Ormsby, and additionally including the participation of Clark Barwick, Elden Elmanto, Saul Glasman, Jeremiah Heller, Denis Nardin, and Jay Shah. It was a pleasure discussing these ideas with you. It is also a pleasure to thank Serge Bouc for useful correspondence concerning Corollary 2.11, and Alexander Duncan for writing and running some GAP code to check an example of Corollary 2.11 The second named author also wishes to thank the Institut Mittag–Leffler for their hospitality during their special program "Algebro-Geometric and Homotopical Methods," where she worked on this paper.

References

[AM86] M. Artin and B. Mazur, *Etale homotopy*, Lecture Notes in Mathematics, vol. 100, Springer-Verlag, Berlin, 1986. Reprint of the 1969 original. MR883959

[Ayo14a] Joseph Ayoub, *La réalisation étale et les opérations de Grothendieck* (French, with English and French summaries), Ann. Sci. Éc. Norm. Supér. (4) **47** (2014), no. 1, 1–145, DOI 10.24033/asens.2210. MR3205601

[Ayo14b] Joseph Ayoub. A guide to (étale) motivic sheaves. In *International Congress of Mathematicians. Vol. II*, pages 1101–1124. Kyung Moon Sa Co. Ltd., Seoul 2014.

[Bar17] Clark Barwick, *Spectral Mackey functors and equivariant algebraic* K*-theory (I)*, Adv. Math. **304** (2017), 646–727, DOI 10.1016/j.aim.2016.08.043. MR3558219

[BGS16] Clark Barwick, *Spectral Mackey functors and equivariant algebraic* K*-theory (I)*, Adv. Math. **304** (2017), 646–727, DOI 10.1016/j.aim.2016.08.043. MR3558219

[Bou00] Serge Bouc, *Burnside rings*, Handbook of algebra, Vol. 2, Handb. Algebr., vol. 2, Elsevier/North-Holland, Amsterdam, 2000, pp. 739–804, DOI 10.1016/S1570-7954(00)80043-1. MR1759611

[CD16] Denis-Charles Cisinski and Frédéric Déglise, *Étale motives*, Compos. Math. **152** (2016), no. 3, 556–666, DOI 10.1112/S0010437X15007459. MR3477640

[Del77] P. Deligne, *Cohomologie étale*, Lecture Notes in Mathematics, Vol. 569, Springer-Verlag, Berlin-New York, 1977. Séminaire de Géométrie Algébrique du Bois-Marie SGA $4\frac{1}{2}$; Avec la collaboration de J. F. Boutot, A. Grothendieck, L. Illusie et J. L. Verdier. MR0463174

[DPSW16] Rachel Davis, Rachel Pries, Vesna Stojanoska, and Kirsten Wickelgren, *The galois action and cohomology of a relative homology group of Fermat curves.* arXiv:1610.02006v1, 2016. To appear in the *Journal of Algebra.*

[Dre69] Andreas Dress, *A characterisation of solvable groups*, Math. Z. **110** (1969), 213–217, DOI 10.1007/BF01110213. MR0248239

[EGH$^+$11] Pavel Etingof, Oleg Golberg, Sebastian Hensel, Tiankai Liu, Alex Schwendner, Dmitry Vaintrob, and Elena Yudovina, *Introduction to representation theory*, Student Mathematical Library, vol. 59, American Mathematical Society, Providence, RI, 2011. With historical interludes by Slava Gerovitch. MR2808160

[Fau08] Halvard Fausk, *Equivariant homotopy theory for pro-spectra*, Geom. Topol. **12** (2008), no. 1, 103–176, DOI 10.2140/gt.2008.12.103. MR2377247

[FK88] Eberhard Freitag and Reinhardt Kiehl, *Étale cohomology and the Weil conjecture*, Ergebnisse der Mathematik und ihrer Grenzgebiete (3) [Results in Mathematics and Related Areas (3)], vol. 13, Springer-Verlag, Berlin, 1988. Translated from the German by Betty S. Waterhouse and William C. Waterhouse; With an historical introduction by J. A. Dieudonné. MR926276

[FLM01] H. Fausk, L. G. Lewis Jr., and J. P. May, *The Picard group of equivariant stable homotopy theory*, Adv. Math. **163** (2001), no. 1, 17–33, DOI 10.1006/aima.2001.1997. MR1867202

[Fri82] Eric M. Friedlander, *Étale homotopy of simplicial schemes*, Annals of Mathematics Studies, vol. 104, Princeton University Press, Princeton, N.J.; University of Tokyo Press, Tokyo, 1982. MR676809

[Glu81] David Gluck, *Idempotent formula for the Burnside algebra with applications to the* p*-subgroup simplicial complex*, Illinois J. Math. **25** (1981), no. 1, 63–67. MR602896

[HO16] J. Heller and K. Ormsby, *Galois equivariance and stable motivic homotopy theory*, Trans. Amer. Math. Soc. **368** (2016), no. 11, 8047–8077, DOI 10.1090/tran6647. MR3546793

[Hoy14] Marc Hoyois, *A quadratic refinement of the Grothendieck-Lefschetz-Verdier trace formula*, Algebr. Geom. Topol. **14** (2014), no. 6, 3603–3658, DOI 10.2140/agt.2014.14.3603. MR3302973

[Ill00] Sören Illman, *Existence and uniqueness of equivariant triangulations of smooth proper* G*-manifolds with some applications to equivariant Whitehead torsion*, J. Reine Angew. Math. **524** (2000), 129–183, DOI 10.1515/crll.2000.054. MR1770606

[Isa04] Daniel C. Isaksen, *Etale realization on the* $\mathbb{A}^1$*-homotopy theory of schemes*, Adv. Math. **184** (2004), no. 1, 37–63, DOI 10.1016/S0001-8708(03)00094-X. MR2047848

[LMSM86] L. G. Lewis Jr., J. P. May, M. Steinberger, and J. E. McClure, *Equivariant stable homotopy theory*, Lecture Notes in Mathematics, vol. 1213, Springer-Verlag, Berlin, 1986. With contributions by J. E. McClure. MR866482

[May96] J. P. May, *Equivariant homotopy and cohomology theory*, CBMS Regional Conference Series in Mathematics, vol. 91, Published for the Conference Board of the Mathematical Sciences, Washington, DC; by the American Mathematical Society, Providence, RI, 1996. With contributions by M. Cole, G. Comezaña, S. Costenoble, A. D. Elmendorf, J. P. C. Greenlees, L. G. Lewis, Jr., R. J. Piacenza, G. Triantafillou, and S. Waner. MR1413302

[MH73] John Milnor and Dale Husemoller, *Symmetric bilinear forms*, Springer-Verlag, New York-Heidelberg, 1973. Ergebnisse der Mathematik und ihrer Grenzgebiete, Band 73. MR0506372

[Mil80] James S. Milne, *Étale cohomology*, Princeton Mathematical Series, vol. 33, Princeton University Press, Princeton, N.J., 1980. MR559531

[Mor12] Fabien Morel, $\mathbb{A}^1$*-algebraic topology over a field*, Lecture Notes in Mathematics, vol. 2052, Springer, Heidelberg, 2012. MR2934577

[MV99] Fabien Morel and Vladimir Voevodsky, $\mathbf{A}^1$*-homotopy theory of schemes*, Inst. Hautes Études Sci. Publ. Math. **90** (1999), 45–143 (2001). MR1813224

[Qui07] Gereon Quick, *Stable étale realization and étale cobordism*, Adv. Math. **214** (2007), no. 2, 730–760, DOI 10.1016/j.aim.2007.03.005. MR2349718

[Qui13] Gereon Quick, *Profinite* G*-spectra*, Homology Homotopy Appl. **15** (2013), no. 1, 151–189, DOI 10.4310/HHA.2013.v15.n1.a9. MR3079203

[Rob85] Mark Roberts, *Equivariant Milnor numbers and invariant Morse approximations*, J. London Math. Soc. (2) **31** (1985), no. 3, 487–500, DOI 10.1112/jlms/s2-31.3.487. MR812778

[Seg] G. B. Segal, *Equivariant stable homotopy theory*, Actes du Congrès International des Mathématiciens (Nice, 1970), Gauthier-Villars, Paris, 1971, pp. 59–63. MR0423340

[sga73] *Théorie des topos et cohomologie étale des schémas. Tome 3.* Lecture Notes in Mathematics, Vol. 305. Springer-Verlag, Berlin, 1973. Séminaire de Géométrie Algébrique du Bois-Marie 1963–1964 (SGA 4), Dirigé par M. Artin, A. Grothendieck et J. L. Verdier. Avec la collaboration de P. Deligne et B. Saint-Donat.

[sga03] *Revêtements étales et groupe fondamental (SGA 1)* (French), Documents Mathématiques (Paris) [Mathematical Documents (Paris)], vol. 3, Société Mathématique de France, Paris, 2003. Séminaire de géométrie algébrique du Bois Marie 1960–61. [Algebraic Geometry Seminar of Bois Marie 1960-61]; Directed by A. Grothendieck; With two papers by M. Raynaud; Updated and annotated reprint of the 1971 original [Lecture Notes in Math., 224, Springer, Berlin; MR0354651 (50 #7129)]. MR2017446

[Sol67] Louis Solomon, *The Burnside algebra of a finite group*, J. Combinatorial Theory **2** (1967), 603–615. MR0214679

[tD79] Tammo tom Dieck, *Transformation groups and representation theory*, Lecture Notes in Mathematics, vol. 766, Springer, Berlin, 1979. MR551743

[Yos83] Tomoyuki Yoshida, *Idempotents of Burnside rings and Dress induction theorem*, J. Algebra **80** (1983), no. 1, 90–105, DOI 10.1016/0021-8693(83)90019-4. MR690705

Current address: J. L. Kass, Dept. of Mathematics, University of South Carolina, 1523 Greene Street, Columbia, SC 29208, United States of America

Email address: kassj@math.sc.edu

URL: http://people.math.sc.edu/kassj/

Current address: K. Wickelgren, School of Mathematics, Georgia Institute of Technology, 686 Cherry Street, Atlanta, GA 30332-0160

Email address: kwickelgren3@math.gatech.edu

URL: http://people.math.gatech.edu/~kwickelgren3/

Contemporary Mathematics
Volume **707**, 2018
http://dx.doi.org/10.1090/conm/707/14252

Multiplicative structure on Real Johnson-Wilson theory

Nitu Kitchloo, Vitaly Lorman, and W. Stephen Wilson

ABSTRACT. We prove that the Real Johnson-Wilson theories $ER(n)$ are homotopy associative and commutative ring spectra up to phantom maps. We further show that the $ER(n)$ represent associatively and commutatively multiplicative cohomology theories on the category of (possibly non-compact) spaces. We also revisit a result of the first and third authors concerning the $MR(n)$-orientation of $\mathrm{MO}[2^{n+1}]$.

1. Introduction

At the prime 2, Johnson-Wilson theory $E(n)$ [**JW73**] is a complex-oriented cohomology theory which has a C_2-equivariant refinement, $\mathbb{E}(n)$ as a genuine C_2-equivariant spectrum, where the action of C_2 stems from complex conjugation. This was first constructed in [**HK01**], and Real Johnson-Wilson theory $ER(n)$ is defined to be the C_2-fixed points of $\mathbb{E}(n)$. The underlying nonequivariant spectrum of $\mathbb{E}(n)$ is Johnson-Wilson theory $E(n)$, and it is a homotopy associative, commutative, and unital ring spectrum. The goal of this note is to investigate whether the same properties hold for $\mathbb{E}(n)$ and $ER(n)$.

Interest in this problem comes from the fact that $ER(n)$ is quickly becoming a useful and computable cohomology theory. For $n = 1$ and 2, it reproduces familiar cohomology theories, $ER(1) = KO_{(2)}$ and $ER(2) = TMF_0(3)$ (the latter after suitable completion, see [**HM17**]). The $ER(n)$-cohomology of a large (and growing) collection of spaces has been computed: real projective spaces and their products [**KW08a, KW08b, Ban13**] (for $n = 2$), complex projective spaces [**Lor16, KLW17**], BO and some of its connective covers [**KW14, KLW17**], and half of all Eilenberg MacLane spaces [**KLW16, KLW17**]. Furthermore, these computations have applications. In [**KW08a, KW08b**], the first and third authors used computations in $ER(2)$-cohomology to prove new nonimmersion results for real projective spaces.

The existence of a multiplicative structure on Real Johnson-Wilson theory has been suggested in a comment in [**HK01**] (Comment 5 following the proof of Theorem 2.28) which claims that $\mathbb{E}(n)^\star(\mathbb{E}(n) \wedge \mathbb{E}(n))$ may be calculated and from this it may be shown that $\mathbb{E}(n)$ is a (homotopy) associative, commutative, and unital ring spectrum. The results in this note were born in the attempt to verify the above

---

2010 *Mathematics Subject Classification.* 55P91, 55N22, 55N91.

The first author is supported in part by the NSF through grant DMS 1307875.

claim. Unfortunately, we were unsuccessful in doing so. However, we show that $\mathbb{E}(n)$ represents an $\mathbb{MU}$-algebra which is homotopy unital, associative, and commutative *up to phantom maps*. By a phantom map, we mean a map $f : X \longrightarrow Y$ which has trivial restriction to any finite CW complex mapping into X. In addition, we show that the $\mathbb{E}(n)$-cohomology of an equivariant topological space is canonically a commutative ring.

THEOREM 1.1. *$\mathbb{E}(n)$ is a homotopy commutative, homotopy associative, unital $\mathbb{MU}$-algebra up to phantom maps. In other words, there exist unit and multiplication maps:*

$$1 : \mathbb{MU} \longrightarrow \mathbb{E}(n), \qquad \hat{\mu} : \mathbb{E}(n) \wedge \mathbb{E}(n) \longrightarrow \mathbb{E}(n),$$

such that all the obstructions to $\hat{\mu}$ being a homotopy associative and homotopy commutative $\mathbb{MU}$-algebra structure are phantom maps. Differently said, all the corresponding structure diagrams commute up to phantom maps. Furthermore, the forgetful map:

$$\rho : \mathbb{E}(n)^0(\mathbb{E}(n) \wedge \mathbb{E}(n)) \longrightarrow E(n)^0(E(n) \wedge E(n)),$$

maps $\hat{\mu}$ to the canonical product μ on the non-equivariant Johnson-Wilson spectrum $E(n)$.

Theorem 1.1 tells us that the Real Johnson-Wilson theory is valued in commutative rings when applied to *finite* CW complexes. Our second result extends this to the category of all spaces.

THEOREM 1.2. *With any choice of multiplication $\hat{\mu}$ as above, the spectrum $\mathbb{E}(n)$ represents a multiplicative cohomology theory on the category of C_2-spaces valued in (bigraded) commutative rings. There are natural transformations of ring-valued cohomology theories $\mathbb{MU}^\star(-) \longrightarrow \mathbb{E}(n)^\star(-)$ and $\mathbb{E}(n)^\star(-) \longrightarrow E(n)^*(-)$.*

The results of this document justify the assumption of commutativity in the computations of the $ER(n)$-cohomology of topological spaces made by the authors in previous work. We conclude by revisiting a result of the first and third authors concerning the $MR(n)$-orientation of $\mathrm{MO}[2^{n+1}]$ to correct an error in the proof of [**KW14**, Theorem 1.4].

In the course of proving Theorem 1.2 we show that the infinite loop space underlying $ER(n)$, $\underline{ER(n)}_0$, is a homotopy commutative, associative, and unital H-ring space (Lemma 5.1). Applying the Bousfield-Kuhn functor shows that the $K(n)$-localization, $L_{K(n)}ER(n)$, is in fact a homotopy commutative, associative and unital ring spectrum (not just up to phantom maps). The authors are grateful to Tyler Lawson for pointing this out.

Hahn and Shi [**HS17**] have recently proved that Johnson-Wilson theory $E(n)$ admits an A_∞ ring structure for which the C_2-action is given by an A_∞ involution (but not necessarily an A_∞ ring automorphism). The question of (homotopy) commutativity is not readily addressed by their techniques. They show that these issues resolve on $K(n)$-localization giving rise to an E_∞ ring spectrum structure on $L_{K(n)}ER(n)$.

The present paper shows a (homotopy) commutative and associative ring structure up to phantom maps, but to the authors' present knowledge, $ER(n)$ is not (yet) known to be a homotopy commutative and associative ring spectrum.

The authors would like to thank Neil Strickland for helpful discussion related to this document and the referee for diligent and valuable comments.

2. Background

In this section, we recall a few background definitions and theorems from [**KW07b**] and [**KLW16**] that we use in subsequent sections.

A genuine C_2-equivariant spectrum $\mathbb{E}$ is a family of C_2-spaces $\underline{\mathbb{E}}_{a+b\alpha}$, indexed over elements $a + b\alpha \in RO(C_2)$ where α denotes the sign representation, together with a compatible system of equivariant homeomorphisms

$$\underline{\mathbb{E}}_{a-r+(b-s)\alpha} \xrightarrow{\simeq} \Omega^{r+s\alpha}\underline{\mathbb{E}}_{a+b\alpha}$$

where the right hand side denotes the space of pointed maps (endowed with the conjugation action) from the one point compactification of the representation $r + s\alpha$. The reader may refer to [**HK01**] for more details on C_2-spectra. We will denote by ER the homotopy fixed point spectrum of the C_2-action on $\mathbb{E}$ and by E the underlying nonequivariant spectrum given by forgetting the C_2-action. An example of a C_2- spectrum of interest to us is $\mathbb{MU}$, whose underlying nonequivariant spectrum is complex cobordism, MU, studied first by Landweber [**Lan68**], Fujii [**Fuj76**], Araki and Murayama [**AM78**], and more recently by Hu-Kriz [**HK01**]. The action of C_2 is induced by the complex conjugation action on the pre-spectrum representing $\mathbb{MU}$ in the usual way (see, e.g. [**HK01**]).

The $p = 2$ Johnson-Wilson theory $E(n)$ lifts to a C_2-spectrum, $\mathbb{E}(n)$, defined as an $\mathbb{MU}$-module by coning off certain equivariant lifts of the Araki generators v_i for $i > n$, and then inverting the lift of v_n. We shall call these equivariant lifts by the same names, v_i. The Real Johnson-Wilson theories, $ER(n)$, are defined as the fixed points of $\mathbb{E}(n)$.

Working with cohomological grading, let Y be a C_2-space and let $\mathbb{E}^{*(1+\alpha)}(Y)$ denote the subgroup of diagonal elements in the equivariant $\mathbb{E}$-cohomology of Y i.e.

$$\mathbb{E}^{*(1+\alpha)}(Y) := \pi_0 \mathrm{Maps}^{C_2}(Y, \underline{\mathbb{E}}_{*(1+\alpha)}).$$

Consider the group homomorphism given by forgetting the C_2-action:

$$\rho : \mathbb{E}^{*(1+\alpha)}(Y) \longrightarrow E^*(Y).$$

Notice that the image of ρ belongs to the graded sub-group of elements in even degree. We recall the following definitions from [**KW07b**].

DEFINITION 2.1. A C_2-space Y is said to have the weak projective property with respect to a C_2-spectrum $\mathbb{E}$ if the map

$$\rho : \mathbb{E}^{*(1+\alpha)}(Y) \longrightarrow E^{2*}(Y),$$

is an isomorphism of graded abelian groups.

Note that in the case that the C_2-spectrum $\mathbb{E}$ is an $\mathbb{MU}$-module spectrum, the map ρ in the definition above is a map of $\mathbb{MU}^{*(1+\alpha)} \cong MU^{2*}$-modules by virtue of the fact that the $\mathbb{MU}$-module structure on $\mathbb{E}$ forgets to an MU-module structure on E.

We will make extensive use of spaces with the weak projective property. In order to recognize such spaces, we need some auxiliary definitions.

DEFINITION 2.2. A pointed C_2-space X is said to be projective if

(1) $H_*(X;\mathbb{Z})$ is of finite type.
(2) X is homeomorphic to $\bigvee_I (\mathbb{C}P^\infty)^{\wedge k_I}$ for some weakly increasing sequence of integers k_I, with the C_2 action given by complex conjugation.

By a C_2-equivariant H-space, we shall mean an H-space whose multiplication map is C_2-equivariant.

DEFINITION 2.3. A C_2-equivariant H-space Y is said to have the projective property if there exists a projective space X, along with a pointed C_2-equivariant map $f : X \longrightarrow Y$, such that $H_*(Y;\mathbb{Z}/2)$ is generated as an algebra by the image of f.

The following theorem, proved in [**KLW16**, Theorem 2.6], establishes that spaces with the projective property have the weak projective property with respect to certain $\mathbb{MU}$-module spectra (in particular, $\mathbb{E}(n)$).

THEOREM 2.4. *Let Y be a C_2-equivariant H-space with the projective property. Let $\mathbb{E}$ denote any complete $\mathbb{MU}$-module spectrum with underlying spectrum E, satisfying the property that the forgetful map: $\rho^* : \mathbb{E}^{*(1+\alpha)} \longrightarrow E^*$, is an isomorphism. Then the space Y has the weak-projective property with respect to $\mathbb{E}$. In other words, the following map is an isomorphism of $\mathbb{MU}^{*(1+\alpha)}$-modules:*

$$\rho : \mathbb{E}^{*(1+\alpha)}(Y) \longrightarrow E^*(Y).$$

REMARK 2.5. The smash product of a finite collection of spaces with the projective property is an example of a space that has the weak projective property with respect to any $\mathbb{E}$ as in Theorem 2.4, but not the projective property (since it is not an H-space). This follows from writing $Y_1 \wedge Y_2 = (Y_1 \times Y_2)/(Y_1 \vee Y_2)$ and the Five Lemma.

There are many examples of spaces with the projective property. The ones of interest in this document will be $\underline{\mathbb{E}(n)}_0$ and its products.

LEMMA 2.6. $\underline{\mathbb{E}(n)}_0^{\times j} \times \underline{\mathbb{MU}}_{i(2^n-1)(1+\alpha)}^{\times s}$ *has the weak projective property with respect to $\mathbb{E}(n)$ for all $i, j, s \geq 1$.*

PROOF. By [**KW13**, Theorem 1-4], the spaces $\underline{\mathbb{E}(n)}_0$ and $\underline{\mathbb{MU}}_{(i(2^n-1)(1+\alpha)}$ have the projective property. Thus $\underline{\mathbb{E}(n)}_0^{\times j} \times \underline{\mathbb{MU}}_{i(2^n-1)(1+\alpha)}^{\times s}$ is a restricted product of a family of spaces with the projective property (i.e. it is the colimit of finite products of spaces with projective property). A finite product of spaces with projective property evidently has the weak projective property, and since $\rho : \mathbb{E}(n)^{*(1+\alpha)}(-) \longrightarrow E(n)^{2*}(-)$ is an isomorphism at each stage, it follows that it is an isomorphism in the limit. □

3. A stable multiplicative structure

We begin with the observation that the spaces $\underline{\mathbb{E}(n)}_0$ are $|v_n| = (2^n-1)(1+\alpha)$-periodic. That is, the adjoint of the multiplication-by-v_n map on the spectrum $\mathbb{E}(n)$ induces, on the 0-space level, an equivalence

$$\underline{\mathbb{E}(n)}_0 \xrightarrow{\simeq} \Omega^{|v_n|}\underline{\mathbb{E}(n)}_0$$

We now express the spectrum $\mathbb{E}(n)$, and its products as colimits of shifted suspension spectra

$$\mathbb{E}(n) = \text{colim}\,_m \Sigma^{-m(2^n-1)(1+\alpha)}\underline{\mathbb{E}(n)}_0, \quad \mathbb{E}(n)^{\wedge k} = \text{colim}\,_m \Sigma^{-mk(2^n-1)(1+\alpha)}\underline{\mathbb{E}(n)}_0^{\wedge k},$$

where the maps are given by successive $v_n^{\wedge k}$ multiplications

$$\Sigma^{-mk|v_n|}\underline{\mathbb{E}(n)}_0^{\wedge k} \longrightarrow \Sigma^{-(m+1)k|v_n|}\underline{\mathbb{E}(n)}_0^{\wedge k}.$$

Applying $\mathbb{E}(n)$ cohomology, Milnor's $\lim^1$-sequence gives us

$$0 \longrightarrow \lim{}^1\mathbb{E}(n)^{-1}(\Sigma^{-mk|v_n|}\underline{\mathbb{E}(n)}_0^{\wedge k}) \longrightarrow \mathbb{E}(n)^0(\mathbb{E}(n)^{\wedge k}) \longrightarrow \lim{}^0\mathbb{E}(n)^0(\underline{\mathbb{E}(n)}_0^{\wedge k}) \longrightarrow 0.$$

Using the $|v_n|$-periodicity of $\mathbb{E}(n)$ noted above to identify $\Sigma^{-|v_n|}\underline{\mathbb{E}(n)}_0$ with $\underline{\mathbb{E}(n)}_0$, we have the sequence

$$0 \longrightarrow \lim{}^1\mathbb{E}(n)^{-1}(\underline{\mathbb{E}(n)}_0^{\wedge k}) \longrightarrow \mathbb{E}(n)^0(\mathbb{E}(n)^{\wedge k}) \longrightarrow \lim{}^0\mathbb{E}(n)^0(\underline{\mathbb{E}(n)}_0^{\wedge k}) \longrightarrow 0.$$

We may now invoke the weak projective property of the spaces $\underline{\mathbb{E}(n)}_0^{\wedge k}$ (see Remark 2.5) and identify the last term with $\lim^0 E(n)^0(\underline{E(n)}_0^{\wedge k})$. The space $\underline{E(n)}_0$ has evenly graded cohomology (since it has the projective property) and homotopy; it follows from the Atiyah-Hirzebruch spectral sequence that the $E(n)$-cohomology of $\underline{E(n)}_0$ is evenly graded as well (as are its k-fold smash products). Thus, in the analogous non-equivariant Milnor sequence

$$0 \longrightarrow \lim{}^1 E(n)^{-1}(\underline{E(n)}_0^{\wedge k}) \longrightarrow E(n)^0(E(n)^{\wedge k}) \longrightarrow \lim{}^0 E(n)^0(\underline{E(n)}_0^{\wedge k}) \longrightarrow 0$$

the $\lim^1$ term vanishes. Consequently, we identify $\lim^0 E(n)^0(\underline{E(n)}_0^{\wedge k})$ with $E(n)^0(\underline{E(n)}_0)$ and obtain our short exact sequence of interest

$$0 \longrightarrow \lim{}^1\mathbb{E}(n)^{-1}(\underline{\mathbb{E}(n)}_0^{\wedge k}) \longrightarrow \mathbb{E}(n)^0(\mathbb{E}(n)^{\wedge k}) \longrightarrow E(n)^0(E(n)^{\wedge k}) \longrightarrow 0, \tag{3.1}$$

with the last map being ρ.

By writing $\mathbb{E}(n)^{\wedge k}$ as a colimit of v_n^k-multiplication maps as above and $\mathbb{MU}^{\wedge s}$ as a colimit of suspension spectra

$$\mathbb{MU}^{\wedge s} = \text{colim}\,_m \Sigma^{\infty - ms|v_n|}\underline{\mathbb{MU}}_{m|v_n|}^{\wedge s}$$

the above proof readily extends to show the existence of a short exact sequence

(3.2)
$$0 \longrightarrow \lim{}^1\mathbb{E}(n)^{-1}(\underline{\mathbb{E}(n)}_0^{\wedge k} \wedge \underline{\mathbb{MU}}_{m|v_n|}^{\wedge s}) \longrightarrow \mathbb{E}(n)^0(\mathbb{E}(n)^{\wedge k} \wedge \mathbb{MU}^{\wedge s})$$
$$\longrightarrow E(n)^0(E(n)^{\wedge k} \wedge MU^{\wedge s}) \longrightarrow 0.$$

where we have used Lemma 2.6 above to identify the right hand term as before.

We are now ready to construct the $\mathbb{MU}$-algebra structure on $\mathbb{E}(n)$ that will be shown to be homotopy commutative and homotopy associative up to phantom maps.

Definition 3.1. Define $\hat{\mu}$ to be any element in $\mathbb{E}(n)^0(\mathbb{E}(n)^{\wedge 2})$ that lifts the canonical ring structure of $E(n)^0(E(n)^{\wedge 2})$ along ρ. Define the unit map $1 : \mathbb{MU} \longrightarrow \mathbb{E}(n)$ to be the canonical map expressing $\mathbb{E}(n)$ as a localized quotient of $\mathbb{MU}$.

As a formal consequence of the short exact sequence constructed above, we obtain Theorem 1.1 from the introduction:

THEOREM 1.1. *The class $\hat{\mu}$ defines a homotopy commutative, homotopy associative, and unital* $\mathbb{MU}$*-algebra structure on* $\mathbb{E}(n)$ *up to phantom maps.*

PROOF. By construction, $\hat{\mu}$ maps to the canonical ring structure on $E(n)$ under the forgetful map, ρ. It follows that any obstruction to the homotopy associativity, commutativity, or unitality of $\hat{\mu}$, viewed as a class in $\mathbb{E}(n)^0(\mathbb{E}(n)^{\wedge k} \wedge \mathbb{MU}^{\wedge s})$, maps to zero under ρ. We claim that the only elements of the kernel of ρ are phantoms.

To see this, recall the short exact sequence (3.2) above:

$$0 \to \lim{}^1 \mathbb{E}(n)^{-1}(\underline{\mathbb{E}(n)}_0^{\wedge k} \wedge \underline{\mathbb{MU}}^{\wedge s}_{m|v_n|}) \to \mathbb{E}(n)^0(\mathbb{E}(n)^{\wedge k} \wedge \mathbb{MU}^{\wedge s}) \to E(n)^0(E(n)^{\wedge k} \wedge MU^{\wedge s}) \to 0$$

where the right hand side was identified via

$$\lim{}^0 \mathbb{E}(n)^0(\underline{\mathbb{E}(n)}_0^{\wedge k} \wedge \underline{\mathbb{MU}}^{\wedge s}_{m|v_n|}) \cong \lim{}^0 E(n)^0(\underline{E(n)}_0^{\wedge k} \wedge \underline{MU}^{\wedge s}_{m|v_n|}) \cong E(n)^0(E(n)^{\wedge k} \wedge MU^{\wedge s}).$$

Notice that by exactness, an element in the kernel of ρ is in the image of the $\lim^1$-term in $\mathbb{E}(n)^0(\mathbb{E}(n)^{\wedge k} \wedge \mathbb{MU}^{\wedge s})$ and so restricts trivially to all the terms $\mathbb{E}(n)^0(\underline{\mathbb{E}(n)}_0^{\wedge k} \wedge \underline{\mathbb{MU}}^{\wedge s}_{m|v_n|})$ in the inverse system. Any map from a finite CW complex into $\mathbb{E}(n)^{\wedge k} \wedge \mathbb{MU}^{\wedge s}$ must factor through a finite stage of the colimit, that is, through $\underline{\mathbb{E}(n)}_0^{\wedge k} \wedge \underline{\mathbb{MU}}^{\wedge s}_{m|v_n|}$ for some m. It follows that any element in the image of the $\lim^1$ term is zero upon restriction to any finite CW-complex. In other words, the kernel of ρ consists entirely of phantoms. □

REMARK 3.2. One may attempt to compute the group of phantom maps $\lim{}^1 \mathbb{E}(n)^{-1}(\underline{\mathbb{E}(n)}_0^{\wedge k})$ explicitly by identifying it with the vector space $E(n)^{-1}(E(n)^{\wedge k}) \otimes \mathbb{Z}/2$ (suitably extended by a $\mathbb{Z}/2$-agebra). This is an open problem, but it appears to the authors that this vector space is trivial for $n = 1$, but may fail to be so for $n > 1$. Hence we at present have no general way of ensuring that the ring structure we have constructed is rigid up to homotopy for the spectra $\mathbb{E}(n)$, $n > 1$.

4. Multiplicative structure on $\mathbb{MU}_{(2)}[v_n^{-1}]$

We would like to show that the multiplication $\hat{\mu}$ constructed above naturally induces a commutative algebra structure on the $\mathbb{E}(n)$-cohomology of *any* (not necessarily finite-dimensional) space. An essential ingredient in our construction will be the multiplication on $\mathbb{MU}_{(2)}[v_n^{-1}]$. We pause to describe it in this section. The ingredient we need is the following proposition, which appears as Proposition 9.15 in [**HHR16**] and is proved in [**HH14**].

PROPOSITION 4.1. [**HH14**, Corollary 4.11] *Let R be a G-equivariant commutative ring with $D \in \pi_\star^G(R)$. If D has the property that for every $H \subset G$, $N_H^G i_H^* D$ divides a power of D, then the spectrum $D^{-1}R$ has a unique commutative algebra structure such that the map $R \longrightarrow D^{-1}R$ is a map of commutative rings.*

We begin by constructing a C_2-equivariant associative and commutative ring (in the highly structured sense) that lifts $\mathbb{E}(n)$. We begin with the fact that $\mathbb{MU}$ has this structure (see [**HHR16**] or [**HK01**]). We localize at $p = 2$. The spectrum $\mathbb{MU}_{(2)}$ is a C_2-equivariant commutative ring, as shown in [**HH14**]. By [**HK01**], the forgetful map

$$\rho : \pi_{*(1+\alpha)}(\mathbb{MU}_{(2)}) \longrightarrow \pi_{2*}(MU_{(2)})$$

is an isomorphism. The classes v_i (Araki, Hazewinkel, or others) in $\pi_{2*}(MU_{(2)})$ may now be lifted via ρ^{-1} to equivariant classes, $\rho^{-1}(v_i)$. While in the rest of the manuscript, we abuse notation by denoting $\rho^{-1}(v_i)$ by v_i, in the following lemma, we will distinguish between the nonequivariant v_i and the equivariant $\rho^{-1}(v_i)$. Note that for $i \leq n$ and the Araki v_i, the images of these lifts $\rho^{-1}(v_i)$ in the coefficients of $\mathbb{E}(n)$ are exactly the equivariant v_i we have been working with throughout.

Our next step is to invert $\rho^{-1}(v_n)$.

LEMMA 4.1. *The spectrum $(\rho^{-1}(v_n))^{-1}\mathbb{MU}_{(2)}$ is a C_2-equivariant commutative ring.*

PROOF. We apply Proposition 4.1 above (quoted from [**HHR16**]). The map i_H^* is exactly ρ, and so $\rho(\rho^{-1}(v_n)) = v_n \in \pi_{2(2^n-1)}(MU_{(2)})$. We need to show that $N_{\{e\}}^{C_2}(v_n)$ divides a power of $\rho^{-1}(v_n)$. In fact, we claim that $N_{\{e\}}^{C_2}(v_n) = -[\rho^{-1}(v_n)]^2$. To see this, we apply the isomorphism ρ to both sides. Let c denote the action of the generator of C_2. Using the fact that $c(v_n) = -v_n$, the double coset formula (see e.g. Proposition 10.9(v) in [**Sch**] or [**May96**]) reduces in our case to

$$\rho \circ N_{\{e\}}^{C_2}(v_n) = v_n \cdot c(v_n) = -v_n^2,$$

which completes the proof. □

REMARK 4.2. Let us again denote the spectrum $(\rho^{-1}(v_n))^{-1}\mathbb{MU}_{(2)}$ by $\mathbb{MU}[v_n^{-1}]$. This spectrum serves as a commutative proxy for $\mathbb{E}(n)$. Indeed, essentially all prior results of the authors that hold for $\mathbb{E}(n)$ extend verbatim to this spectrum.

5. Unstable properties of the multiplicative structure

We now address the question of the (unstable) multiplicative structure on $\mathbb{E}(n)$. We begin with the following lemma:

LEMMA 5.1. *There is a unique (homotopy) commutative, associative, and unital equivariant H-ring structure on the infinite loop space of $\mathbb{E}(n)$ that lifts any fixed H-ring structure on $\underline{E(n)}_0$:*

$$\hat{\mu}_0 : \underline{\mathbb{E}(n)}_0 \times \underline{\mathbb{E}(n)}_0 \longrightarrow \underline{\mathbb{E}(n)}_0.$$

PROOF. Recall that Lemma 2.6 shows that $\underline{\mathbb{E}(n)}_0^{\times j}$ has the weak projective property. As $E(n)$ is a (homotopy) associative and commutative ring spectrum, we may define the map $\hat{\mu}_0 : \underline{\mathbb{E}(n)}_0 \times \underline{\mathbb{E}(n)}_0 \longrightarrow \underline{\mathbb{E}(n)}_0$ as the preimage of the multiplication on $\underline{E(n)}_0$ along the isomorphism

$$\mathbb{E}(n)^0(\underline{\mathbb{E}(n)}_0 \times \underline{\mathbb{E}(n)}_0) \xrightarrow[\cong]{\rho} E(n)^0(\underline{E(n)}_0 \times \underline{E(n)}_0)$$

The unity, commutativity, and associativity of $\hat{\mu}_0$ are similarly verified by applying the isomorphism ρ, as the desired relations all hold in $E(n)$-cohomology. Likewise, given a choice of H-ring structure on $\underline{E(n)}_0$, the uniqueness of the equivariant lift follows from the fact that ρ is an isomorphism. □

REMARK 5.2. The unstable multiplication $\hat{\mu}_0$ we constructed in the previous lemma and the stable multiplication $\hat{\mu}$ we constructed in Definition 3.1 are compatible in the sense that $\hat{\mu}$ restricts to $\hat{\mu}_0$ on the zero space of $\mathbb{E}(n)$. To see this, apply the isomorphism ρ and note that this claim is true nonequivariantly by construction.

We now prove our second main result, Theorem 1.2.

PROOF. (of Theorem 1.2) Let X be a space and consider $f \in \mathbb{E}(n)^V(X)$ and $g \in \mathbb{E}(n)^W(Y)$. We define the product $fg \in \mathbb{E}(n)^{V+W}(X)$ as follows. First, note that since $\mathbb{E}(n)$ is an $\mathbb{MU}[v_n^{-1}]$-module, we may multiply f and g by classes in the coefficients $\mathbb{MU}[v_n^{-1}]_\star$. Let k and l be the minimal integers such that

$$v_n^k f \in \mathbb{E}(n)^{V'}(X), \qquad v_n^l g \in \mathbb{E}(n)^{W'}(X)$$

with $V', W' \leq 0$ (by this we mean that when we express each representation as a combination of irreducibles, each coefficient should be nonpositive). These classes are represented by maps

$$X \xrightarrow{v_n^k f} \underline{\mathbb{E}(n)}_{V'} = \Omega^{-V'} \underline{\mathbb{E}(n)}_0 , \qquad X \xrightarrow{v_n^l g} \underline{\mathbb{E}(n)}_{W'} = \Omega^{-W'} \underline{\mathbb{E}(n)}_0 .$$

We adjoin the loops over to form classes

$$\Sigma^{-V'} X \xrightarrow{v_n^k f} \underline{\mathbb{E}(n)}_0 , \qquad \Sigma^{-W'} X \xrightarrow{v_n^l g} \underline{\mathbb{E}(n)}_0 .$$

Note that since $-V', -W' \geq 0$, these are positive suspensions and so the sources of these maps are *spaces*. We may thus smash them together, precompose with the diagonal on X and postcompose with the multiplication on the zero space (which factors through the smash product) constructed in Lemma 5.1:

$$\Sigma^{-V'-W'} X \xrightarrow{\Delta} \Sigma^{-V'} X \wedge \Sigma^{-W'} X \xrightarrow{v_n^k f \wedge v_n^l g} \underline{\mathbb{E}(n)}_0 \wedge \underline{\mathbb{E}(n)}_0 \xrightarrow{\hat{\mu}_0} \underline{\mathbb{E}(n)}_0$$

This produces a class in $\mathbb{E}(n)^{V'+W'}(X)$. Finally, we multiply by v_n^{-k-l} to define the product $f \cdot g \in \mathbb{E}(n)^{V+W}(X)$.

The unit of this multiplication comes from the unit on $\mathbb{MU}[v_n^{-1}]$. The unity, associativity, and commutativity of this multiplication follow from the corresponding properties of $\underline{\mathbb{E}(n)}_0$.

We have shown that for any space X, $\mathbb{E}(n)^\star(X)$ is a graded associative and commutative ring. It remains to show that this is compatible with the multiplication on $\mathbb{MU}[v_n^{-1}]$-cohomology. If we carry out the above construction to define multiplication on $\mathbb{MU}[v_n^{-1}]^\star(X)$, it is evident that this agrees with the multiplication coming from the ring spectrum structure on $\mathbb{MU}[v_n^{-1}]$. To see that this multiplication agrees with the one defined on $\mathbb{E}(n)$-cohomology, it suffices to check this fact on zero spaces. To see that this diagram

$$\begin{array}{ccc} \underline{\mathbb{MU}}_0 \times \underline{\mathbb{MU}}_0 & \xrightarrow{\hat{\mu}_{\mathbb{MU}}} & \underline{\mathbb{MU}}_0 \\ \downarrow & & \downarrow \\ \underline{\mathbb{E}(n)}_0 \times \underline{\mathbb{E}(n)}_0 & \xrightarrow{\hat{\mu}_{\mathbb{E}(n)}} & \underline{\mathbb{E}(n)}_0 \end{array}$$

commutes, we may use the fact that $\underline{\mathbb{MU}}_0$ and $\underline{\mathbb{MU}}_0 \times \underline{\mathbb{MU}}_0$ are spaces with weak projective properties to map the diagram isomorphically along ρ where its commutativity is apparent. □

6. The MR(n) orientation for MO[2^{n+1}] revisited

Let $n > 0$ and let $\mathrm{MO}[2^{n+1}]$ denote the Thom spectrum for the virtual bundle over BO given by multiplication by 2^{n+1} seen as a self-map of BO. In other words, $\mathrm{MO}[2^{n+1}]$ is the spectrum that represents real vector bundles ξ endowed with an isomorphism $\xi \to 2^{n+1}\zeta$ for some bundle ζ.

In [**KW14**], the first and third authors showed that $\mathrm{MO}[2^{n+1}]$ admits an orientation with respect to $ER(n)$. However, that proof requires the homotopy commutativity of $\mathbb{E}(n)$ which is unclear in light of this document. Consequently, in this section, we reproduce the argument in [**KW14**] to show that $\mathrm{MO}[2^{n+1}]$ admits a canonical orientation with respect to the commutative ring spectrum $\mathrm{MR}(n)$ defined as the homotopy fixed points of the spectrum $\mathbb{MU}[v_n^{-1}]$, where it is understood that $\mathbb{MU}$ is 2-local. This orientation descends to an orientation with respect to $ER(n)$, generalizing the $\hat{A}$-genus for real K-theory. We also take this opportunity to fix an error in the proof of this result given in [**KW14**]; see Remark 6.4.

THEOREM 6.1. *The spectrum* $\mathrm{MO}[2^{n+1}]$ *supports a canonical orientation* u_{n+1} *with respect to* $\mathrm{MR}(n)$. *Furthermore, given a real vector bundle of the form* $2^{n+1}\zeta$, *this orientation is uniquely determined by the property that the image of* u_{n+1} *in* $\mathrm{MU}[v_n^{-1}]$ *is given by*

$$u_{n+1}(2^{n+1}\zeta) = \mu(2^n\zeta \otimes \mathbb{C}) \cup \psi(\zeta \otimes \mathbb{C})^{2^{n-1}},$$

where μ *is the usual Thom class from usual complex orientation, and* $\psi(\zeta \otimes \mathbb{C})$ *is the series in* $\mathrm{MU}[v_n^{-1}]^*(BU)$ *generated from line bundles by:*

$$\psi(x) = \frac{[-1]_{\mathrm{MU}}(x)}{-x}.$$

First, we lay some groundwork. Consider the Real orientation of $\mathbb{MU}$ given by a $\mathbb{Z}/2$-equivariant map:

$$\mu_1 : \mathbb{MU}(1) \longrightarrow \Sigma^{(1+\alpha)}\mathbb{MU}, \tag{6.1}$$

where $\mathbb{MU}(1) \simeq \mathbb{CP}^\infty$ denotes the C_2-space in the usual prespectrum defining $\mathbb{MU}$. We view $MU^*(MU(1))$ as a rank one free module on μ_1 over $MU^*(BU(1))$. We need the following fact regarding the C_2-action on μ_1.

LEMMA 6.2. *Let x denote the first Chern class in $MU^2(BU(1))$ and let $\psi(x)$ denote the series defined in Theorem* 6.1 *above. Then*

$$c(\mu_1) = \mu_1 \frac{[-1]_{\mathrm{MU}}(x)}{-x} = \mu_1\,\psi(x).$$

PROOF. The action of complex conjugation on $\Sigma^{1+\alpha}\mathbb{MU}$ can be identified with $-c$ (the c from $\mathbb{MU}$ and the -1 from the orientation reversing action on the two sphere). On the other hand, complex conjugation on $\mathrm{MU}(1)$ is induced by the (complex anti-linear) self map of the universal line bundle γ_1 over $\mathrm{BU}(1)$ that sends a vector to its complex conjugate. This map can be seen as a (complex linear) isomorphism from $\overline{\gamma_1}$ to γ_1, where $\overline{\gamma_1}$ is the opposite complex structure on the real bundle underlying γ_1. Since $\overline{\gamma}_1$ is isomorphic to the dual bundle γ_1^*, we see that the action of complex conjugation on MU(1) sends the Thom class $\mu_1 \in \mathrm{MU}^2(\mathrm{MU}(1))$ to the class $[-1]_{\mathrm{MU}}(\mu_1)$. MU(1) and BU(1) are homotopy equivalent and the Thom isomorphism is an $\mathrm{MU}^*(\mathrm{BU}(1))$ module map so that $\mu_1^2 = \mu_1 x$, where $x = c_1(\gamma_1)$.

From this we have $\mu_1^k = \mu_1 x^{k-1}$, so any power series $\sum a_i \mu_1^{i+1}$ can be rewriten as $\mu_1 \sum a_i x^i$. Hence

$$[-1]_{\mathrm{MU}}(\mu_1) = \mu_1 \frac{[-1]_{\mathrm{MU}}(x)}{x}.$$

Incorporating this observation into the $\mathbb{Z}/2$ equivariance of μ_1 from Equation (6.1) by computing on the left and the right, this translates to the equality:

$$c(\mu_1) = \mu_1 \frac{[-1]_{\mathrm{MU}}(x)}{-x} = \mu_1 \psi(x).$$

□

Outline of proof: (of Theorem 6.1) Before beginning the proof of Theorem 6.1 in earnest, we give a brief overview. We will make significant use of the Bockstein spectral sequence constructed from the fibration in [**KW07a**, Theorem 1.6]. Though the results in [**KW07a**] are stated for $ER(n)$, since the class v_n has been inverted, they will apply verbatim to $MR(n)$. In particular, the proof of [**KW07a**, Theorem 1.6] shows that there is a Bockstein spectral sequence $E_r(\mathrm{MO}[4])$, starting with the $\mathrm{MU}[v_n^{-1}]$-cohomology of the spectrum MO[4] and converging to the $MR(n)$-cohomology of MO[4]. The proof of Theorem 6.1 is somewhat technical and assumes familiarity with the properties of this spectral sequence, which may be found summarized in [**KW14**, Theorem 2.1]. In particular, note that the spectral sequence collapses at the $E_{2^{n+1}}$ page. The final part of the proof will make use of the structure of the spectral sequence for a point and for BO. These are described in Section 3 and 5 of [**KW14**], respectively.

The orientation u_{n+1} will be constructed inductively, starting with a class $u_2 \in \mathrm{MU}[v_n^{-1}]^*(\mathrm{MO}[4])$. Once the class u_2 is constructed, we will inductively define u_{k+1} in terms of u_k (see Equation 6.2 below) and begin our analysis of the Bockstein spectral sequence differentials. We will show that for $1 < k < n+1$ that the class u_k survives to the E_{2^k-1} page and that $E_{2^k-1}(\mathrm{MO}[2^k])$ is a rank one free module over the $E_{2^k-1}(\mathrm{BO})$ (the E_{2^k-1} page of the spectral sequence for BO) on the distinguished generator u_k. Continuing in this way, we will conclude that u_{n+1} survives to the last stage $E_{2^{n+1}-1}$ where there is one last possible differential left in the spectral sequence. We will show that this differential on u_{n+1} must be zero, and thus $MR(n)^*(\mathrm{MO}[2^{n+1}])$ is a rank one free module over $MR(n)^*(BO)$, which will establish the theorem.

PROOF. (of Theorem 6.1) We start by defining a class u_2. Let ζ denote the universal real vector bundle over BO, $\zeta_{\mathbb{C}}$ the universal complex vector bundle over BU, and $\overline{\zeta_{\mathbb{C}}}$ its conjugate. Consider the composite

$$\mathrm{BU} \xrightarrow{\Delta} \mathrm{BU} \times \mathrm{BU} \longrightarrow \mathrm{BU}$$

classifying the bundle $\zeta_{\mathbb{C}} \oplus \overline{\zeta_{\mathbb{C}}}$ over BU. Precomposing with the complexification map $\mathrm{BO} \longrightarrow \mathrm{BU}$, we have that $\zeta_{\mathbb{C}} \oplus \overline{\zeta_{\mathbb{C}}}$ pulls back to 4ζ. Taking Thom spectra and mapping into $\mathrm{MU}[v_n^{-1}]$ yields the composite

$$\mathrm{MO}[4] \longrightarrow \mathrm{MU}\wedge\mathrm{MU} \longrightarrow \mathrm{MU}[v_n^{-1}]$$

where the second map is induced by the twisted multiplication map (the counit of the norm-forgetful adjunction)

$$m\circ(id\wedge c) : \mathrm{MU}\wedge\mathrm{MU} \longrightarrow \mathrm{MU}\,.$$

This is an $\mathrm{MU}[v_n^{-1}]$-Thom class for 4ζ, and we define it to be $u_2(4\zeta)$. We have

$$u_2(4\zeta) = \mu((\zeta\otimes\mathbb{C})\oplus(\overline{\zeta\otimes\mathbb{C}})) = \mu(\zeta\otimes\mathbb{C})\mu(\overline{\zeta\otimes\mathbb{C}}).$$

Reasoning from line bundles using Lemma 6.2 and applying the splitting principle shows that

$$\mu(\overline{\zeta\otimes\mathbb{C}}) = \mu(\zeta\otimes\mathbb{C})\psi(\zeta\otimes\mathbb{C}).$$

It follows that

$$u_2(4\zeta) = \mu(2\zeta\otimes\mathbb{C})\cup\psi(\zeta\otimes\mathbb{C}))$$

and that u_2 is c-invariant.

We now begin our investigation of the Bockstein spectral sequence. Since u_2 is c-invariant, it follows that $d_1(u_2)=0$ in the Bockstein spectral sequence converging to $\mathrm{MR}(n)^*(\mathrm{MO}[4])$. We also have $d_2(u_2)=0$ for degree reasons, so that $d_r(u_2)=0$ for $r<3$ will begin our induction. If some Thom class u_r (to be defined inductively in Equation 6.2 below) survives to E_{r+1}, notice that E_{r+1} for the Thom space is a rank one free module over $\mathrm{E}_{r+1}(\mathrm{BO})$ on the Thom class. We will show for $1<k<n+1$ that $E_{2^k-1}(\mathrm{MO}[2^k])$ is a rank one free module over $E_{2^k-1}(\mathrm{BO})$ on a distinguished generator u_k.

By induction, assume this is true for $k>1$. Writing $2^{k+1}\zeta$ as $2(2^k\zeta)$, we have a factorization:

$$[2^{k+1}] = ([2^k]+[2^k])\circ\Delta : \mathrm{BO} \longrightarrow \mathrm{BO}\times\mathrm{BO} \longrightarrow \mathrm{BO},$$

which induces a map of Thom spectra:

$$\tau = m\circ\Delta : \mathrm{MO}[2^{k+1}] \longrightarrow \mathrm{MO}[2^k]\wedge\mathrm{MO}[2^k] \longrightarrow \mathrm{MO}[2^k].$$

This induces a map of spectral sequences $\tau^* : \mathrm{E}_{2^k-1}(\mathrm{MO}[2^k]) \longrightarrow \mathrm{E}_{2^k-1}(\mathrm{MO}[2^{k+1}])$. Define

$$u_{k+1} := \tau^*(u_k). \tag{6.2}$$

Observe that since $c(u_{k+1}) = u_{k+1}$ by naturality and the fact that $c(u_k)=u_k$ by induction (the observation that u_2 is c-invariant above started the induction). By the properties of the Bockstein spectral sequence, we have:

$$\begin{aligned} d_{2^k-1}(u_{k+1}) &= \Delta^*\mu^* d_{2^k-1}(u_k) = \Delta^* d_{2^k-1}\mu^*(u_k) \\ &= \Delta^*(d_{2^k-1}(u_k)\wedge u_k + u_k\wedge d_{2^k-1}(u_k)). \end{aligned}$$

But notice that $\Delta : \mathrm{MO}[2^{k+1}] \longrightarrow \mathrm{MO}[2^k]\wedge\mathrm{MO}[2^k]$ is invariant under the swap map on $\mathrm{MO}[2^k]\wedge\mathrm{MO}[2^k]$. Therefore $\Delta^*(d_{2^k-1}(u_k)\wedge u_k) = \Delta^*(u_k\wedge d_{2^k-1}(u_k))$. It follows that $d_{2^k-1}(u_{k+1})$ is a multiple of 2 and must consequently be zero since $\mathrm{E}_{2^k-1}(\mathrm{MO}[2^{k+1}])$ is a $\mathbb{Z}/2$-module for external degrees greater than zero [1]. It follows that u_{k+1} survives to $\mathrm{E}_{2^k}(\mathrm{MO}[2^{k+1}])$.
We now need to show that $d_{2^k+r}(u_{k+1})=0$ by induction on r, for $0\le r<2^k-1$. Here we use a degree argument that relies on the structure of the Bockstein spectral

[1] Proposition 6.3 below shows that $d_{2^k-1}(u_k)$ is in fact non-trivial for $k\le n$

sequence as described in Section 5 of [**KW14**] and we import our notation from there, in particular the 'hatted' classes below. We write the Bockstein spectral sequence for BO as a tensor product of a ring of permanent cycles and the Bockstein spectral sequence for a point, as in [**KW14**]:

$$E_r(\mathrm{BO}) = \hat{\mathrm{MU}}[\hat{v}_n^{-1}]^*(\mathrm{BO}) \otimes_{\hat{\mathrm{MU}}[\hat{v}_n^{-1}]^*} E_r(\mathrm{pt}).$$

By construction, all classes in $\hat{\mathrm{MU}}[v_n^{-1}]^*(\mathrm{BO})$ have internal degree divisible by 2^{n+2}. On the other hand, the differentials longer than d_{2^k-1} have external degree larger than $2^k - 1$, and hence represent elements divisible by x^{2^k-1}.

Using the structure of the spectral sequence for a point, the domain of these differentials is generated by the classes $y, \hat{v}_{i,l} := \hat{v}_i v_n^{l2^{i+1}}$ and $v_n^{\pm 2^k}$ for $i > k-1$, where y is the permanent cycle representing the nilpotent class x. All of these classes have internal degree divisible by 2^{k+1}. Therefore, for dimensional reasons, there can be no differentials in this spectral sequence that land in internal degree between 2^k and 2^{k+1}, until we reach the differential $d_{2^{k+1}-1}$.

Continuing in this way, we notice that u_{n+1} survives until the last stage: $\mathrm{E}_{2^{n+1}-1}(\mathrm{MO}[2^{n+1}])$. Now consider $d_{2^{n+1}-1}(u_{n+1})$ in degree $1+(2^{n+1}-1)\lambda$. The image of this differential lands inside a subquotient of the group $\mathrm{MU}[v_n^{-1}]^*(\mathrm{MO}[2^{n+1}])$, which we henceforth identify (using the Thom isomorphism) with the group $u_{n+1}\,\mathrm{MU}[v_n^{-1}]^*(\mathrm{BO})$.

Furthermore, classes that have survived past E_{2^n-1} must belong to the $\mathbb{Z}/2[\hat{v}_1, \dots, \hat{v}_{n-1}, v_n^{\pm 2^n}]$-submodule of $u_{n+1}\,\mathrm{MU}[v_n^{-1}]^*(\mathrm{BO})$ generated by $u_{n+1}\hat{\mathrm{MU}}[v_n^{-1}](\mathrm{BO})$, modulo previous differentials.

This allows us to express $d_{2^{n+1}-1}(u_{n+1})$ as:

$$d_{2^{n+1}-1}(u_{n+1}) = v_n^{2^n}\, u_{n+1}\, w, \qquad \text{where} \qquad w \in \frac{\mathrm{MU}[v_n^{-1}]^*(\mathrm{BO})}{\langle \hat{v}_0, \dots, \hat{v}_{n-1}\rangle},$$

for some permanent cycle w. Furthermore, we know that $d^2_{2^{n+1}-1}(u_{n+1}) = 0$. Applying this to the above expression and using the derivation property, we see that:

$$v_n^{2^{n+1}} w^2 = v_n^{2^{n+1}-2^{2n}} \hat{v}_n w.$$

Replacing w with $v_n^{-2^{2n}}\hat{w}$ for some (unique) element $\hat{w} \in \mathrm{MU}[v_n^{-1}]^{(1-2^n)(1-\lambda)}(\mathrm{BO})$, we obtain the relation $\hat{w}^2 = \hat{v}_n\hat{w}$. This implies that $\hat{w}(1 - \hat{v}_n^{-1}\hat{w}) = 0$. Since $(1 - \hat{v}_n^{-1}\hat{w})$ is a unit, we see that $\hat{w} = 0$. In other words, u_{n+1} survives the differential $d_{2^{n+1}-1}$. The proof of Theorem 6.1 is complete on observing that the spectral sequence collapses at $\mathrm{E}_{2^{n+1}}$ (by [**KW14**, Theorem 2.1(iv)]. □

The following proposition shows that we cannot expect to do better:

PROPOSITION 6.3. *There exists a vector bundle ζ such that $2^n\zeta$ is not $ER(n)$-orientable. In particular, the differential d_{2^k-1} in the Bockstein spectral sequence converging to $\mathrm{MR}(n)^*(\mathrm{MO}[2^k])$ is nontrivial on the generator u_k for $k \le n$.*

PROOF. Note that if d_{2^k-1} was trivial on u_k for some $k \le n$, the proof of the above theorem would show that $\mathrm{MO}[2^m]$ was $\mathrm{MR}(n)$-orientable for some $m \le n$. Let us demonstrate a contradiction under that hypothesis by showing the existence of a vector bundle ζ such that $2^n\zeta$ is not $ER(n)$-orientable. Let S^α denote the one point compactification of the sign representation of $\mathbb{Z}/2$. Consider the virtual vector bundle ζ over $B\mathbb{Z}/2$ with Thom space given by:

$$Th(\zeta) = E\mathbb{Z}/2_+ \wedge_{\mathbb{Z}/2} S^{(\alpha-1)}.$$

If $2^n\zeta$ were to admit an MR(n)-orientation, then one would obtain a map representing the Thom class μ:

$$\mu : E\mathbb{Z}/2_+ \wedge_{\mathbb{Z}/2} S^{2^n(\alpha-1)} \longrightarrow MR(n).$$

Postcomposing with the inclusion of fixed points map $MR(n) \longrightarrow \mathbb{MU}[v_n^{-1}]$ (where we view $MR(n)$ as a C_2-spectrum with trivial action) and precomposing with the map to the orbits, we have the composite

$$E\mathbb{Z}/2_+ \wedge S^{2^n(\alpha-1)} \longrightarrow E\mathbb{Z}/2_+ \wedge_{\mathbb{Z}/2} S^{2^n(\alpha-1)} \longrightarrow MR(n) \longrightarrow \mathbb{MU}[v_n^{-1}],$$

which is a C_2-equivariant map (with trivial actions on the middle two terms). Taking adjoints, we obtain a map

$$S^{2^n(\alpha-1)} \longrightarrow F(E\mathbb{Z}/2_+, \mathbb{MU}[v_n^{-1}]) \simeq \mathbb{MU}[v_n^{-1}],$$

where we have used the fact that $\mathbb{MU}[v_n^{-1}]$ is cofree. Since it comes from the Thom class, this map must represent a unit in $\pi^{\mathbb{Z}/2}_{2^n(\alpha-1)}\mathbb{MU}[v_n^{-1}]$. However, the computation of the bigraded homotopy of $\mathbb{MU}[v_n^{-1}]$ given in [**HK01**] shows that there is no such class. Hence we obtain a contradiction to the existence of μ. □

REMARK 6.3. As mentioned earlier, Theorem 6.1 descends to an $ER(n)$-orientation for MO$[2^{n+1}]$. However, this is by no means optimal. For example, for $n = 1$ we know that $ER(1)$ is 2-localized real K-theory. Hence a bundle ξ is $ER(1)$-orientable if and only if it is *Spin*. This is equivalent to $w_1(\xi) = w_2(\xi) = 0$. This holds for bundles of the form $\xi = 4\zeta$, but clearly there are Spin bundles that are not divisible by 4. Similarly, for $ER(2)$, the results of [**KS04**] suggest that a bundle ξ is $ER(2)$-orientable if and only if $w_1(\xi) = w_2(\xi) = w_4(\xi) = 0$, which is clearly true for bundles of the form $\xi = 8\zeta$. It is a compelling question to find a nice answer in general for when a bundle is $ER(n)$-orientable, or even to show that an answer to this question may be given in closed form.

REMARK 6.4. The above Theorem 6.1 corrects an error given in the proof of [**KW14**, Theorem 1.4]. The induction process for the construction of u_{n+1} in [**KW14**] began with a class in the Bockstein spectral sequence converging to $ER(n)^*(\mathrm{MO}[2])$. Unfortunately, that class turns out to not be conjugation invariant as required. Our current argument starts with a manifestly invariant class in the Bockstein spectral sequence converging to $ER(n)^*(\mathrm{MO}[4])$ and use that to generate the other permanent cycles. The rest of the argument is essentially the same as in [**KW14**].

References

[AM78] S. Araki and M. Murayama, *τ-cohomology theories*, Japan. J. Math. (N.S.) **4** (1978), no. 2, 363–416. MR528864

[Ban13] R. Banerjee, *On the ER(2)-cohomology of some odd-dimensional projective spaces*, Topology Appl. **160** (2013), no. 12, 1395–1405, DOI 10.1016/j.topol.2013.05.017. MR3072701

[Fuj76] M. Fujii, *Cobordism theory with reality*, Math. J. Okayama Univ. **18** (1975/76), no. 2, 171–188. MR0420597

[HH14] M. A. Hill and M. J. Hopkins, *Equivariant multiplicative closure*, Algebraic topology: applications and new directions, Contemp. Math., vol. 620, Amer. Math. Soc., Providence, RI, 2014, pp. 183–199, DOI 10.1090/conm/620/12372. MR3290092

[HHR16] M. A. Hill, M. J. Hopkins, and D. C. Ravenel, *On the nonexistence of elements of Kervaire invariant one*, Ann. of Math. (2) **184** (2016), no. 1, 1–262, DOI 10.4007/annals.2016.184.1.1. MR3505179

[HK01] P. Hu and I. Kriz, *Real-oriented homotopy theory and an analogue of the Adams-Novikov spectral sequence*, Topology **40** (2001), no. 2, 317–399, DOI 10.1016/S0040-9383(99)00065-8. MR1808224

[HM17] M. Hill and L. Meier, *The C2–spectrum Tmf1(3) and its invertible modules*, Algebr. Geom. Topol. **17** (2017), no. 4, 1953–2011, DOI 10.2140/agt.2017.17.1953. MR3685599

[HS17] Jeremy Hahn and Xiaolin Danny Shi, *Real orientations of Morava e-theories*, arXiv:1707.03413 (2017).

[JW73] D. C. Johnson and W. S. Wilson, *Projective dimension and Brown-Peterson homology*, Topology **12** (1973), 327–353, DOI 10.1016/0040-9383(73)90027-X. MR0334257

[KLW16] Nitu Kitchloo, Vitaly Lorman, and W. Stephen Wilson, *Landweber flat real pairs and $ER(n)$-cohomology*, `https://www.sciencedirect.com/science/article/pii/S0001870816308817?via3Dihub`

[KLW17] Nitu Kitchloo, Vitaly Lorman, and W. Stephen Wilson, *The $ER(2)$-cohomology of $B\mathbb{Z}/(2^q)$ and $\mathbb{C}P^n$*, Canadian Journal of Mathematics (2017).

[KS04] I. Kriz and H. Sati, *M-theory, type IIA superstrings, and elliptic cohomology*, Adv. Theor. Math. Phys. **8** (2004), no. 2, 345–394. MR2086925

[KW07a] N. Kitchloo and W. S. Wilson, *On fibrations related to real spectra*, Proceedings of the Nishida Fest (Kinosaki 2003), Geom. Topol. Monogr., vol. 10, Geom. Topol. Publ., Coventry, 2007, pp. 237–244, DOI 10.2140/gtm.2007.10.237. MR2402788

[KW07b] Nitu Kitchloo, Vitaly Lorman, and W. Stephen Wilson, *On the Hopf ring for $ER(n)$*, Topology Appl. **154** (2007), no. 8, 1608–1640. MR2317067 (2008h:55016)

[KW08a] N. Kitchloo and W. S. Wilson, *The second real Johnson-Wilson theory and nonimmersions of RP^n*, Homology Homotopy Appl. **10** (2008), no. 3, 223–268. MR2475624

[KW08b] N. Kitchloo and W. S. Wilson, *The second real Johnson-Wilson theory and nonimmersions of RP^n. II*, Homology Homotopy Appl. **10** (2008), no. 3, 269–290. MR2475625

[KW13] N. Kitchloo and W. S. Wilson, *Unstable splittings for real spectra*, Algebr. Geom. Topol. **13** (2013), no. 2, 1053–1070, DOI 10.2140/agt.2013.13.1053. MR3044602

[KW14] Nitu Kitchloo, Vitaly Lorman, and W. Stephen Wilson, *The $ER(n)$-cohomology of $BO(q)$, and Real Johnson-Wilson orientations for vector bundles*, arXiv:1409.1281 (2014).

[Lan68] P. S. Landweber, *Conjugations on complex manifolds and equivariant homotopy of MU*, Bull. Amer. Math. Soc. **74** (1968), 271–274, DOI 10.1090/S0002-9904-1968-11917-2. MR0222890

[Lor16] V. Lorman, *The real Johnson-Wilson cohomology of $\mathbb{CP}^\infty$*, Topology Appl. **209** (2016), 367–388, DOI 10.1016/j.topol.2016.06.018. MR3523487

[May96] J. P. May, *Equivariant homotopy and cohomology theory*, CBMS Regional Conference Series in Mathematics, vol. 91, Published for the Conference Board of the Mathematical Sciences, Washington, DC; by the American Mathematical Society, Providence, RI, 1996. With contributions by M. Cole, G. Comezaña, S. Costenoble, A. D. Elmendorf, J. P. C. Greenlees, L. G. Lewis, Jr., R. J. Piacenza, G. Triantafillou, and S. Waner. MR1413302

[Sch] Stefan Schwede, *Lectures on equivariant stable homotopy theory*, Available at `http://www.math.uni-bonn.de/people/schwede/equivariant.pdf`.

Department of Mathematics, Johns Hopkins University, Baltimore, USA
Email address: `nitu@math.jhu.edu`

Department of Mathematics, University of Rochester, Rochester, USA
Email address: `vlorman@math.jhu.edu`

Department of Mathematics, Johns Hopkins University, Baltimore, USA
Email address: `wsw@math.jhu.edu`

Contemporary Mathematics
Volume **707**, 2018
http://dx.doi.org/10.1090/conm/707/14253

The Morita equivalence between parametrized spectra and module spectra

John A. Lind and Cary Malkiewich

Abstract. We give a Quillen equivalence between May and Sigurdsson's model category of parametrized spectra over BG, and Mandell, May, Schwede, and Shipley's model category of modules over the orthogonal ring spectrum $\Sigma^\infty_+ G$, for each topological group G. More generally, for a topological category $\mathcal{C}$ we introduce an "aggregate" model structure on the category of diagrams of spectra indexed by $\mathcal{C}$, and prove that it is Quillen equivalent to spectra over $B\mathcal{C}$. This strengthens earlier results that hold at the level of ∞-categories, and allows us to give a simple argument that the "dualizable" parametrized spectra are precisely the homotopy retracts of the finite cell spectra.

Contents

1. Introduction

The theory of parametrized spectra is of central importance to the geometric applications of homotopy theory, and to the construction of transfer maps and Thom spectra. The homotopy category of parametrized spectra over a fixed base space B was first defined in the work of Clapp and Puppe [**6**, **7**]. May and Sigurdsson constructed a Quillen model category with this homotopy category [**25**], and shortly thereafter Ando, Blumberg, and Gepner gave an elementary construction of its ∞-category [**1**].

The model category of May and Sigurdsson suffers a few deficiencies, most notably that it is not known to be a monoidal model category. Its compact objects

2010 *Mathematics Subject Classification.* Primary 55R70; Secondary 55P43 55P91 55U35.

also evaded a simple geometric characterization, primarily because of difficulties with fibrant replacement. However, since it is a stable model category, the work of Schwede and Shipley [**31**] shows that it is Quillen equivalent to a category of modules over some ring spectrum R. The ring spectrum R produced by their theorem may be rather unwieldy, but it has an easy characterization of its compact objects as retracts of finite cell objects, and when defining monoidal structures on the category R-modules one does not encounter the same difficulties as above. It is therefore desirable to find a simple model for R and to make this "Morita equivalence" explicit, so that we may carry the clear insights from module theory back over to the geometric setting of parametrized spectra. The present paper does exactly that, and although the theorem of Schwede and Shipley is not invoked in our proof, its perspective guides our arguments.

We work with a connected base space, which we may assume is the classifying space BG of a topological group, or more generally a group-like topological monoid. We provide an explicit, highly structured equivalence between spectra parametrized over BG and modules over the suspension spectrum $\Sigma^\infty_+ G$ of G:

THEOREM 1.1. *For each well-based group-like topological monoid G, there is a Quillen equivalence between the category $\mathcal{S}p_{BG}$ of parametrized spectra over BG and the category $\mathcal{M}\mathrm{od}_{\Sigma^\infty_+ G}$ of modules over the ring spectrum $\Sigma^\infty_+ G$.*

More generally, for any topological category $\mathcal{C}$, there is a Quillen equivalence between parametrized spectra over $B\mathcal{C}$ and a certain localization of the category of $\mathcal{C}$-diagrams of spectra, where the fibrant objects are diagrams in which every morphism is a stable equivalence. When $\mathcal{C}^{\mathrm{op}}$ is the category of open sets in a sufficiently nice open cover of a space X, these diagrams are closely related to locally constant homotopy sheaves on X, see §5.

Theorem 1.1 lifts the earlier result that the underlying ∞-categories of $\Sigma^\infty_+ G$-modules and of spectra over BG are equivalent [**1**]. It allows for geometric arguments to be directly transferred from the setting of module spectra to parametrized spectra, e.g. Theorem 1.3 below.

The broader significance of Theorem 1.1 is that it leads to a zig-zag of Quillen equivalences between May and Sigurdsson's model category and a combinatorial model category (using for instance the forgetful functor to symmetric module spectra on simplicial sets, or [**12**, 2.4.2]). It has been pointed out to the authors that by the framework of Dugger [**9**], any equivalence between the underlying ∞-category of the May-Sigurdsson model category and an ∞-category arising from a combinatorial model category can consequently be lifted to a zig-zag of Quillen equivalences. Therefore, in essence, all model categories of parametrized spectra are equivalent. The model category of $\Sigma^\infty_+ BG$-comodule spectra defined by Hess and Shipley is one such example, by Theorem 1.1 and [**13**, 5.4].

Since modules over $\Sigma^\infty_+ G$ are the same thing as spectra with an action of G, Theorem 1.1 is an instantiation of an old meta-theorem: objects with an action of G are equivalent to families of such objects over BG. This meta-theorem has appeared in various forms, e.g. [**1, 10, 32**], but really goes all the way back to the classical theory of fiber bundles. From a different point of view, this result is another instantiation of the philosophy that parametrized objects are equivalent to objects equipped with an action of a category $\mathcal{C}$. This is at the heart of Lurie's straightening/unstraightening theorem [**20**, 2.2.1.2], and indeed the work of Ando-Blumberg-Gepner on the ∞-category of parametrized spectra [**1**] takes this as its starting point. Finally, we note that a parametrized spectrum over BG with a specified fiber M is classified by a map $BG \longrightarrow B\mathrm{hAut}(M)$ by the work of the first author [**18**]. We can think of the classifying map as encoding a homotopy coherent action of G on M. Theorem 1.1 is a more rigid version of the classification theorem that provides a strict action.

Theorem 1.1 arose as part of our recent work [**19**] relating the transfer map in topological Hochschild homology (THH) and the refined Reidemeister trace. It is essential to our applications that the Quillen equivalence of Theorem 1.1 respects the derived base change functors $f_!, f^*, f_*$ on either side of the equivalence. We use the language of indexed symmetric monoidal categories [**27**] to encode this agreement, whose precise statement takes the following form.

THEOREM 1.2. *Let $\mathcal{T}\mathrm{op}_*^{\mathrm{conn}}$ denote the cartesian monoidal category of pointed connected topological spaces. Upon passage to homotopy categories, the right Quillen adjoint of Theorem* 1.1 *gives an equivalence of $\mathcal{T}\mathrm{op}_*^{\mathrm{conn}}$-indexed symmetric monoidal categories* $\mathrm{ho}\,\mathcal{S}p_{(-)} \simeq \mathrm{ho}\,\mathcal{M}\mathrm{od}_{\Sigma^\infty_+\Omega(-)}$

Theorem 1.2 gives a succinct proof of the well-known equivalence $\mathrm{THH}(\Sigma^\infty_+\Omega X) \simeq \Sigma^\infty_+ LX$, where $LX = \mathrm{Map}(S^1, X)$ denotes the free loop space [**21**]. This is because it allows us to identify the derived tensor product

$$\Sigma^\infty_+\Omega X \wedge^{\mathbf{L}}_{\Sigma^\infty_+\Omega X\wedge\Sigma^\infty_+\Omega X^{\mathrm{op}}} \Sigma^\infty_+\Omega X$$

and the suspension spectrum of the homotopy pullback $X \times^h_{X\times X} X \simeq LX$ as instances of the same recipe, described in terms of base-change functors (see Remark 6.7).

Lurking in the background of this discussion is the analogy between the bicategory $\mathcal{B}\mathrm{imod}$ of ring spectra, bimodules, and homotopy classes of bimodule maps, and the bicategory $\mathcal{E}\mathrm{x}$ of spaces, parametrized spectra and homotopy classes of maps of parametrized spectra [**25**, §17]. Another application of our Morita equivalence is to characterize the duality theory internal to the bicategory $\mathcal{E}\mathrm{x}$, known as Costenoble-Waner duality, using the duality theory of bimodules. In particular, we answer a question posed by May and Sigurdsson [**25**, §18.2], characterizing the compact objects in the triangulated category $\mathrm{ho}\,\mathcal{S}p_B$ as homotopy retracts of finite cell spectra.

THEOREM 1.3. *A parametrized spectrum X over B is Costenoble-Waner dualizable precisely when it is a retract in the homotopy category of a finite cell spectrum over B. More generally, a 1-cell X in the bicategory $\mathcal{E}\mathrm{x}$ from A to B is dualizable over B precisely when each derived pullback i_a^*X to a point $a \in A$ is a retract of a finite cell spectrum over B.*

The equivalence of underlying ∞-categories could of course be used to get an analogous result, but the Quillen equivalence of Theorem 1.2 allows for a pleasingly direct argument. Finally, we get as a corollary a new interpretation of Waldhausen's algebraic K-theory of spaces.

COROLLARY 1.4. *Waldhausen's functor $A(B)$ is equivalent to the K-theory of the category of Costenoble-Waner dualizable spectra over B.*

Outline. Our strategy to prove Theorem 1.1 applies with G replaced by a topological category $\mathcal{C}$, and our arguments proceed with that level of generality. In §2, we discuss the philosophy of cell complexes, explain the distinction between *qf-*

and q-cells in parametrized homotopy theory, and recall the May-Sigurdsson model structure on the category of parametrized spectra. In §3, we describe the category $\mathcal{CS}p$ of $\mathcal{C}$-diagrams of spectra, equipped with the projective model structure, and construct a Quillen adjunction (β, Φ) between $\mathcal{C}$-spectra and parametrized spectra over the classifying space $B\mathcal{C}$. In §4, we construct the *aggregate* model structure on $\mathcal{CS}p$ and prove that the adjunction (β, Φ) is a Quillen equivalence between it and the stable model structure on $\mathcal{S}p_{B\mathcal{C}}$. When $\mathcal{C}$ is a groupoid up to homotopy, in particular a group or group-like monoid, the aggregate model structure coincides with the projective model structure, as we prove in §5. We then deduce Theorem 1.1 as a consequence. In §6, we set up the language of indexed monoidal categories and prove Theorem 1.2. Finally, in §7, we derive the results in the introduction regarding duality and finite cell spectra.

Conventions. By a space, we mean a compactly generated topological space, and we denote the category of spaces by $\mathcal{T}$op. In May-Sigurdsson's work on parametrized homotopy theory [**25**], they work more generally with the category $\mathcal{K}$ of k-spaces because the formation of internal-hom objects in the category of parametrized spaces constructs spaces that are not compactly generated. We do not use internal-hom objects in this paper, and we may safely work in $\mathcal{T}$op. Alternatively, this paper may be read as taking place in $\mathcal{K}$ instead.

2. Cell complexes and parametrized homotopy theory

The basic tension in parametrized homotopy theory is that homotopy groups are encoded in the fibers E_b of an object E parametrized over B, but a presentation of the homotopy theory in terms of cellular objects, either via model categories or classical CW methods, must involve total spaces. In this section, we recall the approach to parametrized stable homotopy theory developed by May-Sigurdsson [**25**] in terms of qf-cells and contrast it with the usual Quillen notion of q-cells.

The category $\mathcal{T}$op$/B$ of spaces over B is the comma category of spaces

$$(E, p) = (p \colon E \longrightarrow B)$$

equipped with a map p to B. As with any comma category, we may lift the Quillen-Serre q-model structure on $\mathcal{T}$op to $\mathcal{T}$op$/B$. The weak equivalences are the weak homotopy equivalences on total spaces. The cofibrations, which we call *q-cofibrations*, are the retracts of relative ${}^q I$-cell complexes, where

$${}^q I = \{i(p) \colon (S^{n-1}, p \circ i) \longrightarrow (D^n, p) \mid n \geq 0,\, p \colon D^n \longrightarrow B\}$$

is the set of inclusions of the boundary of the n-disk, considered as a parametrized space via an arbitrary map p to the base space B. We use the term q-cell to mean a cell attached to a space over B via pushout along a map in ${}^q I$. The fibrations, which we call q-fibrations, are maps in $\mathcal{T}$op$/B$ that have the right lifting property against the set

$${}^q J = \{j(p) \colon (D^n, p \circ j) \longrightarrow (D^n \times I, p) \mid n \geq 1,\, p \colon D^n \times I \longrightarrow B\},$$

or equivalently the maps that are Serre fibrations on total spaces. In particular, the q-fibrant spaces over B are precisely the Serre fibrations over B.

The notion of a q-cofibration is in contrast to the notion of an h-cofibration, also known as an Hurewicz cofibration, which is a map $i \colon A \longrightarrow X$ having the homotopy extension property in the category $\mathcal{T}\mathrm{op}$, meaning that every diagram of the form

$$\begin{array}{ccc} A & \longrightarrow & E^I \\ {\scriptstyle i}\big\downarrow & \nearrow & \big\downarrow{\scriptstyle \mathrm{ev}_0} \\ X & \longrightarrow & E \end{array}$$

admits a diagonal lift as indicated. There is a stronger notion of an f-cofibration, which is a map $i\colon A \longrightarrow X$ in the category $\mathcal{T}\mathrm{op}/B$ having the homotopy extension property in the category $\mathcal{T}\mathrm{op}/B$, meaning that a diagram as above of spaces over B admits a lift that is a map of spaces over B.

There is also a pointed version of a parametrized space, which we call an ex-space. The category $\mathcal{T}\mathrm{op}_B$ of ex-spaces is the comma category of parametrized spaces

$$(E, p, s) = (s\colon (B, \mathrm{id}_B) \longrightarrow (E, p))$$

over B equipped with a section $s\colon B \longrightarrow E$. We let $F_B(E, E')$ refer to the space of maps from E to E' respecting both the inclusion of B and the projection into B.

As with any comma category, we may lift the q-model structure on $\mathcal{T}\mathrm{op}/B$ to $\mathcal{T}\mathrm{op}_B$, the result of which we also call the q-model structure. The weak equivalences and fibrations are detected by the forgetful functor to $\mathcal{T}\mathrm{op}/B$. The cofibrations are the retracts of relative ${}^q\mathcal{I}_+$-cell complexes, where the set ${}^q\mathcal{I}_+$ is obtained from ${}^q\mathcal{I}$ by applying the left adjoint $(-)_+\colon \mathcal{T}\mathrm{op}/B \longrightarrow \mathcal{T}\mathrm{op}_B$ of the forgetful functor, which freely adjoins a disjoint section. We also use the term q-cell for a cell attached to an ex-space by pushout along a map in ${}^q\mathcal{I}_+$. The q-model structures on $\mathcal{T}\mathrm{op}/B$ and $\mathcal{T}\mathrm{op}_B$ are compactly generated with generating sets $({}^q\mathcal{I}, {}^q\mathcal{J})$ and $({}^q\mathcal{I}_+, {}^q\mathcal{J}_+)$, in the sense of [**25**, §4.5]. In particular, they are both cofibrantly generated model categories.

Recall that an orthogonal spectrum X is an enriched diagram of based spaces $\{X_V : V \in \mathrm{ob}\,\mathscr{J}\}$, indexed by a category $\mathscr{J}$ that is also enriched in based spaces (see [**23**, 4.4] for its definition).

DEFINITION 2.1. We define a *parametrized spectrum* X to be a diagram of ex-spaces X_V indexed by $\mathscr{J}$. This means that each morphism space $\mathscr{J}(V, W)$ is mapped continuously into the space $F_B(X_V, X_W)$ of maps of ex-spaces. For each $b \in B$, the diagram of fibers $(X_V)_b$ forms an ordinary orthogonal spectrum, which we call the *fiber spectrum* and denote by X_b.

Let $\mathcal{S}p_B$ be the category of parametrized spectra over B. We define a *level equivalence* to be a map of parametrized spectra $X \longrightarrow Y$ that gives an equivalence on each level, i.e. $X_V \longrightarrow Y_V$ is an equivalence of ex-spaces for each object V in $\mathscr{J}$.

Next we define the stable equivalences. Suppose that $R(-)$ is a levelwise q-fibrant replacement functor. This means that it associates to each parametrized spectrum X a parametrized spectrum RX whose levels $(RX)_V \longrightarrow B$ are q-fibrations, and there is a natural level equivalence $X \longrightarrow RX$. Then we say that $X \longrightarrow Y$ is a *stable equivalence* if the map of fibers $(RX)_b \longrightarrow (RY)_b$ is a π_*-isomorphism of orthogonal spectra, for every $b \in B$. We will soon see that stable equivalences may be computed in terms of more general fibrant replacement functors R (see Lemma 2.2, and compare [**25**, 12.3.4]). It is easy to check that the functor $\Sigma^\infty_+\colon \mathcal{T}\mathrm{op}/B \longrightarrow \mathcal{S}p_B$, which adds a disjoint section and then takes the fiberwise suspension spectrum, takes each weak homotopy equivalence of total spaces to a stable equivalence of parametrized spectra.

It is straightforward to build a model structure on $\mathcal{S}p_B$ around the level equivalences. To be specific, let $F_V : \mathcal{T}\mathrm{op}_B \longrightarrow \mathcal{S}p_B$ denote the shift-desuspension functor indexed by a real inner-product space V. By definition, F_V is left adjoint to the functor that evaluates a $\mathscr{J}$-diagram at the object V. Then there is a level q-model structure on $\mathcal{S}p_B$, with the weak equivalences the level equivalences. The generating cofibrations and generating acyclic cofibrations are the maps of the form $F_V i(p)_+$ for $i(p)_+ \in {}^qI_+$ and $F_V j(p)_+$ for $j(p)_+ \in {}^qJ_+$.

However, when expanding the class of weak equivalences to the stable equivalences, one encounters a technical difficulty. The maps $i(p)$ and $i(p)_+$ are not f-cofibrations, meaning maps with the homotopy extension property in the category $\mathcal{T}\mathrm{op}/B$. They are only h-cofibrations, i.e. Hurewicz cofibrations. As a consequence, applications of the gluing lemma that would allow inductive arguments over q-cell complexes fail in the fiberwise setting. In particular, this prohibits the usual method of verifying that relative cell complexes built out of the generating acyclic q-cofibrations in $\mathcal{S}p_B$ are weak equivalences. There is no known "q-"style model structure on $\mathcal{S}p_B$ whose weak equivalences are the stable equivalences.

In order to construct a stable model structure on parametrized spectra, May-Sigurdsson develop an alternative model structure on parametrized spaces by replacing the q-cells with a more restrictive notion of qf-cells [**25**, §6]. Let

$$^{qf}I = \{i(p) \in {}^qI \mid i \text{ is an } f\text{-cofibration}\}$$

be the set consisting of those generating q-cofibrations for which the inclusion $i\colon S^{n-1} \longrightarrow D^n$ is an f-cofibration. We call such a parametrized disk $p\colon D^n \longrightarrow B$ an f-disk over B, and note that D^n has a collar neighborhood of the boundary whose map into B lies entirely in the image of the boundary S^{n-1}. For example, fix $b \in B$ and write $(D^n)^b = (D^n \xrightarrow{r} * \xrightarrow{b} B)$ for the parametrized space consisting of a single copy of D^n as the fiber over b. Then the inclusion

$$i(b \circ r)\colon (S^{n-1})^b \longrightarrow (D^n)^b$$

is an element of ${}^{qf}I$.

By imposing a similar condition on ${}^qJ$ we get a subset ${}^{qf}J \subset {}^qJ$, and we write ${}^{qf}I_+$ and ${}^{qf}J_+$ for the sets of maps obtained by applying the disjoint section functor to ${}^{qf}I$ and ${}^{qf}J$.

There is a compactly generated model structure on $\mathcal{T}\mathrm{op}/B$ (respectively, $\mathcal{T}\mathrm{op}_B$), called the qf-model structure, in which

- the weak equivalences are the weak homotopy equivalences of total spaces,
- the cofibrations are the qf-cofibrations, which are the retracts of relative ${}^{qf}I$-cell complexes (respectively ${}^{qf}I_+$-cell complexes), and
- the fibrations are the qf-fibrations; these are the maps which have the right lifting property with respect to ${}^{qf}J$ (respectively ${}^{qf}J_+$).

Notice that every qf-cofibration is a q-cofibration and that every q-fibration is a qf-fibration. A quasifibration is a map $X \longrightarrow B$ for which the inclusion of each fiber into the corresponding homotopy fiber is a weak homotopy equivalence. It is a crucial fact that every qf-fibration is a quasifibration [**25**, 6.5.1], so we can still make arguments using the five-lemma as we normally would in fibration theory.

Moving to the level of spectra, we first define a new set of generating qf-cofibrations:

$$^{qf}I_{\mathcal{S}p} = \{F_V i(p)_+ \mid i(p) \in {}^{qf}I\}$$

Then there is a compactly generated model structure on $\mathcal{S}p_B$, called the stable model structure, in which the weak equivalences, cofibrations, and fibrations are:

- the stable equivalences as described above,
- the qf-cofibrations, which are retracts of relative ${}^{qf}I_{\mathcal{S}p}$-cell complexes, and
- the qf-fibrations, which have the right lifting property with respect to all qf-cofibrations that are stable equivalences.

As this model structure is cofibrantly generated, it has generating acyclic cofibrations as well, but we will not need their precise definition. We use the term qf-cell to mean any cell attached to a parametrized space, ex-space, or parametrized spectrum by pushout along a map in ${}^{qf}I$, ${}^{qf}I_+$, or ${}^{qf}I_{\mathcal{S}p}$.

We may use qf-fibrations, as well as the weaker notion of ex-quasifibration [**25**, §8.5], to detect stable equivalences of parametrized spectra.

LEMMA 2.2. [**25**, 12.3.4, 12.4.1] *A map $X \longrightarrow Y$ of parametrized spectra is a stable equivalence if and only if every map of fiber spectra $RX_b \longrightarrow RY_b$ is a π_*-isomorphism of orthogonal spectra, where R is a level-wise qf-fibrant approximation functor. The same statement holds if R is a level-wise ex-quasifibrant replacement functor* .

We remark that the fibrant spectra X in the stable model structure have the property that every level X_V is a qf-fibrant space and therefore a quasifibration. It follows that a map of fibrant spectra $X \longrightarrow Y$ is a stable equivalence precisely when each map of fibers $X_b \longrightarrow Y_b$ is an equivalence of orthogonal spectra. Since the fibers of a quasifibration over two points in the same path component of the base are weak homotopy equivalent, it suffices to check this condition only for a representative point $b \in B$ in each path component of B.

3. $\mathcal{C}$-spectra and spectra over $B\mathcal{C}$

As discussed in the introduction, Theorem 1.1 concerns spectra with an action of a topological group G, but we will begin our proof by working more generally with diagrams of spectra indexed by a topological category $\mathcal{C}$. While this scope of generality is not necessary to prove the main theorem, it allows for some interesting philosophical observations (see Remarks 5.4 and 5.5) and presents no more difficulties than in working with a group. The reader is encouraged to keep in mind throughout this section the special case where $\mathcal{C}$ has a single object and morphism space G.

Let $\mathcal{C}$ be any small topological category. We emphasize that we consider $\mathcal{C}$ to be enriched in unbased spaces, not based spaces. For technical ease, we assume that every space $\mathcal{C}(c, d)$ of morphisms is a retract of a CW complex. In particular, the inclusion of the identity map $* \longrightarrow \mathcal{C}(c, c)$ is an h-cofibration of unbased spaces. We write $\mathcal{S}p$ for the category of orthogonal spectra [**23**], equipped with the smash product $\wedge$ and internal function spectra $F(-, -)$ that make it a closed symmetric monoidal category. A $\mathcal{C}$-diagram of spectra, or more briefly a $\mathcal{C}$-spectrum, is a topological functor $X\colon \mathcal{C} \longrightarrow \mathcal{S}p$, where the topological enrichment of $\mathcal{S}p$ is given by

$$\mathcal{S}p(X,Y) = \Omega^\infty F(X,Y).$$

We can also view $\mathcal{C}$ as a category enriched in spectra by taking the suspension spectra $\Sigma^\infty \mathcal{C}(c,d)_+$ of the morphism spaces in $\mathcal{C}$. A $\mathcal{C}$-spectrum may then be equivalently described as a spectral functor from this spectral category to $\mathcal{S}p$. When $\mathcal{C}$ is a topological monoid G, a $\mathcal{C}$-spectrum is the same thing as a spectrum with an action of G, or equivalently a module spectrum over the ring spectrum $\Sigma^\infty_+ G = \Sigma^\infty G_+$.

We write $X(c)_V$ for the V-th level of the spectrum $X(c)$ associated to an object c of $\mathcal{C}$. We will take care to distinguish between *levels* and *objects*, so that a level-wise property of $\mathcal{C}$-spectra is detected by the $\mathcal{C}$-spaces $X(-)_V$ for all V, and an object-wise property is detected by the spectra $X(c)$ for all objects c of $\mathcal{C}$. Let $\mathcal{C}\mathcal{S}p = \mathrm{Fun}(\mathcal{C},\mathcal{S}p)$ denote the category of $\mathcal{C}$-diagrams of spectra and natural transformations of functors. We equip $\mathcal{C}\mathcal{S}p$ with the projective q-model structure, in which weak equivalences and fibrations are detected object-wise in the stable model structure on spectra [**23**, §9]. We recall that:

- a morphism $f\colon X \longrightarrow Y$ of $\mathcal{C}$-spectra is a weak equivalence (respectively, fibration) if for each object $c \in \mathcal{C}$, the map of spectra $f(c)\colon X(c) \longrightarrow Y(c)$ is a stable equivalence (respectively q-fibration of spectra), and
- the cofibrations are retracts of relative cell complexes built by attaching cells of the form

  $$F_V(i\colon S^{n-1}_+ \longrightarrow D^n_+) \wedge \mathcal{C}(c,-)_+$$

  where F_V denotes the shift desuspension functor from based spaces to orthogonal spectra (left adjoint to evaluation at level V).

We use the terms *q-cofibrations* and *q-fibrations* for the cofibrations and fibrations in the projective q-model structure. The definition is arranged so that evaluating at an object c of $\mathcal{C}$ and at a level V preserves q-fibrations.

We will also use the category $\mathcal{C}\mathcal{S}p_B = \mathrm{Fun}(\mathcal{C},\mathcal{S}p_B)$ of $\mathcal{C}$-diagrams of parametrized spectra over B. We equip $\mathcal{C}\mathcal{S}p_B$ with the projective stable model structure, wherein weak equivalences and fibrations are detected in the stable model structure on $\mathcal{S}p_B$ by the evaluation functors $M \longmapsto M(c)$ for each $c \in \operatorname{ob}\mathcal{C}$.

The category $\mathcal{S}p_B$ of parametrized spectra is enriched in $\mathcal{S}p$ via the function spectrum objects $F_B(M,N)$ and is tensored over $\mathcal{S}p$ via the half-parametrized smash product $M \barwedge X$ [**25**, 11.4.10], so that there is an adjunction

$$\mathcal{S}p_B(M \barwedge X, N) \cong \mathcal{S}p(X, F_B(M,N)).$$

Now suppose that X is a (non-parametrized) $\mathcal{C}$-spectrum and M is a $\mathcal{C}^{\mathrm{op}}$-diagram in $\mathcal{S}p_B$. The tensor product of functors $M \barwedge_{\mathcal{C}} X$ is defined to be the coend of the functor

$$M(-) \barwedge X(-)\colon \mathcal{C} \times \mathcal{C}^{\mathrm{op}} \longrightarrow \mathcal{S}p_B.$$

In other words, $M \barwedge_{\mathcal{C}} X$ is the coequalizer of the diagram of parametrized spectra

$$\bigvee_{c,d\in \operatorname{ob}\mathcal{C}} M(d) \barwedge \mathcal{C}(c,d)_+ \wedge X(c) \rightrightarrows \bigvee_{d\in\operatorname{ob}\mathcal{C}} M(d) \barwedge X(d)$$

defined by the covariant and contravariant functorialities of X and M.

LEMMA 3.1. *Let* $i\colon M \longrightarrow N$ *be a map of* $\mathcal{C}^{\mathrm{op}}$*-spectra over* $B\mathcal{C}$*, and let* $j\colon X \longrightarrow Y$ *be a map of* $\mathcal{C}$*-spectra. If* i *is a* qf*-cofibration and* j *is a* q*-cofibration, then the pushout product*

$$i \,\square\, j\colon (N \,\overline{\wedge}_{\mathcal{C}}\, X) \cup_{M \,\overline{\wedge}_{\mathcal{C}}\, X} (M \,\overline{\wedge}_{\mathcal{C}}\, Y) \longrightarrow N \,\overline{\wedge}_{\mathcal{C}}\, Y$$

is a qf*-cofibration of spectra over* $B\mathcal{C}$*, and if either* i *or* j *is acyclic, then* $i \,\square\, j$ *is as well.*

PROOF. Since the coend $\overline{\wedge}_{\mathcal{C}}$ preserves colimits in each variable separately, so does the pushout-product. Thus it suffices to prove the claim for the generating cofibrations and generating acyclic cofibrations. Assume that $M \longrightarrow N$ is of the form $i \,\overline{\wedge}\, \mathcal{C}(-,d)_+$ where $(i\colon A \longrightarrow B) \in {}^{qf}I_{\mathcal{S}p}$ is a generating qf-cofibration in $\mathcal{S}p_{B\mathcal{C}}$, and $X \longrightarrow Y$ is of the form $j \wedge \mathcal{C}(c,-)_+$ where $j\colon C \longrightarrow D$ is a generating q-cofibration of orthogonal spectra. The pushout-product then becomes

$$i \,\square\, j\colon [(B \,\overline{\wedge}\, C) \cup_{A \,\overline{\wedge}\, C} (A \,\overline{\wedge}\, D) \longrightarrow A \,\overline{\wedge}\, D] \wedge \mathcal{C}(c,d)_+$$

and since the spaces $\mathcal{C}(c,d)$ are retracts of CW complexes this reduces to the same claim for the external smash product of a spectrum over $B\mathcal{C}$ with an ordinary spectrum, found in [**25**, 12.6.5]. □

Let $E\mathcal{C}$ denote the $\mathcal{C}^{\mathrm{op}}$-diagram of spaces whose value on an object c of $\mathcal{C}$ is the two-sided bar construction

$$E\mathcal{C}^c = B(*,\mathcal{C},\mathcal{C}(c,-)) = \left|[q] \longmapsto \coprod_{c_0,\dots,c_q \in \mathrm{ob}\,\mathcal{C}} \mathcal{C}(c_{q-1},c_q) \times \cdots \times \mathcal{C}(c_0,c_1) \times \mathcal{C}(c,c_0)\right|.$$

Each of the spaces $E\mathcal{C}^c$ is contractible, by the usual contracting homotopy argument. The collapse map $\mathcal{C}(c,-) \longrightarrow *$ induces a map $\pi\colon E\mathcal{C}^c \longrightarrow B\mathcal{C}$, so we may consider $E\mathcal{C}$ as a $\mathcal{C}^{\mathrm{op}}$-diagram in spaces over $B\mathcal{C}$. Of course, in the special case of a topological group G, the map π is the usual projection $EG \longrightarrow BG$ with a fiberwise G-action.

LEMMA 3.2. $E\mathcal{C}$ *is a* q*-cofibrant diagram in the projective model structure on* $\mathcal{C}^{\mathrm{op}}\mathcal{T}\mathrm{op}/B\mathcal{C}$*.*

PROOF. We recall that the projective model structure for diagrams of unbased spaces $\mathcal{C}^{\mathrm{op}}\mathcal{T}\mathrm{op}$ has a similar definition to the one for spectra given above, except that all the products are Cartesian products of unbased spaces instead of smash products of based spaces. It suffices to show that the latching maps for the above simplicial space are q-cofibrations when regarded as maps in $\mathcal{C}^{\mathrm{op}}\mathcal{T}\mathrm{op}$. The action of $\mathcal{C}$ preserves the coproduct over the tuples of objects $c_0,\dots,c_q \in \mathrm{ob}\,\mathcal{C}$, so we may focus on one such tuple. Arguing as in [**22**, 4.8], the latching map is an iterated pushout-product of the maps $\emptyset \longrightarrow \mathcal{C}(c_{i-1},c_i)$ and $* \longrightarrow \mathcal{C}(c_i,c_i)$ (which are q-cofibrations by our assumption on $\mathcal{C}$), times the free $\mathcal{C}^{\mathrm{op}}$-diagram $\mathcal{C}(-,c_0)$. This gives a product of a q-cofibration of unbased spaces and a free diagram, which is a q-cofibration of diagrams of unbased spaces. □

We warn the reader that in general $E\mathcal{C}$ is *not* a qf-cell complex, and may not be qf-cofibrant. This can be checked by hand for the double cover $E\mathbf{Z}/2 \longrightarrow B\mathbf{Z}/2$. We also warn that the projection π is often not a fibration, or even a quasifibration, when $\mathcal{C}$ is not a topological group.

By adding a disjoint section and taking the fiberwise suspension spectrum, we obtain a cofibrant $\mathcal{C}^{\mathrm{op}}$-diagram of parametrized spectra $\Sigma^\infty_{B\mathcal{C}} E\mathcal{C}_+$, which we allow ourselves to abbreviate to $E\mathcal{C}_+$. We write $E'\mathcal{C}_+ = \Sigma^\infty_{BC} B(*, \mathcal{C}, \mathcal{C})_+^{qf-\mathrm{cof}}$ for a qf-cofibrant approximation of it in the projective stable model structure on $\mathcal{C}^{\mathrm{op}}\mathcal{S}p_{BC}$. The cofibrant approximation may be calculated by first taking a qf-cofibrant approximation of $E\mathcal{C}$ in $\mathcal{C}^{\mathrm{op}}\mathcal{T}\mathrm{op}/B\mathcal{C}$, then applying the disjoint section and fiberwise suspension spectrum functor.

Let $\beta\colon \mathcal{C}\mathcal{S}p \longrightarrow \mathcal{S}p_{B\mathcal{C}}$ denote the functor which takes a $\mathcal{C}$-diagram X to the parametrized spectrum

$$\beta(X) = E'\mathcal{C}_+ \overline{\wedge}_{\mathcal{C}} X,$$

and similarly on morphisms. Note that we have implicitly taken the fiberwise suspension spectrum of $E'\mathcal{C}_+$. Alternatively, we use the space-level variant of $\overline{\wedge}$ to take the coend of the ex-space $E'\mathcal{C}_+$ and each spectrum level of X, and this gives the same result.

We write $E'\mathcal{C}^c_+$ for the parametrized spectrum obtained by evaluating the $\mathcal{C}^{\mathrm{op}}$-diagram $E'\mathcal{C}_+$ at c. The functor β has a right adjoint Φ which takes a parametrized spectrum M over $B\mathcal{C}$ to the $\mathcal{C}$-spectrum

$$c \longmapsto F_{B\mathcal{C}}(E'\mathcal{C}^c_+, M).$$

The notation for the adjunction (β, Φ) is meant to suggest *Borel* construction and homotopy *fiber*. The following is a consequence of Lemma 3.1.

THEOREM 3.3. *The adjunction $(\beta, \Phi)\colon \mathcal{C}\mathcal{S}p \rightleftarrows \mathcal{S}p_{B\mathcal{C}}$ is a Quillen adjunction with respect to the projective q-model structure on $\mathcal{C}\mathcal{S}p$ and the stable model structure on $\mathcal{S}p_{B\mathcal{C}}$.*

In §5, we will prove that (β, Φ) is a Quillen equivalence when $\mathcal{C}$ is a groupoid up to homotopy. In preparation for our later work, we now give a precise meaning to the idea that Φ plays the role of the homotopy fiber.

The identity morphism of an object c of $\mathcal{C}$ determines a basepoint $* \in E\mathcal{C}^c$, which is non-degenerate by our hypotheses on $\mathcal{C}$. Choose a lift of the basepoint along the qf-cofibrant approximation map $\alpha\colon E'\mathcal{C}^c \longrightarrow E\mathcal{C}^c$. Given a parametrized spectrum M over $B\mathcal{C}$, restriction to this basepoint defines a map

$$\rho\colon F_{B\mathcal{C}}(E'\mathcal{C}^c_+, M) \longrightarrow M_{[c]}$$

to the fiber of M over the point $[c] \in B\mathcal{C}$ determined by the object c.

LEMMA 3.4. *If M is a qf-fibrant spectrum over $B\mathcal{C}$, then the restriction map ρ is a stable equivalence of spectra.*

In fact, we will prove that it is a level equivalence of spectra. Evaluating at an indexing space V, the domain is given by the space $\mathrm{Map}_{B\mathcal{C}}(E'\mathcal{C}, M(V))$ of maps of spaces over $B\mathcal{C}$. Thus it suffices to prove the corresponding result for parametrized spaces, which is the second part of the next lemma.

LEMMA 3.5.

(i) *If $Y \longrightarrow B\mathcal{C}$ is an h-fibration, then the map*

$$\tilde{\rho}\colon \mathrm{Map}_{B\mathcal{C}}(E\mathcal{C}^c, Y) \longrightarrow Y_{[c]}$$

induced by restriction to the basepoint of $E\mathcal{C}^c$ is a weak homotopy equivalence of spaces.

(ii) *If $Y \longrightarrow BC$ is a qf-fibration, then the restriction map*

$$\rho\colon \mathrm{Map}_{BC}(E'C^c, Y) \longrightarrow Y_{[c]}$$

is a weak homotopy equivalence of spaces.

PROOF. For part (i) we follow Shulman's proof of [**32**, Lem. 8.3]. Here we use the fact that the inclusion of the basepoint $* \longrightarrow EC^c$ is an h-cofibration. Since the codomain is contractible, the inclusion is also a homotopy equivalence, and thus, together with the h-fibration $Y \longrightarrow BC$, it induces an acyclic h-fibration

$$\mathrm{Map}(EC^c, Y) \longrightarrow \mathrm{Map}(*, Y) \times_{\mathrm{Map}(*,BC)} \mathrm{Map}(EC^c, BC) \cong Y \times_{BC} \mathrm{Map}(EC^c, BC)$$

on mapping spaces. Pulling back along the inclusion $y \mapsto (y, \pi)$ of the fiber $Y_{[c]}$ into $Y \times_{BC} \mathrm{Map}(EC^c, BC)$ gives the restriction map $\tilde{\rho}$, which is therefore an acyclic h-fibration.

We prove part (ii) by reducing to part (i). Let $Y \longrightarrow Y'$ be an approximation of Y by an h-fibration over BC, say by the usual mapping space construction. The approximation map induces the upwards pointing maps in the next commutative diagram

$$\begin{array}{ccc}
\mathrm{Map}_{BC}(EC^c, Y') & & \\
\downarrow \simeq & \overset{\tilde{\rho}}{\searrow} & \\
\mathrm{Map}_{BC}(E'C^c, Y') & \xrightarrow[\rho]{} & Y'_{[c]} \\
\uparrow \simeq & & \uparrow \simeq \\
\mathrm{Map}_{BC}(E'C^c, Y) & \xrightarrow[\rho]{} & Y_{[c]},
\end{array}$$

and they are both weak homotopy equivalences since $F_{BC}(E'C^c, -)$ preserves weak equivalences between qf-fibrant parametrized spaces. The downward pointing arrow is induced by the qf-cofibrant approximation map $E'C^c \longrightarrow EC^c$, and it is a weak homotopy equivalence because Y', being an h-fibration over BC, is also a q-fibration, and EC^c and $E'C^c$ are both q-cofibrant. The diagonal restriction map $\tilde{\rho}$ is a weak homotopy equivalence by part (i), and this concludes the proof. □

4. The aggregate model structure

In this section we introduce the aggregate model structure on the category of $\mathcal{C}$-spectra and prove that it is Quillen equivalent to May-Sigurdsson's stable model structure on parametrized spectra over BC. In order to motivate our definition of aggregate equivalences, we begin with a lemma. Let X be any $\mathcal{C}$-spectrum, not necessarily cofibrant.

LEMMA 4.1. *The parametrized spectrum $EC_+ \overline{\wedge}_{\mathcal{C}} X$ is at level V homeomorphic to the Bousfield-Kan model for the unbased homotopy colimit* $\mathrm{hocolim}_{\mathcal{C}} X(-)_V$. *The inclusion and projection onto BC are identified with the inclusion and projection onto the diagram consisting of just the basepoints of the spaces $X(c)_V$.*

PROOF. This is expected in the case of a group G, since $EG_+ \overline{\wedge}_G X$ is on each fiber over BG homeomorphic to $G_+ \wedge_G X_b$, which is homeomorphic to X_b. For the general case, we first check using the defining adjunction for the half-parametrized smash product of spaces

$$F_B(M \overline{\wedge} X, N) \cong F(X, F_B(M, N))$$

that for a space E over B and a based space X, we have a canonical homeomorphism

$$E_+ \barwedge X \cong (E \times X) \cup_{(E \times *)} (B \times *).$$

Applying this to the coequalizer system for $E\mathcal{C}_+ \barwedge_{\mathcal{C}} X_V$ gives

$$\begin{array}{c} \bigvee_{c,d \in \mathrm{ob}\,\mathcal{C}} (E\mathcal{C}(d) \times \mathcal{C}(c,d) \times X(c)_V) \cup_{E\mathcal{C}(d)\times\mathcal{C}(c,d)\times *} (B\mathcal{C} \times *) \\ \downdownarrows \\ \bigvee_{d \in \mathrm{ob}\,\mathcal{C}} (E\mathcal{C}(d) \times X(d)_V) \cup_{E\mathcal{C}(d)\times *} (B\mathcal{C} \times *). \end{array}$$

As the big wedge signifies coproduct over $B\mathcal{C}$, this rearranges to

$$\begin{array}{c} \left(\coprod_{c,d \in \mathrm{ob}\,\mathcal{C}} E\mathcal{C}(d) \times \mathcal{C}(c,d) \times X(c)_V \right) \cup_{\coprod_{c,d} E\mathcal{C}(d)\times\mathcal{C}(c,d)} B\mathcal{C} \\ \downdownarrows \\ \left(\coprod_{d \in \mathrm{ob}\,\mathcal{C}} E\mathcal{C}(d) \times X(d)_V \right) \cup_{\coprod_d E\mathcal{C}(d)} B\mathcal{C}. \end{array}$$

Since coequalizers commute with pushouts, this rearranges to the pushout

$$E\mathcal{C} \times_{\mathcal{C}} X(-)_V \cup_{E\mathcal{C}\times_{\mathcal{C}} *} B\mathcal{C}$$

which simplifies to the coend of unbased spaces

$$E\mathcal{C} \times_{\mathcal{C}} X(-)_V \cong \left| [q] \longmapsto \coprod_{c_0,\ldots,c_q \in \mathrm{ob}\,\mathcal{C}} \mathcal{C}(c_{q-1}, c_q) \times \cdots \times \mathcal{C}(c_0, c_1) \times X(c_0)_V \right|.$$

This is the Bousfield-Kan model for the unbased homotopy colimit of $X(-)_V$. □

We define a new notion of equivalence of $\mathcal{C}$-spectra by "aggregating" each diagram together into a single parametrized spectrum, then measuring its homotopy type.

Definition 4.2. A map $f\colon X \longrightarrow Y$ of $\mathcal{C}$-spectra is an *aggregate equivalence* if the induced map

$$E\mathcal{C}_+ \barwedge_{\mathcal{C}} X \longrightarrow E\mathcal{C}_+ \barwedge_{\mathcal{C}} Y$$

is a stable equivalence of parametrized spectra over $B\mathcal{C}$.

In other words, the aggregate equivalences are detected by taking (unbased) homotopy colimits over $\mathcal{C}$. We remark that X and Y are not assumed to be cofibrant diagrams in this definition. This will not get us into trouble because this particular model of the homotopy colimit is known to preserve all equivalences of diagrams of spaces. In particular, every level equivalence of $\mathcal{C}$-spectra is an aggregate equivalence.

In order to verify that object-wise stable equivalences of $\mathcal{C}$-spectra are also aggregate equivalences, we may without loss of generality make X and Y cofibrant diagrams of spectra. The following lemma, combined with Lemma 3.1 finishes the verification.

LEMMA 4.3. *If X is a q-cofibrant $\mathcal{C}$-spectrum, then the map induced by the cofibrant approximation of $E\mathcal{C}$*

$$\alpha \overline{\wedge} \mathrm{id}_X : E'\mathcal{C}_+ \overline{\wedge}_{\mathcal{C}} X \longrightarrow E\mathcal{C}_+ \overline{\wedge}_{\mathcal{C}} X$$

is a level-wise equivalence of parametrized spectra over $B\mathcal{C}$. The same statement holds when X is a q-cofibrant based $\mathcal{C}$-space.

PROOF. The first claim is an immediate consequence of the second. For the second, we argue just as in the proof of Lemma 3.1 that the coend

$$- \overline{\wedge}_{\mathcal{C}} - : \mathcal{C}^{\mathrm{op}}\mathcal{T}\mathrm{op}_{B\mathcal{C}} \times \mathcal{C}\mathcal{T}\mathrm{op}_* \longrightarrow \mathcal{T}\mathrm{op}_{B\mathcal{C}}$$

is a Quillen bifunctor for the projective q-model structures on $\mathcal{C}^{\mathrm{op}}\mathcal{T}\mathrm{op}_{B\mathcal{C}}$ and $\mathcal{C}\mathcal{T}\mathrm{op}_*$, and the q-model structure on $\mathcal{T}\mathrm{op}_{B\mathcal{C}}$. Since $E\mathcal{C}$ and $E'\mathcal{C}$ are both projective q-cofibrant, and X is assumed to be projective q-cofibrant, the weak homotopy equivalence α is preserved by by taking the coend with X. □

Now that we know that the aggregate equivalences contain the object-wise stable equivalences, we may construct the left Bousfield localization of the projective q-model structure on $\mathcal{C}$-spectra along the aggregate equivalences. Since the aggregate equivalences are detected by taking homotopy colimits, the resulting model structure is a stable analog of the model structure on certain categories of diagram spaces [**17**, **30**]. When the category $\mathcal{C}$ is discrete, the aggregate model structure was considered by Cisinski [**5**, §2]. The existence of the Bousfield localization exists by Hirschhorn's general criteria [**15**, 4.1.1], and can be constructed by hand using the method axiomatized in [**25**, §4.5].

THEOREM 4.4. *There is a compactly generated model category structure on $\mathcal{C}\mathcal{S}p$, which we call the* aggregate model structure, *whose weak equivalences are the aggregate equivalences and whose cofibrations coincide with the q-cofibrations in the projective model structure. The fibrations are the object-wise q-fibrations $X \longrightarrow Y$ for which every diagram*

$$\begin{array}{ccc} X(c) & \longrightarrow & X(d) \\ \downarrow & & \downarrow \\ Y(c) & \longrightarrow & Y(d) \end{array}$$

induced by a morphism $c \longrightarrow d$ in $\mathcal{C}$ has the property that the map from the initial vertex to the pullback is a weak homotopy equivalence. In particular, the fibrant diagrams are almost constant, *meaning that every map $X(c) \longrightarrow X(d)$ induced by a morphism in $\mathcal{C}$ is a stable equivalence of spectra.*

By Lemma 4.3, the Borel construction $\beta\colon \mathcal{C}\mathcal{S}p \longrightarrow \mathcal{S}p_{B\mathcal{C}}$ takes aggregate equivalences between cofibrant diagrams to stable equivalences of parametrized spectra. Thus the adjunction (β, Φ), which we already know to be Quillen for the projective model structure, is also Quillen for the aggregate model structure. The main theorem of this section is a strengthening of this observation:

THEOREM 4.5. *The adjunction (β, Φ) is a Quillen equivalence between the aggregate model structure on $\mathcal{C}$-spectra and the stable model structure on parametrized spectra over $B\mathcal{C}$.*

In order to prove the theorem, we will use an elementary fact from category theory: if (F, G) is an adjunction of categories, the left adjoint F detects isomorphisms, and the counit $\epsilon\colon FGx \longrightarrow x$ is an isomorphism on every object x, then (F, G) is an equivalence of categories. In our case, we know that the adjunction (β, Φ) is Quillen, so we may apply the fact to the induced adjunction $(\mathbf{L}\beta, \mathbf{R}\Phi)$ of homotopy categories. Note that the left adjoint β detects weak equivalences between q-cofibrant $\mathcal{C}$-spectra by Lemma 4.3, and so the left derived functor $\mathbf{L}\beta$ detects isomorphisms. This means that the proof of Theorem 4.5 is reduced to the proof of:

Proposition 4.6. *For every qf-fibrant parametrized spectrum M over $B\mathcal{C}$, the derived counit*

$$E'\mathcal{C}_+ \barwedge_{\mathcal{C}} F_{B\mathcal{C}}(E'\mathcal{C}_+, M)^{qf-\mathrm{cof}} \longrightarrow E'\mathcal{C}_+ \barwedge_{\mathcal{C}} F_{B\mathcal{C}}(E'\mathcal{C}_+, M) \xrightarrow{\epsilon} M$$

obtained by precomposing ϵ with a cofibrant approximation of the $\mathcal{C}$-spectrum $F_{B\mathcal{C}}(E'\mathcal{C}_+, M)$ is a level equivalence.

To prove this we use the following lemma, which forms the heart of Quillen's proof of his "Theorem B" [**29**, p. 98]. For a proof that applies to topological categories, see the very general account due to Meyer [**26**, 4.4.1]

Lemma 4.7. *Let X be a based $\mathcal{C}$-space for which every map $X(c) \longrightarrow X(d)$ of spaces induced by a morphism in $\mathcal{C}$ is a weak homotopy equivalence. Then the projection map $\pi_X\colon E\mathcal{C}_+ \barwedge_{\mathcal{C}} X \longrightarrow B\mathcal{C}$ is a quasifibration whose fiber over a point $[c] \in B\mathcal{C}$ is naturally isomorphic to $X(c)$.*

Proof of Proposition 4.6. Since cofibrant approximations of diagrams of spectra are level-wise equivalences, it suffices to consider the case where M is a qf-fibrant space. Without loss of generality we may replace it by an h-fibrant space. Consider the following commuting diagram, in which our desired composite is the bottom edge.

$$\begin{array}{ccccc}
E\mathcal{C}_+ \barwedge_{\mathcal{C}} F_{B\mathcal{C}}(E\mathcal{C}_+, M)^{qf-\mathrm{cof}} & \longrightarrow & E\mathcal{C}_+ \barwedge_{\mathcal{C}} F_{B\mathcal{C}}(E\mathcal{C}_+, M) & & \\
\uparrow{\scriptstyle \alpha \barwedge \mathrm{id}} & & \uparrow{\scriptstyle \alpha \barwedge \mathrm{id}} & \searrow^{\epsilon} & \\
E'\mathcal{C}_+ \barwedge_{\mathcal{C}} F_{B\mathcal{C}}(E\mathcal{C}_+, M)^{qf-\mathrm{cof}} & \longrightarrow & E'\mathcal{C}_+ \barwedge_{\mathcal{C}} F_{B\mathcal{C}}(E\mathcal{C}_+, M) & & M \\
\downarrow{\scriptstyle \mathrm{id} \barwedge F(\alpha,\mathrm{id})} & & \downarrow{\scriptstyle \mathrm{id} \barwedge F(\alpha,\mathrm{id})} & \nearrow_{\epsilon} & \\
E'\mathcal{C}_+ \barwedge_{\mathcal{C}} F_{B\mathcal{C}}(E'\mathcal{C}_+, M)^{qf-\mathrm{cof}} & \longrightarrow & E'\mathcal{C}_+ \barwedge_{\mathcal{C}} F_{B\mathcal{C}}(E'\mathcal{C}_+, M) & &
\end{array}$$

It suffices to show that the maps along the top and left edges are equivalences of spaces. For the left edge this follows from the fact that $F(\alpha, \mathrm{id})$ is an equivalence of spaces and Lemma 4.3. For the horizontal map on the top edge this follows from Lemma 4.1. For the remaining map ϵ, by Lemma 4.7 it suffices to check that the map of fibers $\epsilon_{[c]}$ is an equivalence. We check that its composition with the canonical isomorphism of Lemma 4.7 gives the equivalence $\tilde{\rho}$ of Lemma 3.5(i):

$$F_{B\mathcal{C}}(E\mathcal{C}^c_+, M) \xrightarrow{\cong} (E\mathcal{C}_+ \barwedge_{\mathcal{C}} F_{B\mathcal{C}}(E\mathcal{C}_+, M))_{[c]} \xrightarrow{\epsilon_{[c]}} M_{[c]}$$

This finishes the proof. □

REMARK 4.8. The analog of Theorem 4.5 in the case of $\mathcal{C}$-diagrams of simplicial sets is proved by Heuts and Moerdijk [**14**]. Their route is different than ours; instead of directly proving that a localization of the projective model structure on $\mathcal{C}$-diagrams is Quillen equivalent to the Kan-Quillen model structure on the comma category $s\mathrm{Set}/N\mathcal{C}$ of simplicial sets over the nerve, they first construct a model structure on $s\mathrm{Set}/N\mathcal{C}$ related to the Joyal model structure, prove that it is Quillen equivalent to the projective model structure, then deduce the former statement as a consequence.

5. The case where $\mathcal{C}$ is a groupoid up to homotopy

In this section, we assume that the category $\pi_0\mathcal{C}$ obtained from the topological category $\mathcal{C}$ by applying π_0 to each mapping space is a groupoid. In other words, we assume that $\mathcal{C}$ is a groupoid up to homotopy. We will prove the following theorem and then derive some consequences.

THEOREM 5.1. *When $\pi_0\mathcal{C}$ is a groupoid, the aggregate model structure and the projective model structure on the category of $\mathcal{C}$-spectra coincide. Consequently, the adjunction (β, Φ) is a Quillen equivalence between the projective model structure on $\mathcal{C}$-spectra and the stable model structure on parametrized spectra over $B\mathcal{C}$.*

To prove this, we first observe that Lemma 4.7 and our assumption on $\mathcal{C}$ imply the following lemma.

LEMMA 5.2. *Suppose that $\pi_0\mathcal{C}$ is a groupoid and that X is a $\mathcal{C}$-space. Then the canonical projection* $\operatorname{hocolim}_{\mathcal{C}} X \longrightarrow B\mathcal{C}$ *is a quasifibration whose fiber over a point $[c] \in B\mathcal{C}$ is naturally isomorphic to $X(c)$.*

Then the proof of Theorem 5.1 follows from the lemma. Indeed, it suffices to prove that the aggregate equivalences and the object-wise stable equivalences coincide. By definition, a map $f\colon X \longrightarrow Y$ of $\mathcal{C}$-spectra is an aggregate equivalence precisely when the induced map $E\mathcal{C}_+ \overline{\wedge}_{\mathcal{C}} f$ is a stable equivalence of parametrized spectra over $B\mathcal{C}$. By the lemma, the domain and codomain of this map are level-wise quasifibrations over $B\mathcal{C}$. Thus, f is an aggregate equivalence if and only if each map $(E\mathcal{C}_+ \overline{\wedge}_{\mathcal{C}} f)_{[c]}$ of fiber spectra is a stable equivalence. We may identify such a map with the map of spectra $f(c)\colon X(c) \longrightarrow Y(c)$ indexed by the object c of $\mathcal{C}$, so the proof is complete.

COROLLARY 5.3. *Suppose that G is a group-like topological monoid that is a retract of a CW complex. Then (β, Φ) is a Quillen equivalence between the projective q-model structure on the category of spectra with an action of G and the stable model structure on the category of parametrized spectra over BG.*

This is just a more detailed statement of Theorem 1.1 from the introduction, and has now been proved. Although this corollary was the motivation for the present paper, the generality of Theorem 5.1 provides us with many more model categories that are also Quillen equivalent to the category of parametrized spectra. We summarize two interesting and somewhat related special cases in the remarks below.

REMARK 5.4. Let X be a CW complex and let $\mathcal{C}$ be the category of simplices over X, so that $B\mathcal{C}$ is weakly equivalent to X. The objects of $\mathcal{C}$ are contractible spaces, so a fibrant $\mathcal{C}$-diagram in the aggregate model structure may be interpreted as a homotopy functor. By homotopy left Kan extension and restriction, these correspond to the functors on the category of spaces over X that are *linear* in the sense of Goodwillie calculus. Our Quillen equivalence gives more structure to the statement by Goodwillie [**11**, §5] that linear functors on spaces over X correspond to parametrized spectra over X. However, it stops short of giving a direct Quillen equivalence between the two. It would be interesting to see if our techniques may be combined with those of [**2**, **3**] to give such a direct equivalence.

REMARK 5.5. Similar to the previous remark, let $\mathcal{C}$ be a category consisting of contractible open subsets of X, which cover X and are closed under finite intersections, so that the classifying space $B\mathcal{C}^{\mathrm{op}}$ is equivalent to X. One can include $\mathcal{C}$ into the larger category $\mathcal{D}$ of all open subsets of X. In light of remarks such as [**8**, 2.3], it is reasonable to define a *homotopy sheaf of spectra* to be any diagram equivalent to a homotopy right Kan extension of a diagram on $\mathcal{C}^{\mathrm{op}}$ to $\mathcal{D}^{\mathrm{op}}$. Under this interpretation, Theorem 5.1 says that parametrized spectra over X model locally constant homotopy sheaves of spectra over X, lifting an earlier result of Shulman for parametrized spaces [**32**]. Although we don't pursue in this paper the technical issues needed to make this correspondence precise, we consider it an interesting question whether our techniques could in fact be used to relate parametrized spectra to locally-constant sheaves living in existing model structures on presheaves of spectra on $\mathcal{D}$, such as those found in [**4**, **16**].

6. Indexed symmetric monoidal categories

In this section, we check that the equivalence between the homotopy category of G-spectra and the homotopy category of spectra parametrized over BG respects derived base-change functors on either side. We use the language of indexed symmetric monoidal categories [**27**] to describe this agreement of base-change functors.

Let S be a cartesian monoidal category. An S-indexed symmetric monoidal category is a pseudofunctor $\mathcal{M}\colon S^{\mathrm{op}} \longrightarrow \mathrm{SymmMonCat}$ to the 2-category of symmetric monoidal categories, strong symmetric monoidal functors, and monoidal transformations. In other words, for each object A of S there is a symmetric monoidal category $\mathcal{M}^A$, and for each morphism $f\colon A \longrightarrow B$ there is a strong symmetric monoidal functor $f^*\colon \mathcal{M}^B \longrightarrow \mathcal{M}^A$, along with natural monoidal isomorphisms $(g \circ f)^* \cong f^* \circ g^*$ and $(\mathrm{id}_A)^* \cong \mathrm{id}_{\mathcal{M}^A}$ satisfying associativity and unit conditions.

We are interested in the case where the indexing category S is the category $\mathcal{T}\mathrm{op}$ of topological spaces, the category $\mathcal{T}\mathrm{op}\mathcal{G}\mathrm{rp}$ of well-based group-like topological monoids, or the category $\mathcal{T}\mathrm{op}_*^{\mathrm{conn}}$ of based connected topological spaces.

EXAMPLE 6.1. The assignment $B \longmapsto \mathrm{ho}\,\mathcal{S}p_B$, taking the space B to the homotopy category of parametrized spectra over B, along with the derived base-change functors $f^*\colon \mathrm{ho}\,\mathcal{S}p_B \longrightarrow \mathrm{ho}\,\mathcal{S}p_A$, defines a $\mathcal{T}\mathrm{op}$-indexed symmetric monoidal category $\mathrm{ho}\,\mathcal{S}p_{(-)}$.

EXAMPLE 6.2. Similarly, the assignment $G \longmapsto \mathrm{ho}\,\mathcal{S}p_{BG}$ taking G to the homotopy category of parametrized spectra over BG defines a $\mathcal{T}\mathrm{op}\mathcal{G}\mathrm{rp}$-indexed symmetric monoidal category $\mathrm{ho}\,\mathcal{S}p_{B(-)}$. We assume that G is well-based so that equivalences of groups $G \longrightarrow G'$ give equivalences of spaces $BG \longrightarrow BG'$ built using the bar construction.

EXAMPLE 6.3. The assignment $G \longmapsto \mathrm{ho}\,\mathcal{M}\mathrm{od}_{\Sigma^\infty_+ G}$ taking G to the homotopy category of module spectra over the ring spectrum $\Sigma^\infty_+ G$ defines a $\mathcal{T}\mathrm{op}\mathcal{G}\mathrm{rp}$-indexed symmetric monoidal category $\mathrm{ho}\,\mathcal{M}\mathrm{od}_{\Sigma^\infty_+(-)}$. The symmetric monoidal structure is given by the smash product $M \wedge_S N$ over the sphere spectrum, with the diagonal action defined using the diagonal of G. More generally, the assignment $R \mapsto \mathrm{ho}\,\mathcal{M}\mathrm{od}_R$ defines a $\mathcal{H}$opf-indexed symmetric monoidal category, where $\mathcal{H}$opf is the category of Hopf algebras in spectra, although we will not need that level of generality here.

An equivalence $\mathcal{M} \simeq \mathcal{N}$ of S-indexed symmetric monoidal categories consists of a pseudonatural transformation $\varphi\colon \mathcal{M} \longrightarrow \mathcal{N}$ of pseudofunctors, each of whose components $\varphi^A\colon \mathcal{M}^A \longrightarrow \mathcal{N}^A$ is an equivalence of categories. More concretely, φ consist of a collection of strong symmetric monoidal functors $\varphi^A\colon \mathcal{M}^A \longrightarrow \mathcal{N}^A$, each of which is an equivalence of categories, and natural monoidal isomorphisms

$$\begin{array}{ccc} \mathcal{M}^B & \xrightarrow{\varphi^B} & \mathcal{N}^B \\ {\scriptstyle f^*}\downarrow & \Downarrow\cong & \downarrow{\scriptstyle f^*} \\ \mathcal{M}^A & \xrightarrow{\varphi^A} & \mathcal{N}^A \end{array}$$

that are suitably compatible with the coherence isomorphisms for the base-change functors in $\mathcal{M}$ and $\mathcal{N}$. With this language, we may now state precisely how our Quillen equivalences interact with change of groups G.

THEOREM 6.4. *The derived right adjoint* $\mathbf{R}\Phi\colon \mathrm{ho}\,\mathcal{S}p_{BG} \longrightarrow \mathrm{ho}\,\mathcal{M}\mathrm{od}_{\Sigma^\infty_+ G}$ *associated to the Quillen equivalence* (β, Φ) *induces an equivalence*

$$\mathrm{ho}\,\mathcal{S}p_{B(-)} \simeq \mathrm{ho}\,\mathcal{M}\mathrm{od}_{\Sigma^\infty_+(-)}$$

of $\mathcal{T}\mathrm{op}\mathcal{G}\mathrm{rp}$*-indexed symmetric monoidal categories.*

PROOF. When G is the retract of a CW complex, the derived functor $\mathbf{R}\Phi$ is an equivalence of categories by Corollary 5.3. Let $[*]$ denote the canonical basepoint of $BG = |N.G|$. Then the derived functor is computed by applying Φ to a qf-fibrant spectrum, and so by Lemma 3.4, we may lift the strong symmetric monoidal structure on the derived fiber functor $i^*_{[*]}$ [**25**, 13.7.10] to the functor $\mathbf{R}\Phi$.

In order to prove that $\mathbf{R}\Phi$ is compatible with base-change functors, suppose we are given a continuous homomorphism $f\colon H \longrightarrow G$ and consider the induced diagram of spaces

$$\begin{array}{ccc} EH & \xrightarrow{Ef} & EG \\ \downarrow & & \downarrow \\ BH & \xrightarrow{Bf} & BG. \end{array}$$

Let $Ef^{H-\mathrm{cof}}\colon EH^{H-\mathrm{cof}} \longrightarrow EG^{H-\mathrm{cof}}$ be a qf-cofibrant approximation of the homotopy equivalence Ef in the projective model structure on H-spaces over BG. Taking fiberwise suspension spectra, $Ef^{H-\mathrm{cof}}$ induces a stable equivalence of qf-cofibrant H-spectra over BG. Notice that since $F_{BG}(-,-)$ is a Quillen bifunctor, if M is a qf-fibrant spectrum over BG, then the spectrum $F_{BG}(EG_+^{H-\mathrm{cof}}, M)$ is a q-fibrant H-spectrum, and in particular agrees with the value of the derived restriction functor $f^*\colon \mathcal{M}\mathrm{od}_{\Sigma^\infty_+ G} \longrightarrow \mathcal{M}\mathrm{od}_{\Sigma^\infty_+ H}$ on $\Phi M = F_{BG}(EG_+^{G-\mathrm{cof}}, M)$.

We now pass to derived functors and abbreviate $E'H = EH^{H-\mathrm{cof}}$ and $E'G = EG^{G-\mathrm{cof}}$, as in the previous sections. The derived pull-back functor

$$(Bf)^*\colon \mathrm{ho}\,\mathcal{S}p_{BG} \longrightarrow \mathrm{ho}\,\mathcal{S}p_{BH}$$

has a left adjoint $(Bf)_!$ and the map Ef induces a stable equivalence

$$(Bf)_! E'H_+ \simeq E'G_+$$

of spectra over BG. It follows that there is a natural equivalence of derived functors

$$\begin{aligned}(\mathbf{R}\Phi)(Bf)^*(M) &= F_{BH}(E'H_+, (Bf)^*M) \\ &\simeq F_{BG}((Bf)_! E'H_+, M) \\ &\simeq F_{BG}(E'G_+, M) = f^*(\mathbf{R}\Phi)(M),\end{aligned}$$

and it is easily checked that this composite is a monoidal transformation and is compatible with the coherence isomorphisms $(g \circ f)^* \cong f^* \circ g^*$ for the base-change functors on $\mathrm{ho}\,\mathcal{S}p_{B(-)}$ and $\mathrm{ho}\,\mathcal{M}\mathrm{od}_{\Sigma^\infty_+(-)}$.

Finally, we may extend the result to arbitrary well-based group-like topological monoids G by taking an approximation $G' \xrightarrow{\simeq} G$ of G by a group-like topological monoid that is a CW complex and composing with the induced equivalences of homotopy categories $\mathrm{ho}\,\mathcal{M}\mathrm{od}_{\Sigma^\infty_+ G} \simeq \mathrm{ho}\,\mathcal{M}\mathrm{od}_{\Sigma^\infty_+ G'}$ and $\mathrm{ho}\,\mathcal{S}p_{BG} \simeq \mathrm{ho}\,\mathcal{S}p_{BG'}$. This completes the proof. □

This completes our analysis of how spectra over BG for varying G correspond to module spectra. Our final task is to relate this to the larger category of all topological spaces. One can make a module-theoretic description of the $\mathcal{T}$op-indexed symmetric monoidal category $\mathrm{ho}\,\mathcal{S}p_{(-)}$ of parametrized spectra, using modules over "ring spectra with many objects", i.e. spectrally enriched categories. However, here we will be content to restrict attention to the category $\mathcal{T}\mathrm{op}_*^{\mathrm{conn}}$ of pointed connected topological spaces.

Any well-based variant of the Moore loop space defines a functor $\Omega\colon \mathcal{T}\mathrm{op}_*^{\mathrm{conn}} \longrightarrow \mathcal{T}\mathrm{op}\mathcal{G}\mathrm{rp}$. It induces an equivalence of homotopy categories, with inverse the classifying space functor $B\colon \mathcal{T}\mathrm{op}\mathcal{G}\mathrm{rp} \longrightarrow \mathcal{T}\mathrm{op}_*^{\mathrm{conn}}$. There is a map of spaces $\xi\colon B\Omega X \longrightarrow X$ that is a weak homotopy equivalence when X is connected [**24**, 15.4]. The derived base-change functor

$$\xi^*\colon \mathrm{ho}\,\mathcal{S}p_X \longrightarrow \mathrm{ho}\,\mathcal{S}p_{B\Omega X}$$

is an equivalence of homotopy categories, and commutes with the derived base-change functor f^* of parametrized spectra induced by a map $f\colon X \longrightarrow Y$ of pointed connected spaces. This implies the following consequence of Theorem 6.4.

COROLLARY 6.5. *The composite* $\mathbf{R}\Phi \circ \xi^*$ *induces an equivalence*

$$\mathrm{ho}\,\mathcal{S}p_{(-)} \simeq \mathrm{ho}\,\mathcal{M}\mathrm{od}_{\Sigma^\infty_+ \Omega(-)}$$

of $\mathcal{T}\mathrm{op}_*^{\mathrm{conn}}$*-indexed symmetric monoidal categories.*

REMARK 6.6. Instead of working with homotopy categories, one may attempt to strengthen the theorem to the statement that the base-change functors on the actual categories agree up to coherent homotopy. The work of Ando-Blumberg-Gepner [**1**] using the language of ∞-categories is one way of making this idea precise.

REMARK 6.7. By the uniqueness of adjoints, it follows that the equivalence of Corollary 6.5 commutes with the left and right adjoint $f_!$ and f_* of f^*. Let $\Delta\colon X \longrightarrow X \times X$ be the diagonal map of a space X, and consider the fiberwise suspension spectrum $(X, \Delta)_+ = \Sigma^\infty_{X\times X}(X, \Delta)_+$ of X over $X \times X$. Writing $r\colon X \longrightarrow *$ for the projection of X to a point, one can compute that the derived base-change functors $r_! \Delta^*(X, \Delta)_+$ gives the stable homotopy type of the free loop space

$LX = \mathrm{Map}(S^1, X)$. On the other hand, $\mathbf{R}\Phi(X, \Delta)_+$ is a model for the $\Sigma^\infty\Omega(X \times X)_+$-module spectrum $\Sigma^\infty_+\Omega X$. Considering this as a bimodule, one finds that applying the derived base-change functors of module categories gives the THH of the spherical group ring:

$$\begin{aligned} r_!\Delta^*\mathbf{R}\Phi(X, \Delta)_+ &\simeq \Sigma^\infty_+\Omega X \wedge^{\mathbf{L}}_{\Sigma^\infty_+\Omega X \wedge \Sigma^\infty_+\Omega X^{\mathrm{op}}} \Sigma^\infty_+\Omega X \\ &\simeq \mathrm{THH}(\Sigma^\infty_+\Omega X; \Sigma^\infty_+\Omega X). \end{aligned}$$

Therefore, the equivalence of indexed symmetric monoidal categories proved in this paper recovers the equivalence $\mathrm{THH}(\Sigma^\infty_+\Omega X) \simeq LX_+$. In the companion paper [**19**, §5], we generalize this to an equivalence of shadow functors on bicategories. This comparison is a crucial component in our study of the transfer map of free loop spaces.

7. Costenoble-Waner dualizable spectra

In this final section we prove Theorem 1.3. As background, we recall that May and Sigurdsson proved that every retract of a finite qf-cell spectrum in $\mathrm{ho}\,\mathcal{S}p_B$ is Costenoble-Waner dualizable. They also characterized the Costenoble-Waner dualizable spectra as the compact objects of the triangulated category $\mathrm{ho}\,\mathcal{S}p_B$ [**25**, 18.2.2].

However, this left open the converse question of whether every Costenoble-Waner dualizable spectra is in fact a homotopy retract of a finite qf-cell spectrum. We answer this question in the affirmative. As a result, this excludes the possibility of exotic parametrized spectra which are dualizable but do not have finite presentations in terms of cells and attaching maps.

As discussed in [**25**, 18.2.10], the essential difficulty is that in the parametrized setting, one cannot simultaneously make a space both compact and fibrant. We will circumvent this by taking our compact objects in the category of module spectra, before applying the Borel construction to get parametrized spectra.

In fact, our methods work equally well to characterize all dualizable spectra in the bicategory $\mathcal{E}\mathrm{x}$ of spectra over varying base spaces. Suppose that a spectrum X over $A \times B$ has the property that every derived pullback $i_a^* X$ along the inclusion of a point $a \in A$ is a retract of a finite qf-cell spectrum in $\mathrm{ho}\,\mathcal{S}p_B$. The argument of May and Sigurdsson referenced above actually proves that $X \in \mathrm{ho}\,\mathcal{S}p_{A\times B} = \mathcal{E}\mathrm{x}(A, B)$ is dualizable over B as a 1-cell in the bicategory $\mathcal{E}\mathrm{x}$. See the companion paper [**19**, §4, 5] and the references [**25**, **27**] for more details on duality in the bicategories $\mathcal{E}\mathrm{x}$ and $\mathcal{B}\mathrm{imod}$. We will now prove the converse.

Suppose that $X \in \mathcal{E}\mathrm{x}(A, B)$ is dualizable over B. In other words, X is a spectrum over $A \times B$ and we are given another spectrum $Y \in \mathcal{E}\mathrm{x}(B, A) = \mathrm{ho}\,\mathcal{S}p_{B\times A}$ and morphisms

$$\mathrm{coev}\colon U_A \longrightarrow X \odot_B Y \qquad \mathrm{eval}\colon Y \odot_A X \longrightarrow U_B$$

in the homotopy categories of spectra over $A \times A$ and $B \times B$ satisfying the triangle identities for a duality. Here, the bifunctor $X \odot_B Y = \pi_!^B \Delta_B^*(X \overline{\wedge} Y)$ is the horizontal composition in the bicategory $\mathcal{E}\mathrm{x}$ [**25**, §17]. Specializing to the case $A = *$, we have the notion of a Costenoble-Waner dualizable spectrum X in $\mathrm{ho}\,\mathcal{S}p_B$.

We assume without loss of generality that A and B are connected and pointed. For the sake of clarity, let us write

$$\Phi_X\colon \mathrm{ho}\,\mathcal{S}p_X \longrightarrow \mathrm{ho}\,\mathcal{S}p_{B\Omega X} \longrightarrow \mathrm{ho}\,\mathcal{M}\mathrm{od}_{\Sigma^\infty_+\Omega X}$$

for the composite equivalence from Corollary 6.5. The morphisms $\Phi_{A\times A}(\mathrm{coev})$ and $\Phi_{B\times B}(\mathrm{eval})$ exhibit $\Phi_{A\times B}X$ and $\Phi_{B\times A}Y$ as a dual pair in the bicategory $\mathcal{B}$imod. In particular, forgetting the action of ΩA, the spectrum $\Phi_{A\times B}X$ is dualizable as a $\Sigma^\infty_+\Omega B$-module, so it is a retract of a finite cell spectrum in $\mathrm{ho}\,\mathcal{M}\mathrm{od}_{\Sigma^\infty_+\Omega B}$.

It now suffices to prove that the inverse of the equivalence Φ_B takes finite $\Sigma^\infty_+\Omega B$-cell spectra to finite qf-cell spectra over B. Without loss of generality, we may work in $\mathrm{ho}\,\mathcal{S}p_{BG}$, where G is a topological group. In other words, Theorem 1.3 follows from:

PROPOSITION 7.1. *If X is a finite $\Sigma^\infty_+ G$-cell spectrum, then βX is equivalent in* $\mathrm{ho}\,\mathcal{S}p_{BG}$ *to a finite qf-cell spectrum.*

PROOF. After sufficient desuspensions, we may assume that that X is the suspension spectrum of a finite free G-cell complex of pointed spaces, which we also denote by X. By Lemma 4.3, it suffices to prove the result for the functor $X \longmapsto EG_+ \overline{\wedge}_G X$ instead of β. Write X as the colimit of equivariant skeleta $X^{(i)}$ where $X^{(i+1)}$ is obtained from $X^{(i)}$ by attaching a single cell of the form $G \times D^n$. We inductively construct a finite qf-cell complex Y_i over BG with a weak homotopy equivalence $f_i : Y_i \longrightarrow EG_+ \overline{\wedge}_G X^{(i)}$ of ex-spaces over BG. The base case is $Y_0 = BG = EG_+ \overline{\wedge}_G X^{(0)}$. For the inductive step, we apply $EG_+ \overline{\wedge}_G(-)$ to the pushout which constructs $X^{(i+1)}$ from $X^{(i)}$, and build around it the following diagram of ex-spaces.

$$\begin{array}{ccccccc}
 & & & \overset{\alpha}{\dashrightarrow} & & & Y_i \\
 & & & & & & \downarrow{\scriptstyle f_i}\ \simeq \\
S^{n-1}_+ & \underset{(*,\mathrm{id})}{\dashrightarrow} & (EG \times S^{n-1})_+ & \xrightarrow{\cong} & EG_+ \overline{\wedge}_G (G \times S^{n-1})_+ & \longrightarrow & EG_+ \overline{\wedge}_G X^{(i)} \\
 & & \downarrow & & \downarrow & & \downarrow \\
 & & (EG \times D^n)_+ & \xrightarrow{\cong} & EG_+ \overline{\wedge}_G (G \times D^n)_+ & \longrightarrow & EG_+ \overline{\wedge}_G X^{(i+1)}
\end{array}$$

On the left, we include S^{n-1} into $EG \times S^{n-1}$ using the basepoint $* \in EG$. The weak equivalence f_i is provided by the inductive hypothesis. We observe that Y_i need not be fibrant over BG, so f_i may not have an inverse in the homotopy category of spaces over BG, only in the homotopy category of spaces. The latter maneuver gives a map $\alpha\colon S^{n-1} \longrightarrow Y_i$ as in the diagram and a homotopy between the two routes in the upper triangle. This gives a diagram

$$\begin{array}{ccccc}
D^n & \xleftarrow{i} & S^{n-1} & \xrightarrow{\alpha} & Y_i \\
{\scriptstyle (*,\mathrm{id})}\downarrow\ \simeq & & {\scriptstyle (*,\mathrm{id})}\downarrow\ \simeq & & {\scriptstyle f_i}\downarrow\ \simeq \\
EG \times D^n & \xleftarrow{\mathrm{id} \times i} & EG \times S^{n-1} & \longrightarrow & EG_+ \overline{\wedge}_G X^{(i)}
\end{array}$$

in which the left square commutes and the right square commutes up to a given homotopy. This in turn induces a weak equivalence of spaces

$$D^n \cup^h_{S^{n-1}} Y_i \longrightarrow EG_+ \overline{\wedge}_G X^{(i+1)}$$

from the double mapping cylinder of the top row to the pushout of the bottom row. We insist that the induced map uses the given homotopy at double-speed along the first half of the mapping cylinder, then is constant at α for the second half of the mapping cylinder. Let us denote this new map by f_{i+1}, and denote its source by Y_{i+1}. We define the inclusion $BG \longrightarrow Y_{i+1}$ by composing the inclusion $BG \longrightarrow Y_i$ with the inclusion of one of the ends of the mapping cylinder. We define the projection $Y_{i+1} \longrightarrow BG$ by composing f_{i+1} with the projection of $EG_+ \overline{\wedge}_G X^{(i+1)}$. Observe that our requirement on how f_{i+1} is defined in terms of the given homotopy to α makes the cell we have just attached to Y_i a qf-cell. In effect, this amounts to composing the projection map $D^n \longrightarrow BG$ of the new cell for Y_{i+1} with the expansion by two map

$$D^n \longrightarrow D^n$$
$$x \longmapsto \begin{cases} 2x & \text{if } |x| \leq \frac{1}{2} \\ x/|x| & \text{if } |x| > \frac{1}{2}, \end{cases}$$

and this makes the resulting disk an f-disk over BG. The map f_{i+1} is a weak equivalence of ex-spaces with source a finite qf-cell, completing the inductive step of the proof. □

We note that it is a consequence of our proof that an object of $\mathrm{ho}\,\mathcal{S}p_B$ is equivalent to a finite q-cell spectrum if and only if it is equivalent to a finite qf-cell spectrum. Recall that Waldhausen's functor $A(B)$ may be described as the K-theory of the Waldhausen category of retractive spaces over B that are retracts of finite q-cell complexes. Since K-theory is invariant under stabilization of Waldhausen categories, A-theory may equivalently be described as the K-theory of parametrized finite q-cell spectra over B, and thus also as the K-theory of parametrized finite qf-cell spectra over B. Theorem 1.3 then implies the interpretation of Waldhausen's A-theory stated in Corollary 1.4.

Acknowledgements. The authors thank Thomas Nikolaus for a helpful discussion of the consequences of Theorem 1.2 for ∞-categories, and of an alternate approach to Theorem 1.3. J.L. was supported in part by the DFG through SFB1085. C.M. was supported in part by an AMS Simons Travel Grant.

References

[1] M. Ando, A. J. Blumberg, and D. Gepner, *Parametrized spectra, multiplicative Thom spectra, and the twisted Umkehr map*, arXiv:1112.2203 (2011).

[2] G. Biedermann, B. Chorny, and O. Röndigs, *Calculus of functors and model categories*, Adv. Math. **214** (2007), no. 1, 92–115, DOI 10.1016/j.aim.2006.10.009. MR2348024

[3] G. Biedermann and O. Röndigs, *Calculus of functors and model categories, II*, Algebr. Geom. Topol. **14** (2014), no. 5, 2853–2913, DOI 10.2140/agt.2014.14.2853. MR3276850

[4] J. Block and A. Lazarev, *Homotopy theory and generalized duality for spectral sheaves*, Internat. Math. Res. Notices **20** (1996), 983–996, DOI 10.1155/S1073792896000608. MR1422371

[5] D.-C. Cisinski, *Locally constant functors*, Math. Proc. Cambridge Philos. Soc. **147** (2009), no. 3, 593–614, DOI 10.1017/S030500410900262X. MR2557145

[6] M. Clapp, *Duality and transfer for parametrized spectra*, Arch. Math. (Basel) **37** (1981), no. 5, 462–472, DOI 10.1007/BF01234383. MR643290

[7] M. Clapp and D. Puppe, *The homotopy category of parametrized spectra*, Manuscripta Math. **45** (1984), no. 3, 219–247, DOI 10.1007/BF01158038. MR734840

[8] P. Boavida de Brito and M. Weiss, *Manifold calculus and homotopy sheaves*, Homology Homotopy Appl. **15** (2013), no. 2, 361–383, DOI 10.4310/HHA.2013.v15.n2.a20. MR3138384

[9] D. Dugger, *Universal homotopy theories*, Adv. Math. **164** (2001), no. 1, 144–176, DOI 10.1006/aima.2001.2014. MR1870515

[10] E. Dror, W. G. Dwyer, and D. M. Kan, *Equivariant maps which are self homotopy equivalences*, Proc. Amer. Math. Soc. **80** (1980), no. 4, 670–672, DOI 10.2307/2043448. MR587952

[11] T. G. Goodwillie, *Calculus. III. Taylor series*, Geom. Topol. **7** (2003), 645–711, DOI 10.2140/gt.2003.7.645. MR2026544

[12] Y. Harpaz, J. Nuiten, and M. Prasma, *The abstract cotangent complex and Quillen cohomology of enriched categories*, arXiv:1612.02608.

[13] K. Hess and B. Shipley, *Waldhausen K-theory of spaces via comodules*, Adv. Math. **290** (2016), 1079–1137, DOI 10.1016/j.aim.2015.12.019. MR3451948

[14] G. Heuts and I. Moerdijk, *Left fibrations and homotopy colimits*, Math. Z. **279** (2015), no. 3-4, 723–744, DOI 10.1007/s00209-014-1390-7. MR3318247

[15] P. S. Hirschhorn, *Model categories and their localizations*, Mathematical Surveys and Monographs, vol. 99, American Mathematical Society, Providence, RI, 2003. MR1944041

[16] J. F. Jardine, *Stable homotopy theory of simplicial presheaves*, Canad. J. Math. **39** (1987), no. 3, 733–747, DOI 10.4153/CJM-1987-035-8. MR905753

[17] J. A. Lind, *Diagram spaces, diagram spectra and spectra of units*, Algebr. Geom. Topol. **13** (2013), no. 4, 1857–1935, DOI 10.2140/agt.2013.13.1857. MR3073903

[18] J. A. Lind, *Bundles of spectra and algebraic K-theory*, Pacific J. Math. **285** (2016), no. 2, 427–452, DOI 10.2140/pjm.2016.285.427. MR3575574

[19] J.A. Lind and C. Malkiewich, *The transfer map of free loop spaces*, arXiv:1604.03067 (2016).

[20] J. Lurie, *Higher topos theory*, Annals of Mathematics Studies, vol. 170, Princeton University Press, Princeton, NJ, 2009. MR2522659

[21] I. Madsen, *Algebraic K-theory and traces*, Current developments in mathematics, 1995 (Cambridge, MA), Int. Press, Cambridge, MA, 1994, pp. 191–321. MR1474979

[22] C. Malkiewich, *Cyclotomic structure in the topological Hochschild homology of DX*, Algebr. Geom. Topol. **17** (2017), no. 4, 2307–2356, DOI 10.2140/agt.2017.17.2307. MR3686399

[23] M. A. Mandell, J. P. May, S. Schwede, and B. Shipley, *Model categories of diagram spectra*, Proc. London Math. Soc. (3) **82** (2001), no. 2, 441–512.

[24] J. P. May, *Classifying spaces and fibrations*, Mem. Amer. Math. Soc. **1** (1975), no. 1, 155, xiii+98. MR0370579

[25] J. P. May and J. Sigurdsson, *Parametrized homotopy theory*, Mathematical Surveys and Monographs, vol. 132, American Mathematical Society, Providence, RI, 2006. MR2271789

[26] J.-P. Meyer, *Bar and cobar constructions. II*, J. Pure Appl. Algebra **43** (1986), no. 2, 179–210, DOI 10.1016/0022-4049(86)90094-0. MR866618

[27] K. Ponto and M. Shulman, *Duality and traces for indexed monoidal categories*, Theory Appl. Categ. **26** (2012), No. 23, 582–659. MR3065938

[28] K. Ponto and M. Shulman, *The multiplicativity of fixed point invariants*, Algebr. Geom. Topol. **14** (2014), no. 3, 1275–1306, DOI 10.2140/agt.2014.14.1275. MR3190594

[29] D. Quillen, *Higher algebraic K-theory. I*, Algebraic K-theory, I: Higher K-theories (Proc. Conf., Battelle Memorial Inst., Seattle, Wash., 1972), Springer, Berlin, 1973, pp. 85–147. Lecture Notes in Math., Vol. 341. MR0338129

[30] S. Sagave and C. Schlichtkrull, *Diagram spaces and symmetric spectra*, Adv. Math. **231** (2012), no. 3-4, 2116–2193, DOI 10.1016/j.aim.2012.07.013. MR2964635

[31] S. Schwede and B. Shipley, *Stable model categories are categories of modules*, Topology **42** (2003), no. 1, 103–153, DOI 10.1016/S0040-9383(02)00006-X. MR1928647

[32] M. A. Shulman, *Parametrized spaces model locally constant homotopy sheaves*, Topology Appl. **155** (2008), no. 5, 412–432, DOI 10.1016/j.topol.2007.11.001. MR2380927

Reed College, Portland, Oregon 97202
Email address: john.alexander.lind@gmail.com

Binghamton University, Binghamton, New York 13902
Email address: malkiewich@math.binghamton.edu

Contemporary Mathematics
Volume **707**, 2018
http://dx.doi.org/10.1090/conm/707/14254

tmf is not a ring spectrum quotient of string bordism

Carl McTague

ABSTRACT. This paper shows that $\mathrm{tmf}[\frac{1}{6}]$ is not a ring spectrum quotient of $\mathrm{MO}\langle 8\rangle[\frac{1}{6}]$. In fact, for any prime $p > 3$ and any sequence X of homogeneous elements of $\pi_*\mathrm{MO}\langle 8\rangle$, the $\pi_*\mathrm{MO}\langle 8\rangle$-module

$$\pi_*\left(\mathrm{MO}\langle 8\rangle_{(p)}/X\right)$$

is not (even abstractly) isomorphic to $\pi_*\mathrm{tmf}_{(p)}$.

It does so by showing that, for any commutative ring spectrum R and any sequence X of homogeneous elements of $\pi_*(R)$, there is an isomorphism of graded $\mathbf{Q}$-vector spaces

$$\pi_*(R/X)\otimes\mathbf{Q}\cong \mathrm{H}_*(\mathrm{Tot}(\mathrm{K}(X)))\otimes\mathbf{Q},$$

where the right-hand side is the rational homology of the (total) Koszul complex of X, which is strictly bigger than $\pi_*(R)/(X)\otimes\mathbf{Q}$ unless X is a $\pi_*(R)\otimes\mathbf{Q}$-quasi-regular sequence. The result then follows from the fact that the kernel of the p-local Witten genus cannot be generated by a $\pi_*\mathrm{MO}\langle 8\rangle\otimes\mathbf{Q}$-quasi-regular sequence.

1. Introduction

Given a commutative ring spectrum R, an R-module M, and a *sequence* $X = x_1, x_2, \ldots$ of homogeneous elements of $\pi_*(R)$, one can construct the quotient R-module M/XM, by iteratively forming cofibers

(1)

$$M/(x_1,\ldots,x_{n+1})M = \text{cofiber}\left[\Sigma^{\deg x_{n+1}} M/(x_1,\ldots,x_n)M \xrightarrow{x_{n+1}} M/(x_1,\ldots,x_n)M\right]$$

and, if the sequence is infinite, passing to their telescope

$$M/(x_1,x_2,x_3,\ldots)M = \mathrm{hocolim}_n\left[M/(x_1,\ldots,x_n)M\right].$$

When $M = R$, the resulting R-module, denoted R/X, often comes equipped with a homotopy commutative R-algebra structure. For example, if R is a commutative S-algebra with $\pi_i(R) = 0$ for i odd, and if X is a sequence of non zerodivisors such that $\pi_*(R/X)$ is concentrated in degrees congruent to 0 modulo 4, then R/X, regarded as an object in the *homotopy category* of R-modules, admits a unique (up to canonical isomorphism) ring object structure, and it is commutative and associative [**7**, Thm. V.3.2]. To emphasize: although R is required to have a commutative ring spectrum structure, the quotient R/X is only guaranteed to admit a homotopy commutative one.

2010 *Mathematics Subject Classification.* Primary 55N34, 55P43; Secondary 58J26.

If X is regular in the sense that each element x_{n+1} of X is not a zerodivisor for $\pi_*(R)/(x_1, \dots, x_n)$, then the long exact homotopy sequence for each cofiber breaks into short exact sequences, and

$$\pi_*(R/X) \cong \pi_*(R)/(X),$$

where (X) is the *ideal* generated by the elements of X.

This, together with a theory of localization, makes possible a one-page construction (from MU as a commutative ring spectrum, and for $p > 2$)[1] of a rogue's gallery of spectra (listed with coefficient ring),

BP	$\mathbf{Z}_{(p)}[v_1, v_2, \dots]$	$\mathrm{BP}\langle n\rangle$	$\mathbf{Z}_{(p)}[v_1, \dots, v_n]$
$\mathrm{E}(n)$	$\mathbf{Z}_{(p)}[v_1, \dots, v_n, v_n^{-1}]$	$\mathrm{P}(n)$	$\mathbf{F}_p[v_n, v_{n+1}, \dots]$
$\mathrm{B}(n)$	$\mathbf{F}_p[v_n^{-1}, v_n, v_{n+1}, \dots]$	$\mathrm{k}(n)$	$\mathbf{F}_p[v_n]$
$\mathrm{K}(n)$	$\mathbf{F}_p[v_n, v_n^{-1}]$,		

all equipped with unique (up to canonical isomorphism) homotopy commutative MU-algebra structures [**7**, §V.4]. (Previously these spectra had been constructed using Baas-Sullivan theory or the Landweber exact functor theorem, and laboriously shown to admit ring spectrum structures [**2**, **17**–**19**, **22**].)

It also makes possible a one-line construction of Landweber-Ravenel-Stong elliptic cohomology [**14**] (from MSpin as a commutative ring spectrum),

$$\mathrm{ell}\left[\tfrac{1}{2}\right] = \mathrm{MSpin}\left[\tfrac{1}{2}\right]/\big(\text{kernel of the Ochanine elliptic genus}\big),$$

similarly equipped with a unique (up to canonical isomorphism) homotopy commutative MSpin-algebra structure. (In fact, since the kernel of the Ochanine elliptic genus is the ideal consisting of $\mathbf{HP}^2$-bundles, one may write

$$\mathrm{ell}\left[\tfrac{1}{2}\right] = \mathrm{MSpin}\left[\tfrac{1}{2}\right]/\big(\mathbf{HP}^2\text{-bundles}\big)$$

[**13**]; cf. also the Conner-Floyd isomorphism for ell [**10**, **12**].)

By contrast, this paper shows that it is impossible to construct the spectrum tmf of topological modular forms [**6**, **8**] in an analogous way from the string bordism ring spectrum $\mathrm{MO}\langle 8\rangle$. (Even though

$$\begin{aligned}\pi_*\mathrm{tmf}\left[\tfrac{1}{6}\right] &\cong \pi_*(\mathrm{MO}\langle 8\rangle)\left[\tfrac{1}{6}\right]/\big(\text{kernel of the Witten genus}\big)\\ &= \pi_*(\mathrm{MO}\langle 8\rangle)\left[\tfrac{1}{6}\right]/\big(\mathbf{CaP}^2\text{-bundles}\big)\end{aligned}$$

[**16**].)

THEOREM 1. *For any prime $p > 3$ and any sequence X of homogeneous elements of $\pi_*\mathrm{MO}\langle 8\rangle$, the $\pi_*\mathrm{MO}\langle 8\rangle$-module*

$$\pi_*\big(\mathrm{MO}\langle 8\rangle_{(p)}/X\big)$$

is not (even abstractly) isomorphic to $\pi_*\mathrm{tmf}_{(p)}$.

As we shall see, the reason is twofold:

[1] When $p = 2$, $\pi_*(\mathrm{BP})$ is not concentrated in degrees congruent to 0 modulo 4. Nevertheless, BP satisfies weaker conditions [**21**, Thm. 2.7] ensuring that it admits a homotopy commutative $\mathrm{MU}_{(2)}$-algebra structure [**21**, Prop. 2.9]; the same is true of $\mathrm{BP}\langle n\rangle$ and $\mathrm{E}(n)$, provided the generators v_i are chosen carefully [**21**, Prop. 2.10].

(I) For any (homotopy) commutative ring spectrum R and any sequence X of homogeneous elements of $\pi_*(R)$, there is an isomorphism of graded $\mathbf{Q}$-vector spaces

$$\pi_*(R/X) \otimes \mathbf{Q} \cong \mathrm{H}_*(\mathrm{Tot}(\mathrm{K}(X))) \otimes \mathbf{Q},$$

where the right-hand side is the rational homology of the (total) Koszul complex of X (Theorem 3), which is strictly bigger than $\pi_*(R)/(X) \otimes \mathbf{Q}$ unless X is a $\pi_*(R) \otimes \mathbf{Q}$-quasi-regular sequence (Theorem 4).

(II) The kernel of the p-local Witten genus

$$\pi_*\mathrm{MO}\langle 8\rangle_{(p)} \to \pi_*\mathrm{tmf}_{(p)}$$

cannot be generated by a $\pi_*\mathrm{MO}\langle 8\rangle \otimes \mathbf{Q}$-quasi-regular sequence (Proposition 7).

COROLLARY 2. *There do not exist* $\mathrm{MO}\langle 8\rangle$*-module (let alone* $\mathrm{MO}\langle 8\rangle$*-algebra) isomorphisms*

$$\mathrm{MO}\langle 8\rangle_{(p)}/X \to \mathrm{tmf}_{(p)} \qquad\qquad \mathrm{MO}\langle 8\rangle\left[\tfrac{1}{6}\right]/X \to \mathrm{tmf}\left[\tfrac{1}{6}\right].$$

(If such isomorphisms existed, their underlying $\pi_*\mathrm{MO}\langle 8\rangle$-module isomorphisms would contradict Theorem 1.)

This helps to better appreciate the elaborate known constructions of tmf and its orientations, cf. [**1, 6, 9, 15**]. It also raises the question of whether there is a general modification of, or alternative to, the ring spectrum quotient construction, one better suited to non quasi-regular sequences, which produces tmf from $\mathrm{MO}\langle 8\rangle$ in a simple way.

2. The Koszul Complex $\mathrm{K}(X)$

Let A be a graded and commutative ring, E an A-module, and $f : E \to A$ a map of graded A-modules. Recall that the *Koszul complex* $\mathrm{K}(f)$ of f is the exterior algebra $\bigwedge E$ equipped with differential

$$d_k(e_1 \wedge \cdots \wedge e_k) = \sum_{i=1}^{k} (-1)^{i+1} f(e_i)\, e_1 \wedge \cdots \wedge \hat{e}_i \wedge \cdots \wedge e_k,$$

where $\hat{e}_i$ means that the wedge summand e_i is omitted. $\bigwedge E$ is bigraded,

$$\deg(e_1 \wedge \cdots \wedge e_k) = \big(k, \deg(e_1) + \cdots + \deg(e_k)\big),$$

and the differential d_k decreases the first component by 1 and preserves the second, so $\mathrm{K}(f)$ may be thought of as a graded complex of A-modules

$$\cdots \to \wedge^2 E \xrightarrow{d_2} \wedge^1 E \xrightarrow{d_1} A \to 0.$$

A (possibly infinite) sequence $X = x_1, x_2, \ldots$ of homogeneous elements of A defines a map of graded A-modules

$$f : \bigoplus_i A(-\deg x_i) \to A$$
$$f\left(\textstyle\bigoplus_i a_i\right) = \textstyle\sum_i x_i a_i.$$

Write $\mathrm{K}(X)$ for the Koszul complex of this map. (Here and below, $N(m)$ denotes the shift of a graded module N defined by $N(m)_k = N_{k+m}$.)

$\mathrm{K}(f)$ *is functorial:* A map $\phi : E' \to E$ of graded A-modules extends to a map of bigraded A-algebras $\wedge\phi : \wedge E' \to \wedge E$, which defines a map of graded complexes $\mathrm{K}(f \circ \phi) \to \mathrm{K}(f)$.

Also, $\mathrm{K}(f)$ *is exponential:* The sum of graded A-linear maps $f : E \to A$ and $f' : E' \to A$ is a graded A-linear map $f \oplus f' : E \oplus E' \to A$, and the isomorphism of graded A-modules $\bigwedge(E \oplus E') \cong \bigwedge E \otimes \bigwedge E'$ extends to an isomorphism of graded complexes

$$\mathrm{K}(f \oplus f') \cong \mathrm{K}(f) \otimes \mathrm{K}(f').$$

In particular, if $E' = A(m)$, then the isomorphism of graded A-modules

$$\bigwedge^k (E \oplus A(m)) \cong \bigwedge^k E \oplus \bigwedge^{k-1} E(m),$$

induces, for any finite sequence $x_1, \ldots, x_{n+1}$ of homogeneous elements of A, a short exact sequence of graded complexes

$$0 \to \mathrm{K}(x_1, \ldots, x_n) \to \mathrm{K}(x_1, \ldots, x_{n+1}) \to \mathrm{K}(x_1, \ldots, x_n)(-\deg x_{n+1})[-1] \to 0$$

whose connecting homomorphism is induced by multiplication by x_{n+1}. (Here and below, $C[m]$ denotes the shift of a complex (C, d_C) defined by $C[m]_i = C_{i+m}$ and $d_{C[1]} = (-1)^m d_C$.)

Thus $\mathrm{K}(x_1, \ldots, x_{n+1})$ is isomorphic to the mapping cone

$$\mathrm{K}(x_1, \ldots, x_{n+1}) \cong \operatorname{cone}\left[\mathrm{K}(x_1, \ldots, x_n)(-\deg x_{n+1}) \xrightarrow{x_{n+1}} \mathrm{K}(x_1, \ldots, x_n)\right] \tag{2}$$

(compare [**20**, Tag 0628] or [**5**, Cor. 1.6.13]). Observe how similar (2) is to (1).

3. $\pi_*(R/X)$ is Rationally Isomorphic to $\mathrm{H}_*(\mathrm{Tot}(\mathrm{K}(X)))$

THEOREM 3. *For any (homotopy) commutative ring spectrum R and any sequence X of homogeneous elements of $\pi_*(R)$, there is an isomorphism of graded $\mathbf{Q}$-vector spaces*

$$\pi_*(R/X) \otimes \mathbf{Q} \cong \mathrm{H}_*(\mathrm{Tot}(\mathrm{K}(X))) \otimes \mathbf{Q}$$

making the diagram

$$\begin{array}{ccc} \pi_*(R) \otimes \mathbf{Q} & \longrightarrow & \mathrm{H}_0(\mathrm{K}(X)) \otimes \mathbf{Q} \\ \downarrow & & \downarrow \\ \pi_*(R/X) \otimes \mathbf{Q} & \xleftarrow{\cong} & \mathrm{H}_*(\mathrm{Tot}(\mathrm{K}(X))) \otimes \mathbf{Q} \end{array}$$

commute.

REMARK. The right-hand side is the rational homology of the *total* Koszul complex, so the isomorphism determines, for each integer n, an isomorphism of $\mathbf{Q}$-vector spaces

$$\pi_n(R/X) \otimes \mathbf{Q} \cong \bigoplus_{i+j=n} \mathrm{H}_i(\mathrm{K}(X))_j \otimes \mathbf{Q},$$

where $(-)_j$ denotes the degree-j component of the graded module $(-)$.

REMARK. Theorem 3 does not hold integrally. In general, there is a spectral sequence

$$\mathrm{H}_*(\mathrm{K}(X)) \implies \pi_*(R/X)$$

[**18**, p. 185]. (According to Theorem 3, this spectral sequence collapses tensor $\mathbf{Q}$.)

PROOF OF THEOREM 3. Let $X = x_1, x_2, \ldots$ be an infinite sequence. First we use induction to argue that the result holds for each finite subsequence

$$X_n = x_1, \ldots, x_n \quad (n \geq 0).$$

(Base case) The result holds integrally for the empty sequence X_0. Indeed, by definition $R/X_0 = R$ and $\mathrm{K}(X_0) = (\bigwedge \pi_*(R), d) = \pi_*(R)[0]$, so for any integer i,

$$\pi_i(R/X_0) = \pi_i(R) = \mathrm{H}_0(\mathrm{K}(X_0))_i = \mathrm{H}_0(\pi_*(R)[0])_i = \mathrm{H}_i(\mathrm{Tot}(\pi_*(R)[0])) = \mathrm{H}_i(\mathrm{Tot}(\mathrm{K}(X_0))).$$

(Inductive step) Suppose that the conclusion holds for some subsequence X_n ($n \geq 0$). We argue that it then also holds for the subsequence X_{n+1}. Let $m = \deg(x_{n+1})$.

According to (2), there is a distinguished triangle of graded complexes

$$\mathrm{K}(X_n)(-m) \xrightarrow{x_{n+1}} \mathrm{K}(X_n) \longrightarrow \mathrm{K}(X_{n+1}) \longrightarrow \mathrm{K}(X_n)(-m)[-1]$$

and hence a distinguished triangle of total complexes

$$\mathrm{Tot}(\mathrm{K}(X_n))[-m] \xrightarrow{x_{n+1}} \mathrm{Tot}(\mathrm{K}(X_n)) \longrightarrow \mathrm{Tot}(\mathrm{K}(X_{n+1})) \longrightarrow \mathrm{Tot}(\mathrm{K}(X_n))[-1-m].$$

Its long exact homology sequence looks like

$$\cdots \to \mathrm{H}_{i-m}(\mathrm{Tot}(\mathrm{K}(X_n))) \to \mathrm{H}_i(\mathrm{Tot}(\mathrm{K}(X_n))) \to \mathrm{H}_i(\mathrm{Tot}(\mathrm{K}(X_{n+1}))) \to \mathrm{H}_{i-1-m}(\mathrm{Tot}(\mathrm{K}(X_n))) \to \cdots,$$

which looks just like the long exact homotopy sequence

$$\cdots \longrightarrow \pi_{i-m}(R/X_n) \longrightarrow \pi_i(R/X_n) \longrightarrow \pi_i(R/X_{n+1}) \longrightarrow \pi_{i-1-m}(R/X_n) \longrightarrow \cdots$$

for the cofiber sequence (1) above,

$$\Sigma^m R/X_n \xrightarrow{x_{n+1}} R/X_n \longrightarrow R/X_{n+1} \longrightarrow \Sigma^{1+m} R/X_n.$$

Specifically, by inductive hypothesis there are isomorphisms (solid arrows) fitting into the commutative diagram

$$\begin{array}{ccccccccc}
\cdots \to & \mathrm{H}_{i-m}(\mathrm{Tot}(\mathrm{K}(X_n))) \otimes \mathbf{Q} & \to & \mathrm{H}_i(\mathrm{Tot}(\mathrm{K}(X_n))) \otimes \mathbf{Q} & \to & \mathrm{H}_i(\mathrm{Tot}(\mathrm{K}(X_{n+1}))) \otimes \mathbf{Q} & \to & \mathrm{H}_{i-1-m}(\mathrm{Tot}(\mathrm{K}(X_n))) \otimes \mathbf{Q} & \to \cdots \\
& \downarrow \cong & & \downarrow \cong & & \vdots \cong & & \downarrow \cong & \\
\cdots \longrightarrow & \pi_{i-m}(R/X_n) \otimes \mathbf{Q} & \longrightarrow & \pi_i(R/X_n) \otimes \mathbf{Q} & \longrightarrow & \pi_i(R/X_{n+1}) \otimes \mathbf{Q} & \longrightarrow & \pi_{i-1-m}(R/X_n) \otimes \mathbf{Q} & \longrightarrow \cdots .
\end{array}$$

Since exact sequences of vector spaces split, there are isomorphisms (dashed arrows) making the diagram commute.

To see that these dashed isomorphisms fit together to make the diagram in the statement of the theorem commute, sum the middle squares over i, and use the inductive hypothesis and the fact that $\mathrm{K}(X_n) \to \mathrm{K}(X_{n+1})$ is a map of graded complexes to form the commutative diagram

$$\begin{array}{ccccc}
 & \nearrow & \mathrm{H}_0(\mathrm{K}(X_n)) \otimes \mathbf{Q} & \longrightarrow & \mathrm{H}_0(\mathrm{K}(X_{n+1})) \otimes \mathbf{Q} \\
 & & \downarrow & & \downarrow \\
\pi_*(R) \otimes \mathbf{Q} & & \mathrm{H}_*(\mathrm{Tot}(\mathrm{K}(X_n))) \otimes \mathbf{Q} & \longrightarrow & \mathrm{H}_*(\mathrm{Tot}(\mathrm{K}(X_{n+1}))) \otimes \mathbf{Q} \\
 & & \downarrow \cong & & \downarrow \cong \\
 & \searrow & \pi_*(R/X_n) \otimes \mathbf{Q} & \longrightarrow & \pi_*(R/X_{n+1}) \otimes \mathbf{Q}.
\end{array}$$

By induction, then, the result holds for each finite subsequence $X_n (n \geq 0)$.

(Infinite case) The infinite case follows since

$$\begin{aligned}\pi_*(R/X)\otimes\mathbf{Q} &\overset{(1)}{\cong} \pi_*\big(\operatorname{hocolim}_n(R/X_n)\big)\otimes\mathbf{Q}\\ &\overset{(2)}{\cong} \operatorname{colim}_n\big(\pi_*(R/X_n)\otimes\mathbf{Q}\big)\\ &\overset{(3)}{\cong} \operatorname{colim}_n\big(\mathrm{H}_*(\mathrm{Tot}(\mathrm{K}(X_n)))\otimes\mathbf{Q}\big)\\ &\overset{(4)}{\cong} \mathrm{H}_*\big(\operatorname{colim}_n(\mathrm{Tot}(\mathrm{K}(X_n)))\otimes\mathbf{Q}\big)\\ &\overset{(5)}{\cong} \mathrm{H}_*\big(\mathrm{Tot}(\mathrm{K}(X))\otimes\mathbf{Q}\big).\end{aligned}$$

These isomorphisms hold by

(1) definition of R/X,
(2) properties of hocolim and cocontinuity of $-\otimes\mathbf{Q}$ (as it is left adjoint to $\mathrm{Hom}(\mathbf{Q},-)$),
(3) the finite case proved above,
(4) exactness of filtered colimits in the category of A-modules, and
(5) cocontinuity of $\bigwedge(-)$ (as it is left adjoint to the degree-1 functor $(-)_1$). □

4. $\pi_*(R)\to\pi_*(R/X)$ is surjective only if X is a quasi-regular sequence

Theorem 1 could be proved using regular sequences, but it is easier to prove using *quasi*-regular sequences. Let us recall how regular and quasi-regular sequences are related to the homology of the Koszul complex.

Consider a commutative (but not necessarily graded) ring A, an A-module N, and an ideal $\mathfrak{X}$ of A. The $\mathfrak{X}$-adic topology on N is the topology compatible with the group structure of N, for which the submodules $\mathfrak{X}^rN$ $(r\geq 0)$ form a fundamental system of neighborhoods of zero. This topology is separated if and only if

$$\bigcap_{r\geq 0}\mathfrak{X}^r=0.$$

Suppose now that $\mathfrak{X}$ is generated by a (possibly infinite) sequence $X=x_1,x_2,\dots$ of elements of A. Consider the graded A-module $\bigoplus_{r\geq 0}\mathfrak{X}^rN$ and the map of graded A-modules

$$a_N^X : A[X_1,X_2,\dots]\otimes_A N\to\bigoplus_{r\geq 0}\mathfrak{X}^rN$$

defined by

$$a_N^X(P\otimes n)=P(x_1,x_2,\dots)n$$

for n in N and P a homogeneous polynomial in indeterminants $X_1,X_2,\dots$ (not to be confused with the sequence X and its elements $x_1,x_2,\dots$). Let $\mathfrak{d}$ be the ideal of $A[X_1,X_2,\dots]$ generated by the elements $(x_iX_j-x_jX_i)$ for $1\leq i<j$. Then $P(x_1,x_2,\dots)=0$ if $P\in\mathfrak{d}$, so a_N^X determines a map of graded A-modules

$$\alpha_N^X : (A[X_1,X_2,\dots]/\mathfrak{d})\otimes_A N\to\bigoplus_{r\geq 0}\mathfrak{X}^rN.$$

By tensor product with $A/\mathfrak{X}$ one obtains a map of graded A-modules

$$\beta_N^X : (A/\mathfrak{X})[X_1,X_2,\dots]\otimes_A N\to\bigoplus_{r\geq 0}(\mathfrak{X}^rN/\mathfrak{X}^{r+1}N).$$

The maps a_N^X, α_N^X, and β_N^X are surjective.

THEOREM 4 ([**4**, §9, n° 7, Thm. 1]). *Consider the following conditions:*

(i) *X is N-regular.*
(ii) $\mathrm{H}_i(\mathrm{K}(X) \otimes N) = 0$ *for all* $i > 0$.
(iii) $\mathrm{H}_1(\mathrm{K}(X) \otimes N) = 0$.
(iv) *α_N^X is bijective.*
(v) *β_N^X is bijective.*

We then have implications (i) $\Longrightarrow$ (ii) $\Longrightarrow$ (iii) $\Longleftrightarrow$ (iv) $\Longrightarrow$ (v).

If, for $1 \le i \le n$, the A-module $N/(x_1 N + \cdots + x_n N)$ is separated in the $\mathfrak{X}$-adic topology, then conditions (i) and (v) are equivalent.

(The proof of Theorem 4 given in [**4**] is for a finite sequence X but applies, *mutatis mutandis*, to an infinite sequence.)

DEFINITION. The sequence X is called *N-quasi-regular* if it satisfies condition (v) of Theorem 4. An A-quasi-regular sequence is simply called *quasi-regular.*

In short, every N-regular sequence X satisfies $\mathrm{H}_1(\mathrm{K}(X) \otimes N) = 0$, and every sequence X satisfying $\mathrm{H}_1(\mathrm{K}(X) \otimes N) = 0$ is N-quasi-regular.

Theorems 3 and 4 together yield:

COROLLARY 5. *For any (homotopy) commutative ring spectrum R and any sequence X of homogeneous elements of $\pi_*(R)$, the homomorphism*

$$\pi_*(R) \to \pi_*(R/X)$$

is surjective only if X is $\pi_*(R) \otimes \mathbf{Q}$-quasi-regular.

PROOF. If $\pi_*(R) \to \pi_*(R/X)$ is surjective, then $\pi_*(R) \otimes \mathbf{Q} \to \pi_*(R/X) \otimes \mathbf{Q}$ is too. By Theorem 3, then, $\mathrm{H}_1(\mathrm{K}(X)) \otimes \mathbf{Q} = 0$. And by Theorem 4, X must be $\pi_*(R) \otimes \mathbf{Q}$-quasi-regular. □

Note that, since $\mathrm{MO}\langle 8 \rangle$ is connective, Theorem 1 could be proved using regular (rather than quasi-regular) sequences, by the following result.

COROLLARY 6 (of Theorem 4, [**4**, §9, n° 7, Cor. 2]). *Suppose A is a positively graded ring, N is a graded, bounded-below A-module, and X is a sequence of homogeneous elements of A of degree > 0. Then conditions (i) and (v) of Theorem 4 are equivalent.*

(The hypotheses ensure that the $\mathfrak{X}$-adic topology on each of the A-modules $M/(x_1 M + \cdots + x_n M)$ is separated.)

5. The kernel of the p-local Witten genus cannot be generated by a quasi-regular sequence

PROPOSITION 7. *For $p > 3$, the kernel of the p-local Witten genus*

$$\phi_W \otimes \mathbf{Z}_{(p)} : \pi_* \mathrm{MO}\langle 8 \rangle_{(p)} \to \pi_* \mathrm{tmf}_{(p)}$$

cannot be generated by a $\pi_* \mathrm{MO}\langle 8 \rangle \otimes \mathbf{Q}$-quasi-regular sequence. (In particular, it cannot be generated by a quasi-regular sequence.)

PROOF. Denote the kernel $\mathfrak{K}$, and suppose it was generated by a $\pi_*\mathrm{MO}\langle 8\rangle\otimes\mathbf{Q}$-quasi-regular sequence $X = x_1, x_2, \ldots$.

Rationally, the Witten genus is a surjective homomorphism of polynomial rings whose generators may be chosen so that

$$\begin{aligned}\pi_*(\mathrm{MO}\langle 8\rangle)\otimes\mathbf{Q} &\cong \mathbf{Q}[y_2, y_3, y_4, \ldots]\\ \pi_*(\mathrm{tmf})\otimes\mathbf{Q} &\cong \mathbf{Q}[\mathbf{G}_2, \mathbf{G}_3]\\ \ker\phi_W\otimes\mathbf{Q} &= (y_4, y_5, y_6, \cdots)\end{aligned}$$

where $\deg(y_i) = \deg(\mathbf{G}_i) = 4i$ (see [**11**], [**3**, Prop. 4.4], and [**9**, Thm. 6.25]). It follows that

$$(\mathfrak{K}^1/\mathfrak{K}^2)_n\otimes\mathbf{Q} \cong \begin{cases}\mathbf{Q} & \text{if } n = 4i \geq 16\\ 0 & \text{otherwise.}\end{cases}$$

Since X is a $\pi_*\mathrm{MO}\langle 8\rangle\otimes\mathbf{Q}$-quasi-regular sequence generating $\mathfrak{K}\otimes\mathbf{Q}$, the map

$$\underbrace{\big(\pi_*\mathrm{MO}\langle 8\rangle/\mathfrak{K}\big)[X_1, X_2, \ldots]\otimes\mathbf{Q}}_{\mathbf{Q}[\mathbf{G}_2,\mathbf{G}_3][X_1,X_2,\ldots]} \to \bigoplus_{r\geq 0}\left(\mathfrak{K}^r/\mathfrak{K}^{r+1}\right)\otimes\mathbf{Q}$$

induced by the substitution $X_i \mapsto \overline{x}_i \in \mathfrak{K}^1/\mathfrak{K}^2$ is by definition bijective. *It follows that X has precisely one element of degree $4i$, for each $i \geq 4$, and no other elements.*

Let $\mathfrak{A}$ be the augmentation ideal of $\pi_*\mathrm{MO}\langle 8\rangle_{(p)}$. Applying the snake lemma to the commutative diagram of $\pi_*\mathrm{MO}\langle 8\rangle_{(p)}$-modules

$$\begin{array}{ccccccccc} 0 & \longrightarrow & \mathfrak{K}^2 & \longrightarrow & \mathfrak{A}^2 & \longrightarrow & \mathfrak{A}^2/\mathfrak{K}^2 & \longrightarrow & 0\\ & & \downarrow & & \downarrow & & \downarrow & & \\ 0 & \longrightarrow & \mathfrak{K}^1 & \longrightarrow & \mathfrak{A}^1 & \longrightarrow & \mathfrak{A}^1/\mathfrak{K}^1 & \longrightarrow & 0\end{array}$$

(in which the rows are exact) produces an exact sequence

$$\mathfrak{K}^1/\mathfrak{K}^2 \longrightarrow \mathfrak{A}^1/\mathfrak{A}^2 \longrightarrow \mathfrak{A}^1/(\mathfrak{K}^1+\mathfrak{A}^2) \longrightarrow 0.$$

Since

$$\pi_*\mathrm{MO}\langle 8\rangle_{(p)}/\mathfrak{K} \cong \mathbf{Z}_{(p)}[\mathbf{G}_2, \mathbf{G}_3]$$

(again by [**3**, Prop. 4.4] and [**9**, Thm. 6.25], see also [**16**]), it follows that

$$\mathfrak{A}^1/(\mathfrak{K}^1+\mathfrak{A}^2) \cong \mathbf{Z}_{(p)}\{\mathbf{G}_2, \mathbf{G}_3\}$$

and therefore that $\mathfrak{K}^1/\mathfrak{K}^2 \to \mathfrak{A}^1/\mathfrak{A}^2$ is surjective in degrees > 12.

According to Hovey's calculations [**11**] (see also [**16**, Thm. 4])

$$(\mathfrak{A}^1/\mathfrak{A}^2)_n \cong \begin{cases}\mathbf{Z}_{(p)}\oplus\mathbf{Z}/p & \text{if } n = 2(p^i+p^j) \text{ for some } 0<i<j\\ \mathbf{Z}_{(p)} & \text{if } n = 4m \geq 8 \text{ but } n \neq 2(p^i+p^j) \text{ for any } 0<i<j\\ 0 & \text{otherwise.}\end{cases}$$

Since no single element can generate $\mathbf{Z}_{(p)}\oplus\mathbf{Z}/p$ as a $\mathbf{Z}_{(p)}$-module, it follows that the map

$$\underbrace{\big(\pi_*\mathrm{MO}\langle 8\rangle_{(p)}/\mathfrak{K}\big)}_{\mathbf{Z}_{(p)}[\mathbf{G}_2,\mathbf{G}_3]}[X_1, X_2, \ldots] \to \bigoplus_{r\geq 0}\left(\mathfrak{K}^r/\mathfrak{K}^{r+1}\right)$$

induced by the substitution $X_i \to \overline{x}_i \in \mathfrak{K}^1/\mathfrak{K}^2$ cannot be surjective, contradicting the assumption that X generates $\mathfrak{K}$. $\square$

Acknowledgments

Thanks to Jack Morava, John Greenlees, Nitu Kitchloo, Burt Totaro, Doug Ravenel, John Lind, and Doug Haessig for helpful conversations.

References

[1] M. Ando, M. J. Hopkins, and N. P. Strickland, *Elliptic spectra, the Witten genus and the theorem of the cube*, Invent. Math. **146** (2001), no. 3, 595–687, DOI 10.1007/s002220100175. MR1869850

[2] N. A. Baas, *On bordism theory of manifolds with singularities*, Math. Scand. **33** (1973), 279–302 (1974), DOI 10.7146/math.scand.a-11491. MR0346824

[3] T. Bauer, *Computation of the homotopy of the spectrum* tmf, Groups, homotopy and configuration spaces, Geom. Topol. Monogr., vol. 13, Geom. Topol. Publ., Coventry, 2008, pp. 11–40, DOI 10.2140/gtm.2008.13.11. MR2508200

[4] N. Bourbaki, *Éléments de mathématique* (French), Masson, Paris, 1980. Algèbre. Chapitre 10. Algèbre homologique. [Algebra. Chapter 10. Homological algebra]. MR610795

[5] W. Bruns and J. Herzog, *Cohen-Macaulay rings*, Cambridge Studies in Advanced Mathematics, vol. 39, Cambridge University Press, Cambridge, 1993. MR1251956

[6] C. L. Douglas, J. Francis, A. G. Henriques, and M. A. Hill (eds.), *Topological modular forms*, Mathematical Surveys and Monographs, vol. 201, American Mathematical Society, Providence, RI, 2014. MR3223024

[7] A. D. Elmendorf, I. Kriz, M. A. Mandell, and J. P. May, *Rings, modules, and algebras in stable homotopy theory*, Mathematical Surveys and Monographs, vol. 47, American Mathematical Society, Providence, RI, 1997. With an appendix by M. Cole. MR1417719

[8] M. J. Hopkins, *Topological modular forms, the Witten genus, and the theorem of the cube*, Proceedings of the International Congress of Mathematicians, Vol. 1, 2 (Zürich, 1994), Birkhäuser, Basel, 1995, pp. 554–565. MR1403956

[9] M. J. Hopkins, *Algebraic topology and modular forms*, Proceedings of the International Congress of Mathematicians, Vol. I (Beijing, 2002), Higher Ed. Press, Beijing, 2002, pp. 291–317. MR1989190

[10] M. J. Hopkins and M. A. Hovey, *Spin cobordism determines real K-theory*, Math. Z. **210** (1992), no. 2, 181–196, DOI 10.1007/BF02571790. MR1166518

[11] M. Hovey, *The homotopy of $M\mathrm{String}$ and $M\mathrm{U}\langle 6\rangle$ at large primes*, Algebr. Geom. Topol. **8** (2008), no. 4, 2401–2414, DOI 10.2140/agt.2008.8.2401. MR2465746

[12] M. A. Hovey, *Spin bordism and elliptic homology*, Math. Z. **219** (1995), no. 2, 163–170, DOI 10.1007/BF02572356. MR1337212

[13] M. Kreck and S. Stolz, *$\mathbf{HP}^2$-bundles and elliptic homology*, Acta Math. **171** (1993), no. 2, 231–261, DOI 10.1007/BF02392533. MR1251581

[14] P. S. Landweber, D. C. Ravenel, and R. E. Stong, *Periodic cohomology theories defined by elliptic curves*, The Čech centennial (Boston, MA, 1993), Contemp. Math., vol. 181, Amer. Math. Soc., Providence, RI, 1995, pp. 317–337, DOI 10.1090/conm/181/02040. MR1320998

[15] J. Lurie, *A survey of elliptic cohomology*, Algebraic topology, Abel Symp., vol. 4, Springer, Berlin, 2009, pp. 219–277, DOI 10.1007/978-3-642-01200-6_9. MR2597740

[16] C. McTague, *The Cayley plane and string bordism*, Geom. Topol. **18** (2014), no. 4, 2045–2078, DOI 10.2140/gt.2014.18.2045. MR3268773

[17] O. K. Mironov, *Existence of multiplicative structures in the theory of cobordism with singularities* (Russian), Izv. Akad. Nauk SSR Ser. Mat. **39** (1975), no. 5, 1065–1092, 1219. MR0402784

[18] J. Morava, *A product for the odd-primary bordism of manifolds with singularities*, Topology **18** (1979), no. 3, 177–186, DOI 10.1016/0040-9383(79)90001-6. MR546788

[19] N. Shimada and N. Yagita, *Multiplications in the complex bordism theory with singularities*, Publ. Res. Inst. Math. Sci. **12** (1976/77), no. 1, 259–293, DOI 10.2977/prims/1195190968. MR0415642

[20] T. Stacks Project Authors. *Stacks Project*. `http://stacks.math.columbia.edu`, 2017.

[21] N. P. Strickland, *Products on* MU-*modules*, Trans. Amer. Math. Soc. **351** (1999), no. 7, 2569–2606, DOI 10.1090/S0002-9947-99-02436-8. MR1641115

[22] D. Sullivan, *Combinatorial invariants of analytic spaces*, Proceedings of Liverpool Singularities—Symposium, I (1969/70), Springer, Berlin, 1971, pp. 165–168. MR0278333

Mathematics Department, University of Rochester, Rochester, NY 14627, USA
Email address: carl.mctague@rochester.edu

Mathematics Department, Johns Hopkins University, Baltimore, MD 21218, USA
Email address: mctague@math.jhu.edu
URL: http://www.mctague.org/carl

Contemporary Mathematics
Volume **707**, 2018
http://dx.doi.org/10.1090/conm/707/14255

Cocycle schemes and $MU[2k, \infty)$–orientations

Eric Peterson

ABSTRACT. We recall the study of $MU[2k, \infty)$–orientations as elucidated by Ando, Hopkins, and Strickland. Their work prompts us to investigate a particular algebraic moduli which, after 2–localization, we (together with Adam Hughes and JohnMark Lau) fully describe for all values of k. It gives a strikingly good (but imperfect) approximation of our topological motivator.

1. Motivation: Integration in extraordinary cohomology theories

There is a whole industry in algebraic topology dedicated to understanding "orientations" of ring spectra, defined as follows:

DEFINITION 1.1. Suppose that $G(n) \to O(n)$ is a multiplicative system of structure groups:

$$\begin{array}{ccccccc} \cdots & \longrightarrow & G(n) & \longrightarrow & G(n+1) & \longrightarrow & \cdots \\ & & \downarrow & & \downarrow & & \\ \cdots & \longrightarrow & O(n) & \longrightarrow & O(n+1) & \longrightarrow & \cdots . \end{array}$$

A *G–orientation* of a ring spectrum E is a map φ of ring spectra

$$\varphi\colon MG \to E,$$

where MG is the Thom spectrum associated to the system of structure groups.

EXAMPLE 1.2. There are orientations $MO \to H\mathbb{F}_2$ and $MSO \to H\mathbb{Z}$ given by coconnective truncation.

This definition is stanadard, but it may not make plain to an interloper why this is of any special interest. The first observation (and the usual motivation) is that evaluating such a map on a point gives rise to an E_*–valued genus

$$\varphi_*\colon MG_* \to E_*,$$

i.e., a ring homomorphism which converts G–structured bordism classes to elements of E_*.

EXAMPLE 1.3. The induced maps $\pi_*MO \to \pi_*H\mathbb{F}_2 = \mathbb{F}_2$ and $\pi_*MSO \to \pi_*H\mathbb{Z} = \mathbb{Z}$ count the number of (signed) points in a dimension 0 (oriented) manifold.

2010 *Mathematics Subject Classification.* Primary 55N22.

However, this is really just scratching the surface, and there are techniques for getting a lot more out of an orientation than this simple invariant. For example, Kochman [**Koc78**, Section 4] models the G–structured bordism groups $MG_m(X)$ via a chain complex of monoids

$$\begin{aligned} \cdots &\xrightarrow{\partial} \left\{ Z^n \to X \,\middle|\, \begin{array}{c} Z^n \text{ a compact } n\text{–manifold} \\ \text{with } G\text{–structure} \end{array} \right\} \\ &\xrightarrow{\partial} \left\{ Z^{n-1} \to X \,\middle|\, \begin{array}{c} Z^{n-1} \text{ a compact } (n-1)\text{–manifold} \\ \text{with } G\text{–structure} \end{array} \right\} \\ &\xrightarrow{\partial} \cdots, \end{aligned}$$

where each map "∂" is given by restricting to the boundary of the source manifold. Under this definition, we then consider the bordism homology group $MG_{n-m}(\underline{E}_n)$, where $\underline{E}_n$ is the nth space in the Ω–spectrum for the G–oriented ring spectrum E. Using this model, one computes

$$MG_n\underline{E}_{n-m} = \left\{ M^n \xrightarrow{\omega} \underline{E}_{n-m} \,\middle|\, \begin{array}{c} M \text{ a closed } n\text{–dimensional } G\text{–manifold,} \\ \omega \in E^{n-m}(M) \text{ a class of codimension } m \end{array} \right\} \Big/ \sim .$$

In terms of stable homotopy theory, the spectrum

$$MG \wedge \Sigma^{-(n-m)}\Sigma^{\infty}_{+}\underline{E}_{n-m}$$

contains in π_{-m} information about G–structured bordism classes of dimension n which are equipped with m–codimensional E–cohomology classes. Incorporating the orientation φ, one can then build the composite

$$\mathbb{S}^{-m} \xrightarrow{(M,\omega)} MG \wedge \Sigma^{-(n-m)}\Sigma^{\infty}_{+}\underline{E}_{n-m} \xrightarrow{\operatorname{colim}_{n\to\infty}} MG \wedge E \xrightarrow{\varphi\wedge 1} E \wedge E \xrightarrow{\mu} E,$$

which gives a recipe for an element of $\pi_{-m}E$. Classically, this element is identifiable.

LEMMA 1.4. *For $\varphi_{MO}\colon MO \to H\mathbb{F}_2$ and $\varphi_{MSO}\colon MSO \to H\mathbb{Z}$, this assignment is only interesting in the case $m = 0$, where*

$$\varphi_{MO}\colon (M^n, \omega \in H^n(M;\mathbb{F}_2)) \mapsto \int_M \omega \in \mathbb{F}_2 = \pi_0 H\mathbb{F}_2,$$

$$\varphi_{MSO}\colon (M^n \text{ oriented, } \omega \in H^n(M;\mathbb{Z})) \mapsto \int_M \omega \in \mathbb{Z} = \pi_0 H\mathbb{Z}. \quad \square$$

In general, we think of this as a way to extract an integral for E–cohomology classes on G–structured manifolds.

Remembering the Pontryagin–Thom equivalence $\mathbb{S} \simeq M\text{Framed}$, this construction suddenly gains widespread application: to *any* ring spectrum E, the unit map $\eta_E : \mathbb{S} \to E$ associates a theory of integration for E–cohomology classes on framed bordism classes by the orientation

$$M\text{Framed} \simeq \mathbb{S} \xrightarrow{\eta_E} E.$$

This also gives a further interpretation of orientations in general: the sequence of ring maps

$$\mathbb{S} \to MSO \xrightarrow{\varphi} H\mathbb{Z}$$

can then be thought of as a *factorization* which witnesses some overdeterminacy of the naturally occurring framed integral. After all, given a framed manifold one can integrate a top-dimensional integral cohomology class, but actually a mere orientation of the tangent bundle is all that is required to produce an integral.[1]

Recognizing that O and SO form the beginning of the Postnikov tower for O, we are inspired to ask the following:

QUESTION 1.5. Given a ring spectrum $\eta_E\colon \mathbb{S} \to E$, through what stage of the Postnikov filtration does η_E factor?

Here's a somewhat longer list of known orientations of cohomology theories:

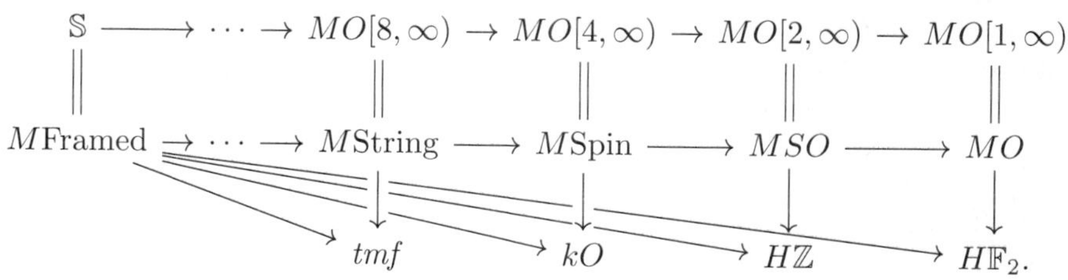

The first two vertical maps are classical; the orientation $M\text{Spin} \to kO$ is due to Atiyah, Bott, and Shapiro [**ABS64, Joa97**]; and the orientation

$$\sigma\colon M\text{String} \to tmf$$

is due to Ando, Hopkins, Rezk, and Strickland [**AHR, AHS01, AS01, HAS99**]. One can see an interesting correspondence beginning to take form: the level of Postnikov filtration on the top is matched (imperfectly) by the chromatic height of the spectrum on the bottom. The precise nature of this correspondence is pretty widely open, and our goal is to shed computational light on one aspect of it, motivated by the construction of the String–orientation of *tmf*. Their first step is to instead consider maps

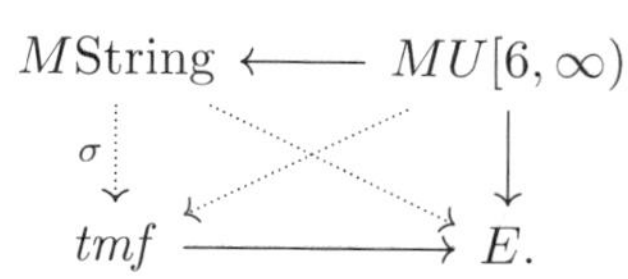

approximating σ on both sides. The replacement of MString by $MU[6, \infty)$ is meant to be simpler because $MU[6, \infty)$ is closer to MU, whose theory of orientations is exceedingly well–understood: this the theory of *complex orientations*. Accordingly, E is chosen to be a complex-orientable spectrum approximating *tmf*. Their game, then, is to scale the Postnikov tower from orientations $MU \to E$ to gain an understanding of orientations $MU[6, \infty) \to E$, then use this knowledge to show that wisely chosen approximations suffice to construct the desired map σ.

2. The algebraic geometry of the Thom construction

Chromatic homotopy theory, the study of complex-oriented cohomology theories, is naturally cast as a marriage of arithmetic geometry and algebraic topology. According to the taste of the author, the exposition can vary widely in which of these two fields is favored, and the version set forward by Ando, Hopkins, and Strickland is one of the more arithmetically dense. Their basic object is:

[1] Take this lightly, as this factorization is also a witness that oriented manifolds which *cannot* be framed support such integrals as well.

DEFINITION 2.1 ([**Str99**, Definition 8.11], [**AHS01**, Section 2.1.2]). For a space X and ring spectrum E, we write X_E for the formal affine scheme

$$X_E = \operatorname{Spf} E^0 X$$

topologized by the lattice of compact subspaces of X.

This visibly requires some niceness — for instance, $E^0 X$ has to be a commutative ring, E has to be sufficiently self-entwined so that "E^0" carries all of the interesting information, and the formal topology on $E^0 X$ has to be handled delicately. Details are available for the interested reader [**Str99**, Section 8.2], but for our purposes it suffices to say that they arrange their situation so that this is the case. Here is the basic sort of theorem you can expect to see in their language:

THEOREM 2.2 ([**Str99**, Section 8.6]). *There is a bijection between ring spectrum maps $MUP \to E$ and isomorphisms between $\mathbb{C}P^\infty_E$ and the affine line $\widehat{\mathbb{A}}^1 =$* $\operatorname{Spf} E^0[\![x]\!]$. □

Before discussing bordism orientations further, we will take the time to translate some other theorems about complex-orientable cohomology theories into this language. Of course, the main point of choosing a complex orientation is to provide Thom isomorphisms for complex vector bundles, so this gives us a wide body of results to work on translating. For instance, the complex-oriented E–cohomology of the projectivization $\mathbb{P}(V)$ of a rank n vector bundle V on a space X has a presentation as

$$E^*\mathbb{P}(V) \cong E^*X[\![t]\!]/c_V(t),$$

where $c_V(t)$ is the total Chern class of V, i.e.,

$$c_V(t) = t^n - c_1(V)t^{n-1} + c_2(V)t^{n-2} - \cdots + (-1)^n c_n(V).$$

In their language, one writes:

LEMMA 2.3 ([**Str99**, Proposition 8.28]). *The formal scheme $\mathbb{P}(V)_E$ is finite (of degree n) and free over X_E, and it forms a closed subscheme — i.e., it is an* effective Weil divisor — *of the curve $\mathbb{C}P^\infty_E \times X_E$ (thought of as an X_E-scheme).* □

COROLLARY 2.4 ([**Str99**, Proposition 8.31]). *The induced map $BU(n)_E \to$ $\operatorname{Div}_n^+ \mathbb{C}P^\infty_E$ is an isomorphism, where $\operatorname{Div}_n^+ \mathbb{C}P^\infty_E$ is the formal scheme classifying effective Weil divisors of degree n.* □

Projectivization leads to another essential theorem in vector bundle geometry: the splitting principle. Fixing V over X, there is a map $f\colon Y \to X$ such that f^*V splits naturally as a sum of line bundles and the induced map E^*f, for any complex-orientable E, is injective.

LEMMA 2.5. *Fixing a vector bundle V over a space X, there is a map $f\colon Y \to X$ such that $Y_E \to X_E$ is finite and faithfully flat, and $f_E^*\mathbb{P}(V)_E = \mathbb{P}(f^*V)_E$ splits as a sum of points.* □

We can also directly interpret the cohomology of the Thom spectrum X^V, although this takes a little more vocabulary. In classical language, a Thom isomorphism is an E^0X–module isomorphism $\widetilde{E}^0 X^V \cong E^0 X$, and the entire theory of modules is recast by algebraic geometers as the theory of quasicoherent sheaves over affine schemes.

LEMMA 2.6 ([**Str99**, Definition 8.33]). *The quasicoherent sheaf* $\mathbb{L}(V)$ *over* X_E *determined by* E^0X^V *is a trivializable line bundle.* □

A remarkable feature of the algebro-geometric language employed in this section is that many things we are first taught to think of as structures in topology — here is a collection of Thom isomorphisms, for instance, or a consequence of a calculation that begins with invoking the existence of some basis — can be easily recast as properties of the associated geometric objects — here are *trivializable invertible sheaves*, without a chosen trivializing section. This allows us to make concise distinctions between facts about a complex-*oriented* cohomology theory and facts about a complex-*orientable* cohomology theory, *without a particular complex orientation taken as reference.* In fact, the theorem at the beginning of this section can be stated without such a reference object:

THEOREM 2.7 ([**Str99**, Section 8.6], [**AHS01**, Example 2.53]). *Let* $\mathcal{L}$ *denote the tautological line bundle on* $\mathbb{C}\mathrm{P}^\infty$. *The sheaf* $\mathbb{L}(\mathcal{L})$ *on* $\mathbb{C}\mathrm{P}^\infty_E$ *is invertible exactly when* E *is complex-orientable, in which case it is trivializable. Trivializations then biject with ring maps* $MUP \to E$.

SKETCH OF PROOF. A trivialization of $\mathbb{L}(\mathcal{L})$ gives rise to a trivialization of

$$\mathbb{L}(\mathcal{L}^{\times n} \downarrow (\mathbb{C}\mathrm{P}^\infty)^{\times n}) \cong \mathbb{L}(\mathcal{L})^{\boxtimes n} \downarrow (\mathbb{C}\mathrm{P}^\infty_E)^{\times n}.$$

Applying the splitting principle to a vector bundle V over a space X to produce a map $f\colon Y \to X$, the resulting sheaf $\mathbb{L}(f^*V)$ pulls back from $\mathbb{L}(\mathcal{L}^{\times n})$, hence receives a trivialization. Finally, trivializations descend along finite flat maps. To see that these trivializations are suitably compatible, write $\zeta\colon X_E \to \mathbb{C}\mathrm{P}^\infty_E \times X_E$ for the zero-section and $\mathcal{I}(\mathbb{P}(V)_E)$ for the ideal sheaf of functions vanishing on $\mathbb{P}(V)_E$. There is then is an equivalence $\zeta^*\mathcal{I}(\mathbb{P}(V)_E) \cong \mathbb{L}(V)$. The projectivization construction converts sums of bundles to sums of divisors, and hence the trivialization of $\mathbb{L}(V \oplus W)$ is the product of the trivializations for $\mathbb{L}(V)$ and $\mathbb{L}(W)$. Applying these constructions to the universal bundles over the spaces $BU(n)$, this data constructs a ring map $MUP \to E$. □

3. Orientations for MSU and $MU[6,\infty)$

We now return to the study of ring maps $MU[2k,\infty) \to E$ for positive k. Because we went through the analysis above, we can see where the crucial pieces of input lie: first we should develop a version of the splitting principle for these structured bundles, and then we should tease apart the resulting mess of formal schemes.

Suppose that V is a vector bundle over X and that $Y \to X$ is the space over X guaranteed to us by the splitting principle for V. Now consider first the case of $k = 1$, where V has been lifted as in

$$\begin{array}{ccc} BU & \xrightarrow{+n\cdot 1} & BU \times \mathbb{Z} \\ & \nwarrow & \uparrow{\scriptstyle V} \\ & & X. \end{array}$$

The analog of the splitting principle for BU–structured bundles then says that over Y as in Lemma 2.5, there is an isomorphism

$$f^*(V - n \cdot 1) \cong \left(\bigoplus_{j=1}^{n} \mathcal{L}_j \right) - n \cdot 1 \cong \bigoplus_{j=1}^{n} (\mathcal{L}_j - 1),$$

where now both the left– and right–hand sides have a natural interpretation as vector bundles of virtual rank 0.

For the case $k = 2$, select a further lift

$$\begin{array}{ccccc} BSU & \longrightarrow & BU & \xrightarrow{+n \cdot 1} & BU \times \mathbb{Z} \\ & \nwarrow^{\widetilde{V}} & \nwarrow & & \uparrow V \\ & & & & X. \end{array}$$

We record the following analogous Lemma and proof:

LEMMA 3.1. *The bundle $f^*\widetilde{V}$ is equivalent (as BSU–classes) to a sum of bundles of the form $(\mathcal{H} - 1)(\mathcal{H}' - 1)$ for $\mathcal{H}$, $\mathcal{H}'$ complex line bundles.*

PROOF. Over Y, we may split two lines off of $V = \mathcal{L}_1 + \mathcal{L}_2 + V'$. Because $(\mathcal{L}_1 - \varepsilon)$ and $(\mathcal{L}_2 - \varepsilon)$ both have natural BU–structures, they can both be interpreted as classes in $kU^2(Y) = [Y, BU]$. It follows that their product $(\mathcal{L}_1 - \varepsilon)(\mathcal{L}_2 - \varepsilon)$ has a natural SU–structure, where we cup the two classes in $kU^2(Y)$ to form a class in $kU^4(Y)$, and hence the sum

$$\widetilde{V} + (\mathcal{L}_1 - \varepsilon)(\mathcal{L}_2 - \varepsilon)$$

has a natural SU–structure as well.[2] The underlying bundle has the form

$$\begin{aligned} (V - n\varepsilon) + (\mathcal{L}_1 - \varepsilon)(\mathcal{L}_2 - \varepsilon) &= (V' + \mathcal{L}_1 + \mathcal{L}_2 - n\varepsilon) + (\mathcal{L}_1 - \varepsilon)(\mathcal{L}_2 - \varepsilon) \\ &= (V' + \mathcal{L}_1 + \mathcal{L}_2 - n\varepsilon) \\ &\quad - (\mathcal{L}_1 - \varepsilon) - (\mathcal{L}_2 - \varepsilon) + (\mathcal{L}_1\mathcal{L}_2 - \varepsilon) \\ &= (V' + \mathcal{L}_1\mathcal{L}_2) - (n-1)\varepsilon. \end{aligned}$$

This is the virtualization of a vector bundle of one rank fewer, so we can induct.

To ground the induction, we need only know that all rank 1 bundles with SU–structure are trivial. This follows because admitting an SU–structure is identical to having vanishing first Chern class, but complex line bundles are determined by their first Chern classes. □

LEMMA 3.2 (Hopkins[3]). *In the case $k = 3$, $BU[6, \infty)$–bundles decompose into sums of $BU[6, \infty)$–bundles of the form $(\mathcal{H} - 1)(\mathcal{H}' - 1)(\mathcal{H}'' - 1)$.* □

In light of these enhanced splitting principles for the structure groups SU and $U[5, \infty)$, we are moved to study the universal maps

---

[2] In general, the ring structure gives a lift $(\mathbb{C}\mathrm{P}^\infty)^{\times k} \to BU[2k, \infty)$ of the k–fold external product of the reduced tautological bundle. This lift does not actually depend on the ring structure: it also exists because the bottom several Chern classes of the external product vanish, and it is unique because there are no odd-degree classes in $kU^*(\mathbb{C}\mathrm{P}^\infty)^{\times k}$.

[3] This has been claimed without proof [**Hop95**, pg. 558]. However, I do not know an elementary proof of this, along the lines of Lemma 3.1.

$$(\mathbb{C}\mathrm{P}^\infty_E)^{\times 2} \xrightarrow{(\mathcal{L}_1-1)(\mathcal{L}_2-1)} BSU_E,$$

$$(\mathbb{C}\mathrm{P}^\infty_E)^{\times 3} \xrightarrow{(\mathcal{L}_1-1)(\mathcal{L}_2-1)(\mathcal{L}_3-1)} BU[6,\infty)_E.$$

These maps have some evident properties:

(1) They are *symmetric*: trading $\mathcal{L}_j$ and $\mathcal{L}_k$ gives an isomorphic bundle.
(2) They are *rigid*: replacing $\mathcal{L}_j$ by the trivial line gives the zero bundle.
(3) They are $\operatorname{Div}\mathbb{C}\mathrm{P}^\infty_E$*–linear*: one can act by a line bundle $\mathcal{H} \in kU^0\mathbb{C}\mathrm{P}^\infty$ using the product map

$$\underline{kU}_0 \times \underline{kU}_6 \xrightarrow{\mu} \underline{kU}_6$$

which gives simultaneous decompositions

$$\begin{aligned}(\mathcal{L}_1-1)\mathcal{H}(\mathcal{L}_2-1) &= (\mathcal{L}_1-1)(\mathcal{H}\mathcal{L}_2-\mathcal{H}) = (\mathcal{L}_1-1)(\mathcal{H}\mathcal{L}_2-1)-(\mathcal{H}-1)(\mathcal{L}_1-1)\\ &= (\mathcal{H}\mathcal{L}_1-\mathcal{H})(\mathcal{L}_2-1) = (\mathcal{H}\mathcal{L}_1-1)(\mathcal{L}_2-1)-(\mathcal{H}-1)(\mathcal{L}_2-1).\end{aligned}$$

Equating these last two expressions gives a kind of 2–cocycle condition. With these observations in hand, one is led to investigate the following functions:

DEFINITION 3.3 ([**AHS01**, Definition 2.42]). A *k–variate symmetric 2–cocycle for E* is a trivialization of the line bundle $\mathbb{L}(\bigotimes_{j=1}^k(\mathcal{L}_j-1))$ on $(\mathbb{C}\mathrm{P}^\infty)^{\times k}_E$ satisfying a symmetry condition, a rigidity condition, and the linearity condition[4]

$$\frac{f(t, x +_{\mathbb{C}\mathrm{P}^\infty_E} y, \ldots)}{f(t,x,\ldots)} = \frac{f(t +_{\mathbb{C}\mathrm{P}^\infty_E} x, y, \ldots)}{f(x,y,\ldots)}.$$

We write $C^k(\mathbb{C}\mathrm{P}^\infty_E; \mathcal{I}(0))$ for the set of such.

These are the sections which one "expects" to come from $BU[2k,\infty)$–structures. The main theorem of Ando, Hopkins, and Strickland is that these considerations are sufficient:

THEOREM 3.4 ([**AHS01**, Corollary 2.52]). *For E a complex-orientable ring spectrum and $k \le 3$ an integer, such a trivializing section determines a ring map $MU[2k,\infty) \to E$.*

NOTES ON PROOF. They reduce from general E to $E = MU$, then from there to $E = Hk$ for k a prime field. They then perform the brutal calculation of the available such sections in the algebraic model and compare with $\operatorname{Spec} Hk_*MU[2k,\infty)$, made accessible by Singer's calculation below. They find that the topological and algebraic objects have the same graded ranks and that one surjects onto the other. □

In the case of $k = 0$ and $k = 1$, this degenerates to exactly the classical analysis. In the case where E is an *elliptic spectrum*, they further show that the associated elliptic curve begets a canonical section in the $k = 3$ case.[5] This unicity is the main punchline in the construction of the (complex, nonparametrized) σ–orientation. It's also possible, by placing more hypotheses on E, to extend this same mode of analysis to ring maps $M\mathrm{String} \to E$ [**HAS99**].

---

[4]To explain the notation in this last condition, $\mathbb{C}\mathrm{P}^\infty_E$ is a smooth formal group, owing to the tensor product of line bundles, and we denote its group operation by "$+_{\mathbb{C}\mathrm{P}^\infty_E}$".

[5]Remarkably, attaching elliptic data to E does *not* give a canonical section for $k < 3$!

4. Orientations for $MU[2k,\infty)$?

Several things go wrong in trying to extend this analysis to $MU[2k,\infty)$ for $k \geq 4$, which forms the real subject of this note. In fact, something immediately obstructs any attempt to continue the algebro-geometric analysis described. The following theorem is key to their result:

THEOREM 4.1 ([**Sin68, Sto63**]). *There are classes* $\theta_{2j} \equiv c_j$ (mod decomposables) *such that*

$$H\mathbb{F}_2{}^*(BU[2k,\infty)) \cong \frac{H\mathbb{F}_2{}^*(BU)}{\langle \theta_{2j} \mid \sigma_2(j-1) < k-1 \rangle} \otimes \mathrm{Op}[\mathrm{Sq}^3 \iota_{2k-3}],$$

where $\mathrm{Op}[\mathrm{Sq}^3 \iota_{2k-3}]$ *is the Steenrod–Hopf sub-algebra of* $H\mathbb{F}_2{}^*K(\mathbb{Z}, 2k-3)$ *generated by* $\mathrm{Sq}^3 \iota_{2k-3}$. □

COROLLARY 4.2 ([**HLP13**, Remark 8.1]). *There are odd classes in* $H\mathbb{F}_{2*}BU[2k,\infty)$ *for* $k \geq 4$.

PROOF. The class $\mathrm{Sq}^7 \mathrm{Sq}^3 \iota_{2k-3} \in \mathrm{Op}[\mathrm{Sq}^3 \iota_{2k-3}]$ is non-zero once $k \geq 4$. □

By consequence, there are also odd classes in $H\mathbb{F}_{2*}MU[2k,\infty)$, and there are similar statements at primes $p \geq 3$. This means that we're not allowed to form the affine scheme "$\mathrm{Spec}\, H\mathbb{F}_{p*}MU[2k,\infty)$" for $k \geq 4$, and so the Ando–Hopkins–Strickland program is immediately stymied.

Nonetheless, we can still perform a purely algebraic analysis. In order to find more concrete footing, consider the following result of Ando, Hopkins, and Strickland:

LEMMA 4.3 ([**AHS01**, Theorem 2.50]). *For $k \leq 3$, the Thom diagonal map*

$$E_*MU[2k,\infty) \otimes_{E_*} E_*BU[2k,\infty) \leftarrow E_*MU[2k,\infty)$$

models the tensor product map

$$\begin{aligned} C^k(\mathbb{C}\mathrm{P}^\infty_E; \mathcal{I}(0)) \times C^k(\mathbb{C}\mathrm{P}^\infty_E; \mathcal{O}) &\to C^k(\mathbb{C}\mathrm{P}^\infty_E; \mathcal{I}(0) \otimes \mathcal{O}) \\ &\xrightarrow{\cong} C^k(\mathbb{C}\mathrm{P}^\infty_E; \mathcal{I}(0)), \end{aligned}$$

where $C^k(\mathbb{C}\mathrm{P}^\infty_E; \mathcal{O})$ denotes the set of k–variate $\mathbb{G}_m$–valued functions satisfying the linearity, symmetry, and 2–cocycle conditions. □

This presents $C^k(\mathbb{C}\mathrm{P}^\infty_E; \mathcal{I}(0))$ as a torsor for $C^k(\mathbb{C}\mathrm{P}^\infty_E; \mathcal{O})$. In particular, there is a noncanonical isomorphism between them, and we may as well study the untwisted version, which is a considerable gain in concreteness. Moreover, our case of interest is $E = H\mathbb{F}_2$, so that $\mathbb{C}\mathrm{P}^\infty_{H\mathbb{F}_2} = \widehat{\mathbb{G}}_a$ over $\mathrm{Spec}\,\mathbb{F}_2$, where we too can perform the brutal calculation analogous to the proof of Theorem 3.4:

THEOREM 4.4 ([**HLP13**, Theorem 7.7]). *There is a presentation of $\mathbb{F}_2$–algebras*

$$\begin{aligned} \mathcal{O}(\underline{C}^k(\widehat{\mathbb{G}}_a; \mathcal{O})) = \mathbb{F}_2[z_n \mid \nu_2\varphi(n,k) \leq \nu_2 n] \otimes \\ \otimes \Gamma[b_{n,\gamma_2(n,k)} \mid \nu_2\varphi(n,k) > \nu_2 n] \otimes \\ \otimes \mathbb{F}_2[b_{n,i} \mid \gamma_2(n,k) < i < D_{n,k}]/\langle b_{n,i}^2 \rangle, \end{aligned}$$

where $n \geq k$ ranges over integers, $D_{n,k}$ is the coefficient of the generating function

$$\prod_{i=0}^{\infty} \frac{1}{1 - tx^{2^i}} = \sum_{n,k} D_{n,k} x^n t^k,$$

and $\gamma_p(n,k) = \max\{0, \min\{k - \sigma_p(n), \nu_p(n)\}\}$ *counts the number of divided power classes introduced already.*

The same construction as the universal map in Section 3 begets a ring map

$$\mathcal{O}(\underline{C}^k(\widehat{\mathbb{G}}_a; \mathcal{O})) \to H\mathbb{F}_{2*}BU[2k, \infty).$$

More than this, this map is equivariant for the coaction of $\mathcal{O}(\operatorname{Aut} \widehat{\mathbb{G}}_a)$, i.e., it is a map of comodules for the dual Steenrod algebra. Accordingly, it factors

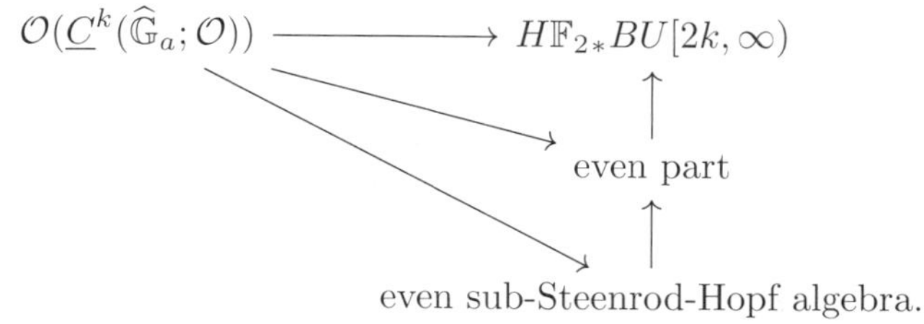

CONJECTURE 4.5 ([**HLP13**, Remark 8.2]). The longest map is an isomorphism.

EVIDENCE. The Poincaré series of the source of the isomorphism is accessible from Theorem 4.4. Singer's method gives partial information about the coaction of the dual Steenrod algebra on the topological side, which is enough to make an educated guess about the Poincaré series of the restricted algebraic target. As far out as we can check (through some thousands of bidegrees) these series agree. □

This means that the algebraic approximation is actually pretty good — much, much closer than you might initially think! This begets a natural question, if we want to use this calculation in homotopy theory:

QUESTION 4.6 ([**HLP13**, Remark 8.5]). Is there a space $X(k)$ with $\operatorname{Spec} E_* X(k) \cong \underline{C}^k(\widehat{\mathbb{G}}_E; \mathcal{O})$ for E a complex-orientable cohomology theory? In particular, is there a space whose $H\mathbb{F}_2$-homology realizes this particular subset of the $H\mathbb{F}_2$-homology of $BU[2k, \infty)$?

These spaces should probably have some other nice properties if they're to be truly useful. However, because these hypothetical spaces are so close to such already-famous spaces, there are many dangerous theorems to steer around.

LEMMA 4.7. *The spaces* $X(k)$ *cannot collectively form the spaces of an* Ω*-spectrum.*

PROOF. The results of Adams and Priddy [**AP76**] show that any connective spectrum delooping BSU is p-locally equivalent to $kU[4, \infty)$. Since we have $X(2) = BSU$, this would force $X(k) = BU[2k, \infty)$ for all k. However, we know that $BU[8, \infty)$ does not model $X(4)$. □

Nonetheless, the putative spaces $X(k)$ may individually be infinite loopspaces, with associated spectra $x(k)$. These, too, are prohibited from being too interesting. Recall that part of the Ando–Hopkins–Strickland program was to study a particular lifting problem, which $K(2)$-locally looks like

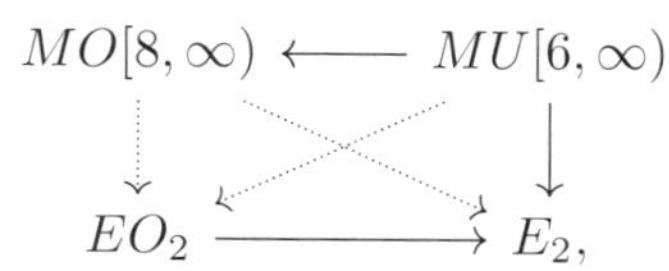

where E_2 is a Morava E–theory associated to a supersingular elliptic curve and EO_2 is its homotopy fixed points under the action of a maximal finite subgroup of the automorphisms of the curve. These sorts of factorizations are largely prohibited at higher chromatic heights.

LEMMA 4.8. *There cannot exist a diagram of ring spectra*

$$\begin{array}{ccccccc} \mathbb{S} & \longrightarrow & MO[j,\infty) & \longrightarrow & xo(k) & \longrightarrow & EO_{p-1} \\ & & \downarrow & & \downarrow & & \downarrow \\ & & MU[j,\infty) & \longrightarrow & x(k) & \longrightarrow & E_{p-1}. \end{array}$$

for any prime $p \geq 5$ and for any *choice of k and j.*

PROOF. Already the "outer" rectangle is prohibited from existing, with the $xo(k) \to x(k)$ column removed. Namely, Hovey has shown [**Hov97**, Proposition 2.3.2] that there cannot exist a map of ring spectra $MO[j,\infty) \to EO_{p-1}$ because the homotopy element $\alpha_{1/1}$ is visible in $\pi_* EO_{p-1}$ but not in $\pi_* MO[j,\infty)$. □

This means that, even if $x(k) \to E_{p-1}$ does exist in any sense, there cannot exist a map $xo(k) \to EO_{p-1}$ factoring it which participates at all naturally with orientations by connective real bordism theories.

In spite of these results, there is an interesting sequence of infinite loopspaces that overlaps with the well–understood part of this analysis.

DEFINITION 4.9. A *Wilson space* is a connected p–local H–space which is indecomposable and which has $\mathbb{Z}_{(p)}$–free homotopy and homology.

THEOREM 4.10 ([**Wil75**, Corollary 6.8]). *Wilson spaces* all *appear as $\underline{BP\langle n\rangle}_k$ for varying values of n and k.* □

At $p = 2$, the first few Wilson spaces are indicated in the following table:

k	1	2	3	4	$\cdots$
Y_k	BU	BSU	$BU[6,\infty)$	$\underline{BP\langle 2\rangle}_8$	$\cdots$.

Their ordinary homology is known as a Hopf ring, computed by Sinkinson [**Sin76**] following along the lines of Ravenel and Wilson's analysis [**RW77**] of $H\mathbb{F}_{p*}\underline{MU}_2$, and the result doesn't look very much like what we described above. Nonetheless, the primacy of Wilson spaces and their overlap with the Ando–Hopkins–Strickland results is encouraging. In light of our summary here, perhaps some connection would be uncovered if we could answer the following question:

QUESTION 4.11. Is there a splitting principle for $\underline{BP\langle 2\rangle}_8$–classes?

References

[ABS64] M. F. Atiyah, R. Bott, and A. Shapiro, *Clifford modules*, Topology **3** (1964), no. suppl. 1, 3–38, DOI 10.1016/0040-9383(64)90003-5. MR0167985

[AHR] M. Ando, M. J. Hopkins, and C. Rezk, *Multiplicative orientations of KO-theory and of the spectrum of topological modular forms*, `http://www.math.illinois.edu/~mando/papers/koandtmf.pdf`, Accessed: 2016-09-13.

[AHS01] M. Ando, M. J. Hopkins, and N. P. Strickland, *Elliptic spectra, the Witten genus and the theorem of the cube*, Invent. Math. **146** (2001), no. 3, 595–687, DOI 10.1007/s002220100175. MR1869850

[AP76] J. F. Adams and S. B. Priddy, *Uniqueness of* BSO, Math. Proc. Cambridge Philos. Soc. **80** (1976), no. 3, 475–509, DOI 10.1017/S0305004100053111. MR0431152

[AS01] M. Ando and N. P. Strickland, *Weil pairings and Morava K-theory*, Topology **40** (2001), no. 1, 127–156, DOI 10.1016/S0040-9383(99)00057-9. MR1791270

[HAS99] M. Hopkins, M. Ando, and N. Strickland, *Elliptic cohomology of $BO\langle 8\rangle$ in positive characteristic*, Unpublished preprint.

[HLP13] A. Hughes, J. Lau, and E. Peterson, *Multiplicative 2-cocycles at the prime 2*, J. Pure Appl. Algebra **217** (2013), no. 3, 393–408, DOI 10.1016/j.jpaa.2012.06.031. MR2974221

[Hop95] M. J. Hopkins, *Topological modular forms, the Witten genus, and the theorem of the cube*, Proceedings of the International Congress of Mathematicians, Vol. 1, 2 (Zürich, 1994), Birkhäuser, Basel, 1995, pp. 554–565. MR1403956

[Hov97] M. A. Hovey, *v_n-elements in ring spectra and applications to bordism theory*, Duke Math. J. **88** (1997), no. 2, 327–356, DOI 10.1215/S0012-7094-97-08813-X. MR1455523

[Joa97] M. Joachim, *The twisted atiyah orientation and manifolds whose universal cover is spin*, ProQuest LLC, Ann Arbor, MI, 1997. Thesis (Ph.D.)–University of Notre Dame. MR2695871

[Koc78] S. O. Kochman, *A chain functor for bordism*, Trans. Amer. Math. Soc. **239** (1978), 167–196, DOI 10.2307/1997852. MR0488031

[RW77] D. C. Ravenel and W. S. Wilson, *The Hopf ring for complex cobordism*, J. Pure Appl. Algebra **9** (1976/77), no. 3, 241–280, DOI 10.1016/0022-4049(77)90070-6. MR0448337

[Sin68] W. M. Singer, *Connective fiberings over* BU *and* U, Topology **7** (1968), 271–303, DOI 10.1016/0040-9383(68)90006-2. MR0232392

[Sin76] K. Sinkinson, *The cohomology of certain spectra associated with the Brown-Peterson spectrum*, Duke Math. J. **43** (1976), no. 3, 605–622. MR0410725

[Sto63] R. E. Stong, *Determination of $H^*(\mathrm{BO}(k,\cdots,\infty),Z_2)$ and $H^*(\mathrm{BU}(k,\cdots,\infty),Z_2)$*, Trans. Amer. Math. Soc. **107** (1963), 526–544, DOI 10.2307/1993817. MR0151963

[Str99] N. P. Strickland, *Formal schemes and formal groups*, Homotopy invariant algebraic structures (Baltimore, MD, 1998), Contemp. Math., vol. 239, Amer. Math. Soc., Providence, RI, 1999, pp. 263–352, DOI 10.1090/conm/239/03608. MR1718087

[Wil75] W. S. Wilson, *The Ω-spectrum for Brown-Peterson cohomology. II*, Amer. J. Math. **97** (1975), 101–123, DOI 10.2307/2373662. MR0383390

Department of Mathematics, Harvard University, 1 Oxford Street, Cambridge, MA 02138

Email address: ecp@math.harvard.edu

Contemporary Mathematics
Volume **707**, 2018
http://dx.doi.org/10.1090/conm/707/14256

The linearity of fixed point invariants

Kate Ponto and Michael Shulman

ABSTRACT. We prove two general decomposition theorems for fixed-point invariants: one for the Lefschetz number and one for the Reidemeister trace. These theorems imply the familiar additivity results for these invariants. Moreover, the proofs of these theorems are essentially formal, taking place in the abstract context of bicategorical traces. This makes it straightforward to generalize the theory to analogous invariants in other contexts, such as equivariant and fiberwise homotopy theory.

CONTENTS

2010 *Mathematics Subject Classification.* 18D05, 18D10, 55M20.
Key words and phrases. trace, additivity, derivators, monoidal model category.
The first author was partially supported by NSF grant DMS-1207670. The second author was partially supported by an NSF postdoctoral fellowship and NSF grant DMS-1128155, and appreciates the hospitality of the University of Kentucky. Any opinions, findings, and conclusions or recommendations expressed in this material are those of the authors and do not necessarily reflect the views of the National Science Foundation.

1. Introduction

In topological fixed-point theory, one of the most basic invariants is the *Lefschetz number* $L(f)$ of an endomorphism f of a finite CW complex X, which vanishes whenever f is homotopic to a map with no fixed points. If f is the identity map, then $L(f)$ is the *Euler characteristic* $\chi(X)$ of the complex X.

Two of the most important tools for calculating Lefschetz numbers (and hence also Euler characteristics) are *additivity* and *multiplicativity*. The classical additivity theorem states that if $X \subset Y$ is a subcomplex and $f\colon Y \to Y$ takes X into itself, then

$$L(f) = L(f|_X) + L(f/X) \tag{1.1}$$

where $f|_X$ denotes the restriction of f to X and f/X denotes the induced map on the quotient Y/X. The classical multiplicativity theorem is that if $p\colon E \to B$ is a fibration with fiber F and simply connected base B, and $f\colon E \to E$ and $f_B\colon B \to B$ are endomorphisms satisfying $p \circ f = f_B \circ p$, then

$$L(f) = L(f_B) \cdot L(f|_F). \tag{1.2}$$

This paper is part of a project to prove similar theorems for generalizations of the Lefschetz number, such as the Reidemeister trace (which refines the Lefschetz number with information appropriate to non-simply-connected spaces), as well as versions of the Lefschetz number and Reidemeister trace for parametrized or equivariant endomorphisms [**Pon10, Pon15**]. These refined invariants are relatively incompatible with the classical approaches to additivity and multiplicativity, so we instead approach these results by embracing a greater level of abstraction.

In [**DP80**], Dold and Puppe observed that the Lefschetz number is an example of the *trace* in a symmetric monoidal category. This generalizes the familiar trace for linear transformations. (See [**PS13b**] for an overview of traces in symmetric monoidal categories.) They also noted that the trace in an equivariant or parameterized stable homotopy category is the equivariant or parameterized Lefschetz number, and that the Lefschetz fixed-point theorem is a consequence of the fact that symmetric monoidal functors preserve traces.

In [**Pon10, PS13a**], the first author observed that the Reidemeister trace can be described in a similar way by generalizing from symmetric monoidal traces to *bicategorical traces*. Here a bicategory is regarded as a "many-object" version of a monoidal category, where the objects are (in the simplest version) groups or rings that encode the actions of fundamental groups. This generalization yields definitions of equivariant and parameterized Reidemeister traces, and the Reidemeister-trace version of the Lefschetz fixed-point theorem follows from the preservation of bicategorical traces by suitable functors of bicategories.

From this perspective, (1.1) and (1.2) have very suggestive parallels with the trace of a linear transformation: the additivity of the Lefschetz number parallels the additivity of the trace of linear transformations on short exact sequences, and the multiplicativity of the Lefschetz number parallels the compatibility between the trace and tensor product of linear transformations. This suggests that additivity and multiplicativity results for fixed-point invariants can usefully be situated in the larger project of establishing such results for the symmetric monoidal and bicategorical traces.

In fact, two special cases of additivity and multiplicativity are immediate consequences of the identification of fixed-point invariants as traces. Traces in a closed symmetric monoidal category are automatically compatible with coproducts and with the monoidal product itself, yielding respectively the "trivial" case of (1.1) when X is the inclusion of a disjoint summand and the "trivial" case of (1.2) when $E = B \times F$ and p is the product projection.

In [**May01**] May extended these compatibility arguments to the general case of (1.1) for Lefschetz numbers, using triangulated categories with compatible monoidal structures, and in [**GPS14**] we reformulated his proof using stable derivators. The proof is much more complicated than that of the trivial case, but the basic idea is the same: the quotient Y/X is a kind of colimit, and the tensor product preserves colimits in each variable, yielding a compatibility relation for traces from which (1.1) can be extracted.

By contrast, the nontrivial case of multiplicativity requires a change of perspective. The hypothesis of simple connectivity is unnatural from an abstract point of view. But in its absence, the analogue of (1.2) must involve not only the Lefschetz number of f_B but its Reidemeister trace, and this pushes us into the world of bicategorical trace. This led us in [**PS12**] to study traces in bicategories that can contain arbitrary fibrations $p : E \to B$, such as those of parametrized spaces and spectra from [**MS06**].

Then in [**PS14**] we gave an abstract proof of multiplicativity: not only (1.2) but also a more refined version that eliminates the hypothesis of simple connectivity, as well as a corresponding result for the Reidemeister trace, all of which can easily be generalized to other contexts including the parametrized and equivariant worlds. The idea is to view the multiplication in (1.2) as the product of two 1×1 matrices, and then generalize it to the product of a $1 \times n$ matrix and an $n \times 1$ matrix. The dimension n encodes $\pi_1(B)$, and the matrix product is a "decategorification" of the composition in a bicategory of parametrized spaces or spectra.

In this paper and its companion [**PS16**], we apply a similar perspective to prove the additivity formula (1.1) and significant generalizations thereof. The most familiar of these generalizations is the following additivity formula for the Reidemeister trace.

THEOREM 1.3. *If X and Y are closed smooth manifolds with $X \subset Y$, $f\colon Y \to Y$ is a continuous map such that $f(X) \subset X$, $\Lambda^f Y$ is the twisted loop space, and $i\colon \Lambda^{f|_X} X \to \Lambda^f Y$ is the inclusion, then*

$$R(f) - i(R(f|_X)) = R_{Y|X}(f).$$

The invariant $R_{Y|X}(f)$ is the relative Reidemeister trace *of f [**Pon11**]. If $X \subset Y$ is a cofibration, $R_{Y|X}(f)$ is a refinement of the Reidemeister trace of $f/X\colon Y/X \to Y/X$ that takes values in $\pi_0^s(\Lambda^f(Y))$.*

While this particular result is a small generalization of [**Fer96**, Theorem 3.2.1], our method is dramatically different from Ferrario's. Most significantly, our method can be applied directly to produce the analogous results for parametrized and equivariant Lefschetz numbers and Reidemeister traces.

The insight leading to our proof of additivity is to rewrite (1.1) as

$$L(f/X) = L(f) - L(f|_X). \tag{1.4}$$

In this form, it can be viewed as a calculation of the invariant associated to a *colimit*, namely the quotient Y/X, in terms of the corresponding invariants associated to the input data (X and Y). We now view the right-hand side of (1.4) as the product of the 1×2 matrix $[L(f), L(f|_X)]$ and the 2×1 matrix $\begin{bmatrix} 1 \\ -1 \end{bmatrix}$. The dimension 2 and the latter matrix are invariants determined by the type of colimit under consideration, and the matrix product is a decategorification of the composition in a suitable bicategory (namely, a bicategory of profunctors). In [**PS16**] we develop this idea into a general method for calculating trace invariants associated to colimits of diagrams; while in this paper we specialize these results to diagrams of spaces, recovering (1.4) by specializing further to quotients.

There is a strong similarity to the results of [**PS14**] in both perspective and technique: in both cases we build on a generalization of the invariance of trace under change of basis that is proven by abstract diagram chases, and we describe the trace of a morphism in terms of traces of constituent pieces. And indeed, from an ∞-categorical point of view, the total space of a fibration can be identified with a colimit whose "diagram shape" is the base space regarded as an ∞-groupoid.

However, because we take a more concrete topological approach, there are also major technical differences between the two theories. The main one is that when dealing with *diagrams* in homotopy theory (which we did not have to do in [**PS14**]), there is a constant need for *fibrant and cofibrant replacement* in order to ensure that various maps that ought to exist really do exist. To avoid this complication, we use a technical device called a *derivator bicategory*, which encapsulates all the fibrant and cofibrant replacement in an abstract structure so that we never have to think about it again.

Although this abstract structure is very convenient for proving the abstract theorems, the topologically inclined reader will naturally want a more concrete description of what is going on. This is the reason for the division between this paper and its companion [**PS16**]: in [**PS16**] we develop the general theory abstractly, while in this paper we apply that theory more concretely to the cases of topological interest. Formally speaking, the latter simply means verifying that the topological context gives rise to a derivator bicategory; we will do that in Part 2 of this paper. However, since many readers may not be interested in that level of abstraction, in Part 1 we give a more concrete translation of the theorems in purely topological language. We encourage the reader with a purely topological interest to focus on Part 1 and treat Part 2 and [**PS16**] as "black boxes". Although we sketch some proofs in Part 1 to give an idea of what is going on, formally speaking the real work is done in [**PS16**] and Part 2.

We begin Part 1 in §2 by explaining how to view the Lefschetz number as a trace, and the implications of this viewpoint (for instance, it immediately implies multiplicativity for trivial bundles). In §3 we then generalize this to traces of *diagrams*, such as the map $X \to Y$ of which we considered the quotient above. The invariant associated to an endomorphism of a diagram includes the Lefschetz number of all the induced endomorphisms of the component spaces. The central theorem then says that the Lefschetz number of the colimit (such as Y/X) can be calculated as a linear combination of these component Lefschetz numbers (such as (1.4)); thus we call it a *linearity formula*.

In §4 we move on to the Reidemeister trace, explaining how to view it as a trace. This involves introducing *parametrized spaces* as in [**MS06**], but we try to remain as explicit as possible. Finally, in §5 we combine §3 and §4 to deduce a linearity formula for traces of diagrams of parametrized spaces, and in §6 we use this theory to deduce the additivity of Reidemeister trace.

Part 2 begins in §7 by recalling the definition of a derivator bicategory from [**PS16**] and the statements of some of the results we used in §5 in the greater generality of a derivator bicategory. Then in §8 we describe a naturally occurring structure on parametrized spaces that we use to construct the derivator bicategory of parametrized spectra, generalizing *indexed monoidal categories* (see [**PS12**]) to *indexed monoidal derivators*. In §9 we describe closed structures on indexed monoidal derivators, and hence on the derivator bicategories they give rise to; these are used in [**PS16**] to make defining and constructing dual pairs easier. Finally, in §10 we explain how to construct an indexed monoidal derivator from an *indexed monoidal model category*. This allows us to produce the desired derivator bicategory of parametrized spectra, starting from an indexed monoidal model category of parametrized spectra (which is essentially what was constructed in [**MS06**]).

Part 1. Topological fixed-point invariants

In these sections we give an explicit, topological description of the linearity results for fixed point theory to the extent that it is illuminating. In practice, the approach we are using is clear when working in this specific context, but there are a few important results whose proofs become unpleasant and relatively intractable when interpreted in particular examples and are far more manageable in greater generality. For these results we provide an indication of the proof, but leave the formal proof to Part 2 and [**PS16**].

It is important to note that while we state the results in this section in terms of classical stable homotopy theory, they apply in equivariant and fiberwise stable homotopy as well. Generalizing further, the approach of §2 and §3 applies in any "homotopy category", such as those arising from model categories, and the approach of §4 to §6 applies to "indexed monoidal model categories" as defined in §10.

2. The Lefschetz number as a trace

We begin by recalling how the Lefschetz number can be regarded as a trace. Let M be a based topological space. We say that M is n-**dualizable** if there is a based space DM and maps

$$\eta\colon S^n \longrightarrow M\wedge DM$$
$$\varepsilon\colon DM\wedge M \longrightarrow S^n$$

such that the composites

$$S^n\wedge M \xrightarrow{\eta\wedge\mathrm{id}} M\wedge DM\wedge M \xrightarrow{\mathrm{id}\wedge\varepsilon} M\wedge S^n$$

$$DM\wedge S^n \xrightarrow{\mathrm{id}\wedge\eta} DM\wedge M\wedge DM \xrightarrow{\varepsilon\wedge\mathrm{id}} S^n\wedge DM$$

become homotopic to transposition maps after smashing with some S^m (in this case one says they are *stably homotopic* to transpositions). We refer to η as the **coevaluation** and ε as the **evaluation** for the duality.

An unbased space M is **n-dualizable** if M with a disjoint basepoint, written M_+, is n-dualizable in the above sense. It is well-known (see [**Ati61, LMSM86**]) that every closed smooth manifold M is n-dualizable. We may take n to be the dimension of a Euclidean space in which M embeds, and DM the Thom space of the normal bundle of the embedding.

If M is an n-dualizable based space and $f\colon M \to M$ is a based endomorphism, we define its **trace** $\mathrm{tr}(f)$ to be the composite map

$$S^n \xrightarrow{\eta} M \wedge DM \xrightarrow{f\wedge \mathrm{id}} M \wedge DM \xrightarrow{\cong} DM \wedge M \xrightarrow{\varepsilon} S^n \tag{2.1}$$

and its **Lefschetz number** $L(f)$ to be the degree of this trace. The **Euler characteristic** of M, $\chi(M)$, is the Lefschetz number of the identity map of M. We apply all these notions to unbased spaces and maps by adjoining disjoint basepoints. These definitions are clearly homotopy invariant, and are known to agree with all other definitions of Lefschetz number and Euler characteristic; see [**DP80**].

There are many reasons why this formulation of Lefschetz number and Euler characteristic is useful, but for us the most important is that it makes it easy to prove the multiplicativity theorem for trivial bundles. Specifically, if M and N are both n-dualizable based spaces, then it is easy and formal to prove that $M \wedge N$ is also n-dualizable; its dual is $DN \wedge DM$. Moreover, if $f\colon M \to M$ and $g\colon N \to N$ are endomorphisms, we can prove by formal manipulation that $\mathrm{tr}(g \wedge f) \sim \mathrm{tr}(g) \circ \mathrm{tr}(f)$ as maps $S^n \to S^n$, hence $L(g \wedge f) = L(g) \cdot L(f)$. Of course, if M, N, f, and g are unbased, then $M_+ \wedge N_+ \cong (M \times N)_+$ and $f_+ \wedge g_+ \cong (f \times g)_+$, so we obtain the multiplicativity theorem for trivial bundles.

3. Linearity of the Lefschetz number

We now generalize this by considering *diagrams* of spaces. Let A and B be small categories; by a **profunctor** from A to B, we will mean a functor from $A \times B^{\mathrm{op}}$ to the category $\mathscr{T}$ of based spaces. For example, for any small category A and any based space X, we have a profunctor $X \wedge A$ from A to A, defined by

$$(X \wedge A)(a, a') = X \wedge \hom_A(a', a)_+,$$

where $\hom_A(a', a)$ has the discrete topology. More generally, if M is a profunctor from A to B and X is a based space, we have a profunctor $X \wedge M$ defined by $(X \wedge M)(a, b) = X \wedge M(a, b)$.

If $M\colon A \times B^{\mathrm{op}} \to \mathscr{T}$ and $N\colon B \times C^{\mathrm{op}} \to \mathscr{T}$ are profunctors, we define their **composite** $M \wedge_B N$ to be the profunctor from A to C where $(M \wedge_B N)(a, c)$ is the *homotopy coend* of the $(B^{\mathrm{op}} \times B)$-indexed diagram $(b, b') \mapsto M(a, b) \wedge N(b', c)$. In good situations (such as when M and N are nondegenerately based and have the homotopy type of CW complexes), this can be constructed as the geometric realization of its *simplicial bar construction*, which is a simplicial space:

$$\cdots \;\substack{\longrightarrow\\ \longleftarrow\\ \longrightarrow\\ \longleftarrow\\ \longrightarrow}\; \textstyle\coprod_{b,b'} M(a,b) \wedge \hom_B(b', b)_+ \wedge N(b', c) \;\substack{\longrightarrow\\ \longleftarrow\\ \longrightarrow}\; \coprod_b M(a,b) \wedge N(b,c)$$

See e.g. [**May75, Mey84, Shu06**] for more details. We have the following natural equivalences for any profunctor $M\colon A \times B^{\mathrm{op}} \to \mathscr{T}$ and any based space X.

$$(X \wedge A) \wedge_A M \simeq X \wedge M \qquad M \wedge_B (X \wedge B) \simeq X \wedge M$$

We are particularly interested in profunctors either to or from the terminal category $\mathbb{1}$ (with one object and only its identity morphism). A profunctor M from A to $\mathbb{1}$ is the same as an A-shaped diagram in $\mathscr{T}$, while a profunctor Φ from $\mathbb{1}$ to A

is the same as an A^{op}-shaped diagram in $\mathscr{T}$. Their composite $\Phi \wedge_A M$ is called the **Φ-weighted homotopy colimit of** M and denoted $\mathrm{colim}^{\Phi}(M)$. In particular, if Φ is constant at S^0, then $\mathrm{colim}^{\Phi}(M)$ is homotopy equivalent to the ordinary homotopy colimit of M. The general framework for linearity will tell us how to compute traces of endomaps of $\Phi \wedge_A M$ that are induced by endo-natural-transformations of M.

In order to state such a result, we need an appropriate notion of duality. We would like to say that a profunctor M from A to B is **right n-dualizable** if there is a profunctor $D_r M$ from B to A and natural transformations $\eta\colon S^n \wedge A \to M \wedge_B D_r M$ and $\epsilon\colon D_r M \wedge_A M \to S^n \wedge B$ such that the composites

$$\begin{array}{ccccc}
S^n \wedge M & & & & S^n \wedge M \\
\big\downarrow\wr & & & & \wr\big\uparrow \\
(S^n \wedge A) \wedge_A M & \xrightarrow{\eta\wedge\mathrm{id}} & M \wedge_B D_r M \wedge_A M & \xrightarrow{\mathrm{id}\wedge\epsilon} & M \wedge_B (S^n \wedge B)
\end{array}$$

$$\begin{array}{ccccc}
D_r M \wedge_A (S^n \wedge A) & \xrightarrow{\mathrm{id}\wedge\eta} & D_r M \wedge_A M \wedge_B D_r M & \xrightarrow{\epsilon\wedge\mathrm{id}} & (S^n \wedge B) \wedge_B D_r M \\
\big\uparrow\wr & & & & \wr\big\downarrow \\
S^n \wedge D_r M & & & & S^n \wedge D_r M
\end{array}$$

become naturally homotopic to identity maps[1] after smashing with some S^m.

However, here there is an additional wrinkle: we need to allow maps η and ϵ where the domains have been modified. Formally, we allow them to be *cofibrantly replaced* with respect to the *projective model structure* [**Hir03**, §11.6]. We think of this as replacing the given diagram by a *cell complex* whose cells are *freely generated* at some object of the indexing category. We will not need to be more specific about this here, since in many cases of interest (see Example 3.7 below) a very simple modification suffices. In the more general cases, [**PS16**] allows us to avoid the issue by using derivators, which "package" the fibrant and cofibrant replacements into an abstract structure so that we don't have to think about them.

The asymmetry between the categories A and B in the above definition means that it incorporates two very different types of duality, which can be best understood by restricting to the case when either A or B is the terminal category $\mathbb{1}$. On the one hand, if M is an A-shaped diagram, we can think of it as a profunctor from A to $\mathbb{1}$. If the resulting profunctor is right n-dualizable, we say that M is **pointwise dualizable**. On the other hand, if Φ is a B^{op}-shaped diagram, we can regard it as a profunctor from $\mathbb{1}$ to B. If this profunctor is right n-dualizable, we say that Φ is **absolute**.

The reason for the name "pointwise dualizable" is the following lemma.

LEMMA 3.1 ([**PS16**, Lemma 3.5]). *A functor $M\colon A \to \mathscr{T}$ is pointwise dualizable if and only if each space $M(a)$ is n-dualizable in $\mathscr{T}$.*

Absoluteness is a much stronger condition than pointwise dualizability. However, the *stability* built into the definition of n-duality allows us to exhibit a few examples.

EXAMPLE 3.2. Let $B = \emptyset$ be the empty category, and $\Phi\colon \emptyset^{\mathrm{op}} \to \mathscr{T}$ the unique functor, regarded as a profunctor from $\mathbb{1}$ to $\emptyset$. Then we can take $D_r\Phi$ to also be the

---

[1] In §2 we said instead that the corresponding composites should be homotopic to transpositions such as $S^n \wedge M \simeq M \wedge S^n$. We have now incorporated these transpositions into the equivalences at the beginning and end of the displayed composites.

unique functor $\emptyset \to \mathscr{T}$, now regarded as a profunctor from $\emptyset$ to $\mathbb{1}$. The composite $D_r\Phi \wedge_{\mathbb{1}} \Phi$ is again the unique functor $\emptyset \times \emptyset^{\mathrm{op}} \to \mathscr{T}$, while the composite $\Phi \wedge_\emptyset D_r\Phi$ is the functor $\mathbb{1} \times \mathbb{1}^{\mathrm{op}} \to \mathscr{T}$ sending the unique object to $\star$ (the one-point based space). The maps $\eta\colon S^0 \wedge \mathbb{1} \to \Phi \wedge_\emptyset D_r\Phi$ and $\epsilon\colon D_r\Phi \wedge_{\mathbb{1}} \Phi \to S^0 \wedge \emptyset$ are the only possible ones (no cofibrant replacement is necessary), and the required identities hold vacuously. Thus, Φ is absolute.

EXAMPLE 3.3. Let $B = \mathbb{1} \sqcup \mathbb{1}$ be the category with two objects and only identity morphisms, and $\Phi\colon (\mathbb{1} \sqcup \mathbb{1})^{\mathrm{op}} \to \mathscr{T}$ the functor sending each object to S^0, regarded as a profunctor from $\mathbb{1}$ to $\mathbb{1} \sqcup \mathbb{1}$. Then we can take $D_r\Phi$ to be the functor $\mathbb{1} \sqcup \mathbb{1} \to \mathscr{T}$ sending each object to S^1, regarded as a profunctor from $\mathbb{1} \sqcup \mathbb{1}$ to $\mathbb{1}$. The composite $D_r\Phi \wedge_{\mathbb{1}} \Phi$ is the functor $(\mathbb{1} \sqcup \mathbb{1}) \times (\mathbb{1} \sqcup \mathbb{1})^{\mathrm{op}} \to \mathscr{T}$ sending each object to S^1, while the composite $\Phi \wedge_B D_r\Phi$ is the functor $\mathbb{1} \times \mathbb{1}^{\mathrm{op}} \to \mathscr{T}$ sending the unique object to $S^1 \vee S^1$ (the coproduct in $\mathscr{T}$ of two copies of S^1). The coevaluation $\eta\colon S^1 \wedge \mathbb{1} \to \Phi \wedge_B D_r\Phi$ is the "pinch" map that wraps S^1 once around each summand of $S^1 \vee S^1$, while the evaluation $\epsilon\colon D_r\Phi \wedge_{\mathbb{1}} \Phi \to S^1 \wedge (\mathbb{1} \sqcup \mathbb{1})$ is the identity on S^1 where possible, and otherwise the zero map. Again, no cofibrant replacement is necessary, and the required identities are easy to check. Thus, Φ is absolute.

The interaction of pointwise duality and absoluteness is central to the subject. For instance, we have the following theorem.

THEOREM 3.4. *If Φ is absolute and M is pointwise dualizable, then the Φ-weighted homotopy colimit* $\mathrm{colim}^{\Phi}(M) = \Phi \wedge_A M$ *is n-dualizable. (More generally, if X and Y are right n-dualizable profunctors from A to B and from B to C respectively, then their composite $X \wedge_B Y$ is also right n-dualizable.)*

Perhaps surprisingly, the proof of this theorem is quite trivial: the dual of $X \wedge_B Y$ is $D_rY \wedge_B D_rX$ and the rest is diagram chases.

Note that the Φ of Examples 3.2 and 3.3 are of the "constant at S^0" sort, so that in both cases Φ-weighted homotopy colimits can be identified with ordinary homotopy colimits. Thus, in both cases, homotopy colimits of pointwise dualizable diagrams are dualizable.

EXAMPLE 3.5. If $B = \emptyset$, there is only one diagram $B \to \mathscr{T}$, which is trivially pointwise dualizable. Thus, Theorem 3.4 implies that its (homotopy) colimit, which is the one-point based space $\star$, is n-dualizable.

EXAMPLE 3.6. If $B = \mathbb{1} \sqcup \mathbb{1}$, then a diagram $B \to \mathscr{T}$ is just a pair of based spaces X and Y, and by Lemma 3.1 it is pointwise dualizable just when X and Y are n-dualizable. Thus, Theorem 3.4 implies that the homotopy colimit of such a diagram, which (at least if X and Y are nondegenerately based) is just the wedge $X \vee Y$, is also n-dualizable.

These conclusions (the initial object is dualizable, and coproducts of dualizable spaces are dualizable) are easy to prove by other means. Somewhat less obvious (though also well-known) is the following example, in which the weight is non-constant.

EXAMPLE 3.7. Let B be the arrow category, with two objects a and b, and one nonidentity morphism $\alpha\colon a \to b$. Let $\Phi\colon B^{\mathrm{op}} \to \mathscr{T}$ be defined by $\Phi(b) = S^0$ and $\Phi(a) = \star$, with $\Phi(\alpha)\colon S^0 \to \star$ the unique map. Then we can take $D_r\Phi\colon B \to \mathscr{T}$ to have $D_r\Phi(a) = S^0$ and $D_r\Phi(b) = \star$, with $D_r\Phi(\alpha)\colon S^0 \to \star$ again the unique map.

Their composite $\Phi \wedge_B D_r\Phi$ is the homotopy pushout of the diagram

$$\begin{array}{ccc} S^0 \wedge S^0 & \longrightarrow & S^0 \wedge \star \\ \downarrow & & \\ \star \wedge S^1 & & \end{array} \qquad \text{or equivalently} \qquad \begin{array}{ccc} S^0 & \longrightarrow & \star \\ \downarrow & & \\ \star & & \end{array}$$

which is the suspension of S^0, namely S^1. Technically by this we mean the profunctor $S^1 \wedge \mathbb{1}$; thus we can take $n = 1$ and let the coevaluation $\eta\colon S^1 \wedge \mathbb{1} \to \Phi \wedge_B D_r\Phi$ be the isomorphism.

Their composite $D_r\Phi \wedge_{\mathbb{1}} \Phi$ in the other order is the $(B \times B^{\mathrm{op}})$-indexed diagram that is $\star$ everywhere except that $(D_r\Phi \wedge_{\mathbb{1}} \Phi)(a, b) = S^0$, which can be drawn like the square on the left below (with the B direction horizontal and the B^{op} direction vertical).

$$\begin{array}{ccc} \star & \longrightarrow & \star \\ \uparrow & & \uparrow \\ S^0 & \longrightarrow & \star \end{array} \qquad\qquad \begin{array}{ccc} S^1 & \longrightarrow & S^1 \\ \uparrow & & \uparrow \\ \star & \longrightarrow & S^1 \end{array}$$

Similarly, the $(B \times B^{\mathrm{op}})$-indexed diagram $S^1 \wedge B$ can be drawn like the square on the right above. Here we do need a cofibrant replacement in order to have a nontrivial map $D_r\Phi \wedge_{\mathbb{1}} \Phi \to S^1 \wedge B$, and as remarked previously, in this case we can use a *Reedy model structure* for the diagram shape $B \times B^{\mathrm{op}}$. The resulting cofibrant replacement of $D_r\Phi \wedge_{\mathbb{1}} \Phi$ looks like this:

$$\begin{array}{ccc} D^1 & \longrightarrow & D^2 \\ \uparrow & & \uparrow \\ S^0 & \longrightarrow & D^1 \end{array}$$

and the evaluation $\epsilon\colon D_r\Phi \wedge_{\mathbb{1}} \Phi \to S^1 \wedge B$ wraps each D^1 around the corresponding S^1. The required identities hold, and so Φ is absolute.

Now a diagram $B \to \mathscr{T}$ is just a pair of based spaces and a map between them, $i\colon Y \to X$, and by Lemma 3.1 it is pointwise dualizable just when X and Y are n-dualizable. The Φ-weighted colimit of such a diagram is the homotopy pushout of

$$\begin{array}{ccc} S^0 \wedge Y & \longrightarrow & S^0 \wedge X \\ \downarrow & & \\ \star \wedge Y & & \end{array} \qquad \text{or equivalently} \qquad \begin{array}{ccc} Y & \longrightarrow & X \\ \downarrow & & \\ \star & & \end{array}$$

which is to say the *mapping cone* of i. Thus, Theorem 3.4 implies that the cone of any map between n-dualizable spaces is n-dualizable.

Next we define traces for endomorphisms of dualizable profunctors. However, there is a problem: the isomorphism $M \wedge D_rM \cong D_rM \wedge M$ in (2.1) doesn't even make sense for profunctors, since $M \wedge_B D_rM$ and $D_rM \wedge_A M$ are not even diagrams of the same shape. For this purpose we introduce a "quotienting" operation that makes them both into ordinary spaces, following [**Pon10, PS13a**].

Specifically, if M is any profunctor from A to A, then we define its **shadow**, denoted $\langle\!\langle M \rangle\!\rangle_A$, to be its homotopy coend. If M and N are profunctors from A to B and from B to A respectively, then we have a natural equivalence

$$\langle\!\langle M \wedge_B N \rangle\!\rangle_A \simeq \langle\!\langle N \wedge_A M \rangle\!\rangle_B.$$

If $M = S^n \wedge A$, then its shadow is the n-fold suspension of a space representing the "Hochschild homology" of the category A. We write $\langle\!\langle A\rangle\!\rangle_A$ for $\langle\!\langle S^0 \wedge A\rangle\!\rangle_A$.

The stable maps $S^n \to \langle\!\langle S^n \wedge A\rangle\!\rangle_A$ play an essential role in this section, so it will be necessary to have a more explicit understanding of them. First observe that smashing with S^n commutes with the homotopy coend, so it is enough to consider stable maps $S^0 \to \langle\!\langle A\rangle\!\rangle_A$, i.e. elements of the 0th stable homotopy group $\pi_0^s(\langle\!\langle A\rangle\!\rangle_A)$ of $\langle\!\langle A\rangle\!\rangle_A$. Now for an unbased space Y, there is an integer n so that $\pi_0^s(Y_+) \cong \pi_n(\Sigma^n(Y_+))$. Since $\tilde{H}_n(\Sigma^n(Y_+)) \cong H_0(Y)$, the Hurewicz theorem implies that $\pi_0^s(Y_+) \cong H_0(Y) \cong \mathbb{Z}\pi_0(Y)$ is the free abelian group on the set of connected components of Y. Therefore, to compute $\pi_0^s(\langle\!\langle A\rangle\!\rangle_A)$, it will suffice to show that $\langle\!\langle A\rangle\!\rangle_A$ has a disjoint basepoint and to count the other components.

For this purpose, recall that $(S^0 \wedge A)(a, a') = \hom_A(a', a)_+$. Homotopy coends commute with adjoining disjoint basepoints, so the homotopy coend of $S^0 \wedge A$ is the homotopy coend of the diagram $\hom_A$ of unbased spaces, with a disjoint basepoint. This homotopy coend, in turn, is the geometric realization of the simplicial set

$$\cdots \substack{\longrightarrow\\[-2pt]\longrightarrow\\[-2pt]\longrightarrow} \coprod_{a,a'} \hom_A(a, a') \times \hom_A(a', a) \substack{\longrightarrow\\[-2pt]\longleftarrow\\[-2pt]\longrightarrow} \coprod_a \hom_A(a, a)$$

Therefore, its set of connected components is the set of endomorphisms $\alpha\colon a \to a$ in A, modulo the equivalence relation generated by $\alpha\beta \sim \beta\alpha$ whenever $\alpha\colon a \to b$ and $\beta\colon b \to a$ are morphisms that are composable in both orders. We call these **conjugacy classes** of A, since when A is a group G regarded as a one-object groupoid, they are precisely the conjugacy classes of G. Thus, $\pi_0^s(\langle\!\langle A\rangle\!\rangle_A)$ is the free abelian group on the conjugacy classes of A. We write the conjugacy class of $\alpha\colon a \to a$ as $[\alpha]$.

We can now define traces for dualizable profunctors. If M is a right dualizable profunctor from A to B, the **trace** of a map $f\colon M \to M$ is the composite

$$\begin{array}{ccccc}
\langle\!\langle S^n \wedge A\rangle\!\rangle_A & & & & \langle\!\langle S^n \wedge B\rangle\!\rangle_B \\
\big\downarrow{\scriptstyle \eta} & & & & \big\uparrow{\scriptstyle \epsilon} \\
\langle\!\langle M \wedge_B D_r M\rangle\!\rangle_A & \xrightarrow{f \wedge \mathrm{id}} & \langle\!\langle M \wedge_B D_r M\rangle\!\rangle_A & \xrightarrow{\cong} & \langle\!\langle D_r M \wedge_A M\rangle\!\rangle_B
\end{array}$$

The following is the central theorem of the subject: the *origin of all linearity formulas*. It says that in the situation of Theorem 3.4, we can calculate the trace of an induced map on the colimit from the trace of an endomap of a diagram. Like Theorem 3.4, its proof is essentially entirely formal manipulation.

THEOREM 3.8 ([**PS13a**, Prop. 7.5]). *If $M\colon A \to \mathscr{T}$ is pointwise dualizable and $\Phi\colon A^{\mathrm{op}} \to \mathscr{T}$ is absolute, then for any $f\colon M \to M$ and $g\colon \Phi \to \Phi$ we have*

$$\mathrm{tr}(g \wedge_A f) = \mathrm{tr}(f) \circ \mathrm{tr}(g).$$

(More generally, if X and Y are right n-dualizable profunctors from A to B and from B to C respectively, then for any $f\colon Y \to Y$ and $g\colon X \to X$ we have $\mathrm{tr}(g \wedge_B f) = \mathrm{tr}(f) \circ \mathrm{tr}(g)$.)

In our applications, we will often take $g = \mathrm{id}_\Phi$ (although in the proof of Lemma 3.10 we will need the more general version). In this case, $\mathrm{id}_\Phi \wedge_A f$ is just the endomorphism of $\mathrm{colim}^\Phi(M)$ induced by f, which we denote $\mathrm{colim}^\Phi(f)$; thus we have

$$\mathrm{tr}(\mathrm{colim}^\Phi(f)) = \mathrm{tr}(f) \circ \mathrm{tr}(\mathrm{id}_\Phi). \tag{3.9}$$

In order to make practical use of this, we need to be able to identify $\mathrm{tr}(f)$ and $\mathrm{tr}(\mathrm{id}_\Phi)$. Note that $\mathrm{tr}(f)$ is a stable map $\langle\!\langle A\rangle\!\rangle_A \to S^0$, i.e. an element of the stable cohomotopy of $\langle\!\langle A\rangle\!\rangle_A$. This could be hard to compute completely. However, for (3.9) all that matters is its composite with the stable map $\mathrm{tr}(\mathrm{id}_\Phi)\colon S^0 \to \langle\!\langle A\rangle\!\rangle_A$, which is an element of $\pi_0^s(\langle\!\langle A\rangle\!\rangle_A)$. Thus, it will be enough to know the composites of $\mathrm{tr}(f)$ with the generators of $\pi_0^s(\langle\!\langle A\rangle\!\rangle_A)$, which as we have seen are induced by conjugacy classes in A. This is what the following lemma provides.

LEMMA 3.10 ([**PS16**, Lemma 5.11]). *If $M\colon A \to \mathscr{T}$ is pointwise dualizable and $f\colon M \to M$, then for any conjugacy class $[a \xrightarrow{\alpha} a]$ in A, the composite*

$$S^n \xrightarrow{[\alpha]} S^n \wedge \langle\!\langle A\rangle\!\rangle_A \xrightarrow{\mathrm{tr}(f)} S^n$$

is stably equal to the trace (in $\mathscr{T}$) of the composite

$$M_a \xrightarrow{M_\alpha} M_a \xrightarrow{f_a} M_a \tag{3.11}$$

where M_a is the value of M on a.

SKETCH OF PROOF. Define a profunctor Φ_a from $\mathbb{1}$ to A by

$$\Phi_a(a') = \hom_A(a',a)_+.$$

This is absolute; its right dual is defined by

$$(D_r\Phi_a)(a') = \hom_A(a,a')_+.$$

The evaluation and coevaluation are defined by the inclusion of the identity map and the composition of morphisms. Moreover, the Φ_a-weighted colimit of M is just the object M_a. We also have an endomorphism $\omega_\alpha\colon \Phi_a \to \Phi_a$ induced by composition with α, and $\omega_\alpha \wedge_A f$ is the composite $f_a \circ M_\alpha$ appearing in (3.11). Finally, $\mathrm{tr}(\omega_\alpha)$ can be shown to be $[\alpha]\colon S^n \to S^n \wedge \langle\!\langle A\rangle\!\rangle_A$, so Theorem 3.8 implies the desired statement. □

Thus, in order to apply Theorem 3.8 for any particular weight Φ, it suffices to show that Φ is absolute, and calculate $\mathrm{tr}(\mathrm{id}_\Phi)$. We refer to $\mathrm{tr}(\mathrm{id}_\Phi)$ as the **coefficient vector** of Φ. Recall that it is an element of the 0th stable homotopy group of $\langle\!\langle A\rangle\!\rangle_A$, hence a linear combination $\sum_{[\alpha]} \phi_{[\alpha]} \cdot [\alpha]$ of the generators $[\alpha]$ corresponding to conjugacy classes in A, with integer coefficients $\phi_{[\alpha]}$. Combining this with Lemma 3.10, we obtain:

THEOREM 3.12. *If $M\colon A \to \mathscr{T}$ is pointwise dualizable, $\Phi\colon A^{\mathrm{op}} \to \mathscr{T}$ is absolute, and $f\colon M \to M$, then we have*

$$\mathrm{tr}(\mathrm{colim}\, f) = \sum_{[\alpha]} \phi_{[\alpha]}\, \mathrm{tr}(f_a \circ M_\alpha)$$

where $\phi_{[\alpha]}$ are the coefficients of Φ.

EXAMPLE 3.13. If A is the empty category there is a unique functor $M\colon A \to \mathscr{T}$, which is trivially pointwise dualizable; hence its colimit, which is the one-point space $\star$, is dualizable. The shadow of A is also $\star$, so the trace of the unique endomorphism of $\star$ is the composite $S^0 \to \star \to S^0$, i.e. the zero endomorphism of S^0.

EXAMPLE 3.14. If $A = \mathbb{1} \sqcup \mathbb{1}$ then we have $\pi_0^s(\langle\!\langle A \rangle\!\rangle) \cong \mathbb{Z} \oplus \mathbb{Z}$. For a pointwise dualizable $M\colon \mathbb{1} \sqcup \mathbb{1} \to \mathscr{T}$ and $f\colon M \to M$, Lemma 3.10 implies that the map $\pi_0^s(\langle\!\langle A \rangle\!\rangle_A) \to \pi_0^s(S^0)$ induced by $\operatorname{tr}(f)$ is the row vector composed of $\operatorname{tr}(f_a)$ and $\operatorname{tr}(f_b)$. Thus, we have

$$\operatorname{tr}(f_a \vee f_b) = \phi_a \cdot \operatorname{tr}(f_a) + \phi_b \cdot \operatorname{tr}(f_b)$$

for some endomorphisms ϕ_a, ϕ_b of S^n. Knowing that such ϕ_a and ϕ_b exist, and are the same for all M and f, enables us to calculate them easily. Namely, let $M_a = \star$ and $M_b = S^0$ and let f be the identity. Then $\operatorname{tr}(f_a) = 0$ by the previous example, and $\operatorname{tr}(f_b) = 1$ since it is the identity; while $M_a \vee M_b \cong S^0$ and $f_a \vee f_b = \mathrm{id}_{S^0}$, so that $\operatorname{tr}(\operatorname{colim}^\Phi f) = 1$ as well. Thus, $1 = \phi_a \cdot 0 + \phi_b \cdot 1$, so $\phi_b = 1$. Similarly, $\phi_a = 1$, so our linearity formula is

$$\operatorname{tr}(f_a \vee f_b) = \operatorname{tr}(f_a) + \operatorname{tr}(f_b).$$

This is, of course, well-known and easy to prove by other means.

Finally, applying this to Example 3.7, we obtain the linearity of Lefschetz numbers.

THEOREM 3.15. *Suppose X and Y are closed smooth manifolds, and we have continuous maps $i\colon Y \to X$, $f\colon X \to X$ and $g\colon Y \to Y$ so that $f \circ i = i \circ g$. Then*

$$L(f) - L(g) = L(h)$$

where h is the endomorphism of the mapping cone of i induced by f and g.

As a special case of this theorem, if $A \subset X$ are closed smooth manifolds and $f\colon X \to X$ is a continuous map so that $f(A) \subset A$ then

$$L(f) - L(f|A) = L(f/A)$$

since the mapping cone on the inclusion is equivalent to the quotient space X/A.

PROOF. Let B be the the category with two objects a and b and one nonidentity morphism $\alpha\colon a \to b$. In Example 3.7, we saw that the functor $\Phi\colon B^{\mathrm{op}} \to \mathscr{T}$ defined by $\Phi(b) = S^0$ and $\Phi(a) = \star$ is absolute, and that Φ-weighted colimits are mapping cones.

Now B has two conjugacy classes, $[\mathrm{id}_a]$ and $[\mathrm{id}_b]$, so $\pi_0^s(\langle\!\langle B \rangle\!\rangle_B) \cong \mathbb{Z} \oplus \mathbb{Z}$. Thus, $\operatorname{tr}(\mathrm{id}_\Phi)$ is determined by two integers ϕ_a and ϕ_b, and for any pointwise dualizable profunctor $M\colon B \to \mathscr{T}$ and endomorphism $f\colon M \to M$ we have a linearity formula

$$\operatorname{tr}(\operatorname{colim}^\Phi(f)) = \phi_a \cdot \operatorname{tr}(f_a) + \phi_b \cdot \operatorname{tr}(f_b).$$

As in Example 3.14, now that we know that ϕ_a and ϕ_b exist, we can calculate them by considering some very special cases. On one hand, if M is the diagram $(S^n \to S^n)$, then its cofiber is $\star$. If f is the identity, we have

$$0 = \phi_a \cdot 1 + \phi_b \cdot 1.$$

On the other hand, if M is the diagram $(\star \to S^n)$, then its cofiber is S^n. If f is again the identity, we have

$$1 = \phi_a \cdot 0 + \phi_b \cdot 1.$$

Solving these equations, we obtain $\phi_b = 1$ and $\phi_a = -1$.

Thus, Theorem 3.12 tells us that if $M\colon B \to \mathscr{T}$ is pointwise dualizable and $f\colon M \to M$, then

$$\operatorname{tr}(\operatorname{colim}(f)) = \operatorname{tr}(f_b) - \operatorname{tr}(f_a).$$

If we let $M_a = Y$, $M_b = X$, $f_a = g$ and $f_b = f$ then we recover the statement of the theorem. □

It is important to note that this approach is not tied to this particular diagram shape. The same proof can be applied to any small category B and functor $\Phi\colon B^{\mathrm{op}} \to \mathscr{T}$ *as long as we can show that the functor* Φ *is absolute.* This is a significant restriction, since verifying absoluteness can be difficult, but it also does allow for significant generalizations; a number of examples can be found in [**PS16**].

The approach is also not tied to the particular context of ordinary spaces: it works in any "homotopy theory". In [**PS16**] we describe it in the general context of *derivators*; the proof of Theorem 3.15 then applies in any *stable* derivator, i.e. one where the suspension functor is an equivalence. (To put based spaces in such a context, one works instead with spectra.) A similar general form of Theorem 3.15 was proven in [**May01**] using triangulated categories and stable model categories. By [**Gro13**], any stable derivator has an underlying triangulated category, and any stable model category gives rise to a stable derivator; thus our general form of Theorem 3.15 has roughly the same generality as that of [**May01**]. However, as remarked above, our method also applies to more general diagram shapes. And, as we now proceed to explain, it also applies to the Reidemeister trace.

4. Parametrized spaces and the Reidemeister trace

We now introduce the context of parametrized spaces and ex-fibrations following [**MS06**], which will enable us to view the Reidemeister trace as a trace in a way similar to the perspective on the Lefschetz number described above.

For us, *fibration* will always mean *Hurewicz fibration.* If $p_1\colon E_1 \to X$ and $p_2\colon E_2 \to X$ are fibrations, a **fiberwise map** is a map $f\colon E_1 \to E_2$ such that $p_2 f = p_1$. Similarly, a **fiberwise homotopy** is a map $H\colon E_1 \times [0,1] \to E_2$ such that $p_2(H(e,t)) = p_1(e)$ for all $e \in E$ and $t \in [0,1]$. We denote fiberwise homotopy equivalences by $\simeq$.

If, on the other hand, we have fibrations $p_1\colon E_1 \to X_1$ and $p_2\colon E_2 \to X_2$ over (possibly) different base spaces, and $\overline{f}\colon X_1 \to X_2$ is a map, then by a **(fiberwise) map over** $\overline{f}$ we mean a map $f\colon E_1 \to E_2$ such that $p_2 f = \overline{f} p_1$. Thus, a "fiberwise map" without qualification is always over the identity id_X.

Two basic constructions on fibrations are *pullback* and *pushforward* along a continuous map $g\colon Y \to X$. Firstly, given a fibration $p\colon E \to X$, we have a pullback fibration $g^*E \to Y$. For any other fibration $E' \to Y$, there is a natural bijection between maps $E' \to E$ over g and fiberwise maps $E' \to g^*E$ (over id_Y). Secondly, given a fibration $q\colon E \to Y$, the composite $E \xrightarrow{q} Y \xrightarrow{g} X$ is not in general a fibration (though it is if g is also a fibration), but up to homotopy we can replace it by one. We denote the result by $g_! E \to X$.

A particularly important construction built out of these operations is the following. Given fibrations $M \to Y \times X$ and $N \to X \times Z$, we can form the product fibration $M \times N \to Y \times X \times X \times Z$, pull back along the diagonal $Y \times X \times Z \to Y \times X \times X \times Z$, then compose with the projection $Y \times X \times Z \to Y \times Z$ (which is a fibration). This yields a fibration $M \times_X N \to Y \times Z$ whose total space is the pullback of M and N over X. (We could have defined this more directly, but the description given above generalizes more easily.)

The fibrational version of a based space is a *sectioned* fibration, i.e. a fibration $p\colon E \to X$ equipped with a continuous section $s\colon X \to E$ (so that $ps = \mathrm{id}_X$). We usually assume that the section is a fiberwise closed cofibration, in which case we call p an **ex-fibration** (see [**MS06**] for details). We say that a fiberwise map between ex-fibrations is an **ex-map** if it also commutes with the specified sections. Note that if $p\colon E \to X$ is an ex-fibration, then each fiber $p^{-1}(b)$ is a nondegenerately based space.

If $E \to X$ is any fibration, then $E \sqcup X \to X$ is an ex-fibration, called the result of *adjoining a disjoint section* to E; we denote it by E_{+X}. Additionally, for any ex-fibration $E \to X$ and any (nondegenerately) based space W, we have an induced ex-fibration $W \wedge E \to X$, whose fiber over $b \in X$ is the ordinary smash product $W \wedge p^{-1}(b)$. Ex-fibrations can be pulled back along a continuous map $g\colon Y \to X$ in a straightforward way. They can also be pushed forwards, by composing and then pushing out along the section (then replacing by a fibration, if necessary).

Two ex-fibrations $p\colon M \to Y$ and $q\colon N \to X$ have an **external smash product** $M \mathbin{\overline{\wedge}} N \to Y \times X$, whose fiber over (a, b) is $p^{-1}(a) \wedge q^{-1}(b)$. If we have ex-fibrations $p\colon M \to Y \times X$ and $q\colon N \to X \times Z$ the **smash pullback** of M and N is the result of pulling back $M \mathbin{\overline{\wedge}} N$ along $Y \times X \times Z \to Y \times X \times X \times Z$, then pushing forward to $Y \times Z$. This yields an ex-fibration $M \odot N \to Y \times Z$; its fiber over (a, c) consists of all the smash products $p^{-1}(a, b) \wedge q^{-1}(b, c)$ for all $b \in X$, with their basepoints identified (and with a suitable quotient topology which relates these products for different b). The smash pullback is associative.

REMARK 4.1. If we are given a fibration $M \to X$, we may regard it either as a fibration $M \to X \times \star$ or as a fibration $M \to \star \times X$. We will only need the second and we will denote it by $\breve{M}$.

Because we will have another use for subscripts on $\odot$ in §5, we will not indicate in the notation the space that is "canceled" in the smash pullback. Instead we will rely on the understanding that $M \odot N$ is a space over the left-hand space of M and the right-hand space of N, as above.

For any space X, let PX denote the space of paths $\gamma\colon [0, 1] \to X$, and define the fibration $PX \to X \times X$ by evaluation at the endpoints, $\gamma \mapsto (\gamma(0), \gamma(1))$. Then $(PX)_{+X\times X}$ is a two-sided homotopy unit for the smash pullback; we denote these units by $U_X \to X \times X$. For any (nondegenerately) based space W and ex-fibration $p\colon M \to Y \times X$ we have natural equivalences

$$(W \wedge U_Y) \odot M \simeq W \wedge M \qquad M \odot (W \wedge U_X) \simeq W \wedge M$$

Given a map $g\colon Y \to X$ of base spaces, we write X_g for the ex-fibration $((\mathrm{id} \times g)^* PX)_{+(X\times Y)}$ and ${}_gX$ for the ex fibration $((g \times \mathrm{id})^* PX)_{+(Y\times X)}$. The ex-fibrations X_g and ${}_gX$ are called **base change objects**, and have the following important properties:

- For any ex-fibration $M \to Z \times X$, we have $M \odot X_g \simeq (\mathrm{id} \times g)^* M$.
- For any ex-fibration $M \to X \times Z$, we have ${}_gX \odot M \simeq (g \times \mathrm{id})^* M$.
- For any ex-fibration $M \to Z \times Y$, we have $M \odot {}_gX \simeq (\mathrm{id} \times g)_! M$.
- For any ex-fibration $M \to Y \times Z$, we have $X_g \odot M \simeq (g \times \mathrm{id})_! M$.

If $M \to Y \times Y$ is an ex-fibration, we denote by $\langle\!\langle M \rangle\!\rangle$ its **shadow**, which is the based space defined by pulling back along the diagonal $Y \to Y \times Y$, then quotienting out the section (i.e. pushing forward along the unique map $Y \to \star$).

For ex-fibrations $M \to Y \times X$ and $N \to X \times Y$, we have a canonical equivalence

$$\langle\!\langle M \odot N \rangle\!\rangle \simeq \langle\!\langle N \odot M \rangle\!\rangle.$$

For a based space X and an ex-fibration $M \to Y \times Y$, we have $\langle\!\langle X \wedge M \rangle\!\rangle \simeq X \wedge \langle\!\langle M \rangle\!\rangle$.

The shadow of the unit ex-fibration U_Y is $(\Lambda Y)_+$, the free loop space of Y with a disjoint basepoint. Similarly, for an endomorphism $g\colon Y \to Y$, the shadow of the base change object Y_g is $(\Lambda^g Y)_+$, where $\Lambda^g Y$ denotes the **twisted free loop space**: its points are pairs consisting of a point $a \in Y$ and a path $a \rightsquigarrow g(a)$. Likewise, the shadow of ${}_gY$ is the space of paths $g(a) \rightsquigarrow a$ (with a disjoint basepoint), which is homeomorphic to $(\Lambda^g Y)_+ \cong \langle\!\langle Y_g \rangle\!\rangle$.

We say that an ex-fibration $M \to Y \times X$ is **(right) n-dualizable** if there is an ex-fibration $D_r M \to X \times Y$ and fiberwise ex-maps

$$\begin{aligned} \eta\colon S^n \wedge U_Y &\longrightarrow M \odot D_r M \\ \varepsilon\colon D_r M \odot M &\longrightarrow S^n \wedge U_X \end{aligned}$$

such that the composites

$$\begin{array}{ccccc} S^n \wedge M & & & & S^n \wedge M \\ \downarrow \wr & & & & \wr \uparrow \\ S^n \wedge U_Y \odot M & \xrightarrow{\eta \odot \mathrm{id}} & M \odot D_r M \odot M & \xrightarrow{\mathrm{id} \odot \varepsilon} & M \odot S^n \wedge U_X \end{array}$$

$$\begin{array}{ccccc} D_r M \odot S^n \wedge U_Y & \xrightarrow{\mathrm{id} \odot \eta} & D_r M \odot M \odot D_r M & \xrightarrow{\varepsilon \odot \mathrm{id}} & S^n \wedge U_X \odot D_r M \\ \uparrow \wr & & & & \wr \downarrow \\ S^n \wedge D_r M & & & & S^n \wedge D_r M \end{array}$$

become fiberwise homotopy equivalent to identity maps after smashing with some S^m. We say that an unsectioned fibration $M \to Y \times X$ is n-dualizable if $M_{+(Y \times X)}$ is so. If $Y = X = \star$, this reduces to the original notion of n-duality from §2.

The following theorem is proven exactly like Theorem 3.4 (in fact, they are both special cases of a single theorem about bicategorical duality).

THEOREM 4.2 ([**MS06**, Theorem 16.5.1]). *If $M \to Y \times X$ and $N \to X \times Z$ are right n-dualizable ex-fibrations, then so is $M \odot N \to Y \times Z$.*

The proof of the following theorem is also formal, when put in the right context.

THEOREM 4.3 ([**MS06**, Theorem 17.3.1], [**Shu08**, Prop. 5.3]). *For any $g\colon Y \to X$, the base change object ${}_gX$ is right n-dualizable with dual X_g.*

The next theorem does require a significant amount of work, but fortunately it has already been done for us by [**MS06**].

THEOREM 4.4 ([**MS06**, Theorem 18.5.2]). *Suppose M is a closed smooth manifold and there is an embedding of M in $\mathbb{R}^n$. For any fibration $M \to X$ the ex-fibration $\widetilde{M_{+X}}$ is right n-dualizable. In particular, this applies to the identity fibration $M \to M$.*

We write $S_M \to M$ for the ex-fibration $M_{+M} \simeq M \sqcup M \simeq S^0 \times M$ over M. According to Remark 4.1, we write $\widetilde{S_M}$ for S_M regarded as a fibration over $\star \times M$. This theorem implies $\widetilde{S_M}$ is dualizable if M is a closed smooth manifold.

If $M \to Y \times X$ is n-dualizable and we have ex-fibrations $Q \to Y \times Y$ and $P \to X \times X$ and a fiberwise ex-map $f\colon Q \odot M \to M \odot P$ (over $Y \times X$) we define the **trace** of f to be the composite

$$\begin{array}{ccccc} \langle\!\langle S^n \wedge Q \rangle\!\rangle & & & & \langle\!\langle S^n \wedge P \rangle\!\rangle \\ \downarrow{\scriptstyle \eta} & & & & \uparrow{\scriptstyle \varepsilon} \\ \langle\!\langle Q \odot M \odot D_r M \rangle\!\rangle & \xrightarrow{f \odot \mathrm{id}} & \langle\!\langle M \odot P \odot D_r M \rangle\!\rangle & \xrightarrow{\simeq} & \langle\!\langle D_r M \odot M \odot P \rangle\!\rangle \end{array}$$

REMARK 4.5. This generalizes the trace in §2 by replacing spaces by ex-fibrations, and also by allowing for *twisting* of both the domain and codomain. The codomain twisting is used in the description of the Reidemeister trace, while the domain twisting will play a role in the proof of its additivity in Theorem 6.3.

We will use several instances of this trace in the next sections. The first example we will need is the Reidemeister trace.

THEOREM 4.6 ([**Pon16**]). *Suppose that $\widetilde{S_X}$ is right n-dualizable, and let $g\colon X \to X$ be any continuous map. Then the trace of the induced map $\widetilde{S_X} \to \widetilde{S_X} \odot_X X_g$ induces a map on homology:*

$$\mathbb{Z} \cong H_0(\Lambda\star) \to H_0(\Lambda^g X)$$

that picks out the Reidemeister trace $R(g)$.

For purposes of this paper, the reader is free to consider this theorem a definition of the Reidemeister trace.

We also have a version of Theorem 3.8 in this context.

THEOREM 4.7 ([**PS13a**, Prop. 7.5]). *Let $M \to Y \times X$ and $N \to X \times Z$ be n-dualizable, let $Q \to Y \times Y$, $P \to X \times X$, and $R \to Z \times Z$ be ex-fibrations, and let $f\colon Q \odot M \to M \odot P$ and $g\colon P \odot N \to N \odot R$ be fiberwise ex-maps. Then the following triangle commutes up to stable homotopy.*

$$\begin{array}{ccc} S^n \wedge \langle\!\langle Q \rangle\!\rangle & \xrightarrow{\mathrm{tr}((\mathrm{id}_M \odot g)\circ(f \odot \mathrm{id}_N))} & S^n \wedge \langle\!\langle R \rangle\!\rangle \\ & {\scriptstyle \mathrm{tr}(f)} \searrow \quad \nearrow {\scriptstyle \mathrm{tr}(g)} & \\ & S^n \wedge \langle\!\langle P \rangle\!\rangle & \end{array}$$

There is one other general fact about traces that we will need in the following sections. Suppose that

$$\begin{array}{ccc} Y & \xrightarrow{g} & X \\ {\scriptstyle f_1}\downarrow & \alpha & \downarrow{\scriptstyle f_2} \\ Y & \xrightarrow[g]{} & X \end{array}$$

is a homotopy commutative square, with specified homotopy $\alpha\colon gf_1 \sim f_2 g$. Then ${}_gX$ is right n-dualizable by Theorem 4.3.

THEOREM 4.8. [**PS14**, Prop. 3.3, Theorem 5.7] *There is an induced map*

$$h\colon Y_{f_1} \odot {}_gX \longrightarrow {}_gX \odot X_{f_2}.$$

whose trace is homotopic to the map

$$\Lambda^{f_1} Y_+ \to \Lambda^{f_2} X_+$$

which sends a path $\gamma\colon a \rightsquigarrow f_1(a)$ *in* Y *to the path* $g(\gamma)\cdot\alpha(a)$.

5. Parametrized profunctors

In order to generalize the proof of the additivity of the Lefschetz number to the Reidemeister trace, we will consider diagrams of ex-fibrations. If A and B are small categories and Y and X are topological spaces, a **parametrized profunctor** from (A,Y) to (B,X) is an $(A\times B^{\mathrm{op}})$-indexed diagram of ex-fibrations over $Y\times X$. Given parametrized profunctors $M\colon (A,Y)\to(B,X)$ and $N\colon (B,X)\to(C,Z)$ we define their **composite** $M\odot_B N$ to be a parametrized profunctor from (A,Y) to (C,Z) where $(M\odot_B N)(a,c)$ is the homotopy coend of the $(B^{\mathrm{op}}\times B)$-indexed diagram

$$(b,b')\mapsto M(a,b)\odot N(b',c).$$

For a (nondegenerately) based topological space W and a parametrized profunctor M from (A,Y) to (B,X) we can define a parametrized profunctor $W\wedge M$ from (A,Y) to (B,X) by taking $(W\wedge M)(a,b)$ to be the ex-fibration $W\wedge(M(a,b))$ over $Y\times X$ as in the previous section. In particular, for a small category A and a space Y we can define a parametrized profunctor $U_{(A,Y)}$ from (A,Y) to (A,Y) by $U_{(A,Y)}(a,a')=\hom_A(a,a')_+\wedge U_Y$. We also have natural equivalences

$$(W\wedge U_{(A,Y)})\odot_A M\simeq W\wedge M \qquad M\odot_B(W\wedge U_{(B,X)})\simeq W\wedge M$$

for any (nondegenerately) based topological space W.

REMARK 5.1. As in §3, we use a subscript on $\odot$ to indicate that we are taking a composite of profunctors.

We say that a parametrized profunctor M from (A,Y) to (B,X) is **right n-dualizable** if there is a parametrized profunctor D_rM from (B,X) to (A,Y) and natural transformations $\eta\colon S^n\wedge(A,Y)\to M\odot_B D_rM$ and $\epsilon\colon D_rM\odot_A M\to S^n\wedge(B,X)$ (possibly with cofibrantly replaced domains) such that the composites

$$\begin{array}{ccccc} S^n\wedge M & & & & S^n\wedge M\\ \downarrow\wr & & & & \wr\uparrow\\ S^n\wedge U_{(A,Y)}\odot_A M & \xrightarrow{\eta\odot\mathrm{id}} & M\odot_B D_rM\odot_A M & \xrightarrow{\mathrm{id}\odot\varepsilon} & M\odot_B S^n\wedge U_{(B,X)} \end{array}$$

$$\begin{array}{ccccc} D_rM\odot_A S^n\wedge U_{(A,Y)} & \xrightarrow{\mathrm{id}\odot\eta} & D_rM\odot_A M\odot_B D_rM & \xrightarrow{\varepsilon\odot\mathrm{id}} & S^n\wedge U_{(B,X)}\odot_B D_rM\\ \uparrow\wr & & & & \wr\downarrow\\ S^n\wedge D_rM & & & & S^n\wedge D_rM \end{array}$$

are naturally fiberwise homotopic to identity maps after smashing with S^m for some m. If we take X and Y to be a point $\star$ this reduces to the duality considered in §3. If we take A and B to be the category $\mathbb{1}$ with one object and no nonidentity morphisms, this reduces to the duality in §4.

As in §3, we say that a right dualizable parametrized profunctor M from (A,Y) to $(\mathbb{1},X)$ is **pointwise dualizable**, and a right dualizable parametrized profunctor Φ from $(\mathbb{1},Y)$ to (B,X) is **absolute**. If Φ and M are parametrized profunctors from $(\mathbb{1},Y)$ to (A,X) and from (A,X) to $(\mathbb{1},Z)$ respectively, then their composite $\Phi\odot_A M$

is the **Φ-weighted homotopy colimit** of M, denoted $\mathrm{colim}^\Phi(M)$. Analogously to Lemma 3.1, we have:

LEMMA 5.2 ([**PS16**, Lemma 3.5]). *A parametrized profunctor M from (A, Y) to $(\mathbb{1}, X)$ is right dualizable if and only if the ex-fibration $M(a) \to Y \times X$ is right dualizable as an ex-fibration for each $a \in A$.*

For a parametrized profunctor M from (A, Y) to (A, Y), its **shadow**, $\langle\!\langle M \rangle\!\rangle_A$, is the homotopy coend of the $(A^{\mathrm{op}} \times A)$-indexed diagram $(a, a') \mapsto \langle\!\langle M(a, a') \rangle\!\rangle$. Using the compatibility between homotopy coends and the shadow for ex-fibrations we have an isomorphism

$$\langle\!\langle M \odot_B N \rangle\!\rangle_A \cong \langle\!\langle N \odot_A M \rangle\!\rangle_B$$

analogous to those in the previous sections, whenever both composites make sense. As before, we write $\langle\!\langle (A, Y) \rangle\!\rangle_A$ for $\langle\!\langle U_{(A,Y)} \rangle\!\rangle_A$.

If M is a right n-dualizable parametrized profunctor from (A, Y) to (B, X), Q is a parametrized profunctor from (A, Y) to (A, Y), and P is a parametrized profunctor from (B, X) to (B, X), we define the **trace** of a natural transformation $f\colon Q \odot_A M \to M \odot_B P$ to be the composite

$$\begin{array}{ccccc}
\langle\!\langle Q \wedge S^n \rangle\!\rangle_A & & & & \langle\!\langle S^n \wedge P \rangle\!\rangle_B \\
\downarrow{\scriptstyle \eta} & & & & \uparrow{\scriptstyle \varepsilon} \\
\langle\!\langle Q \odot_A M \odot_B D_r M \rangle\!\rangle_A & \xrightarrow{f \odot \mathrm{id}} & \langle\!\langle M \odot_B P \odot_B D_r M \rangle\!\rangle_A & \xrightarrow{\simeq} & \langle\!\langle D_r M \odot_A M \odot_B P \rangle\!\rangle_B
\end{array}$$

We now have the following theorem generalizing Theorem 3.4 and Theorem 3.8; we state it for morphisms that are additionally *twisted* in the codomain, since this is what we need for the Reidemeister trace. It is again an instance of [**PS13a**, Prop. 7.5], but now occurring in the "bicategory of diagrams in the derivator bicategory of parametrized spectra". Since the latter derivator bicategory is constructed in Theorem 10.4 below, here is where we really begin relying on the work of Part 2.

THEOREM 5.3. *Let N be a parametrized profunctor from (A, Y) to (B, X) and M be a parametrized profunctor from (B, X) to (C, Z). If M and N are right dualizable then so is $M \odot N$.*

If P is a parametrized profunctor from (C, Z) to (C, Z) and $g\colon N \to N$ and $f\colon M \to M \odot_B P$ are natural transformations, then

$$\mathrm{tr}(g \odot_B f) = \mathrm{tr}(f) \circ \mathrm{tr}(g).$$

In particular, if we take N to be an absolute parametrized profunctor Φ and $g = \mathrm{id}_\Phi$ we have

$$\mathrm{tr}(\mathrm{colim}^\Phi(f)) = \mathrm{tr}(f) \circ \mathrm{tr}(\mathrm{id}_\Phi).$$

As before, in order to be able to use this theorem we need to be able to describe $\mathrm{tr}(f)$ and $\mathrm{tr}(\mathrm{id}_\Phi)$. Our approach will be very similar to the method used in §3, so we start with a generalization of Lemma 3.10.

LEMMA 5.4. *Let M be a right dualizable parametrized profunctor from (A, Y) to (B, X), P be a parametrized profunctor from (B, X) to (B, X), and $f\colon M \to M \odot_B P$ be a natural transformation. For any conjugacy class $[a \xrightarrow{\alpha} a]$ in A, the* *stable homotopy class of the composite*

$$S^n \xrightarrow{[\alpha]} S^n \wedge \langle\!\langle (A,Y) \rangle\!\rangle_A \xrightarrow{\mathrm{tr}(f)} \langle\!\langle P \rangle\!\rangle_B$$

is equal to the trace of the composite

$${}_aM \xrightarrow{{}_\alpha M} {}_aM \xrightarrow{{}_a f} {}_a(M \odot_B P).$$

Here ${}_aM$ denotes the functor $B^{\mathrm{op}} \to \mathscr{T}$ defined by ${}_aM(b) = M(a,b)$.

Formally, this result follows by applying [**PS16**, Lemma 11.15] to the derivator bicategory of parametrized spectra constructed in Theorem 10.4. However, for topological intuition we sketch what the proof looks like in this case.

SKETCH OF PROOF. For each $a \in A$ we can define a parametrized profunctor Φ_a from $(\mathbb{1}, Y)$ to (A, Y) by

$$\Phi_a(a') = \hom_A(a', a)_+ \wedge U_Y.$$

As in Lemma 3.10, this is absolute; its right dual is defined by

$$(D_r\Phi_a)(a') = \hom_A(a, a')_+ \wedge U_Y.$$

The coevaluation and evaluation extend the maps defined in Lemma 3.10.

The value of $\Phi_a \odot_A M$ on an element $b \in B$ is the homotopy coend of the $A \times A^{\mathrm{op}}$ diagram

$$(x, x') \mapsto \Phi_a(x) \odot M(x', b) := A(a, x)_+ \wedge U_Y \odot M(x', b) \cong A(a, x)_+ \wedge M(x', b)$$

This is equivalent to the functor ${}_aM$. In the same way we can identify

$$\alpha \odot_A \mathrm{id}_M \colon \Phi_a \odot_A M \to \Phi_a \odot_A M$$

with the natural transformation ${}_\alpha M \colon {}_aM \to {}_aM$ that is induced by the action of α, and

$$\mathrm{id}_{\Phi_a} \odot_A f \colon \Phi_a \odot_A M \to \Phi_a \odot_A M \odot_B P$$

with ${}_a f \colon {}_aM \to {}_a(M \odot_B P)$.

Now a representative α of a conjugacy class in $A(a,a)$ defines a map $\omega_\alpha \colon \Phi_a \to \Phi_a$ by composition. Explicit descriptions of the coevaluation and evaluation above show that the trace of ω_α is the map $[\alpha] \colon S^0 \to S^0 \wedge \langle\!\langle A, Y \rangle\!\rangle_A$ that takes the non-basepoint of S^0 to α. Thus, Theorem 5.3 implies the desired equality. □

6. Linearity of the Reidemeister traces

We can now use Theorem 5.3 and Lemma 5.4 to generalize Theorem 3.15 to the Reidemeister trace. Recall that in Example 3.7 we showed that if B is the category with two objects and one nonidentity morphism $\alpha \colon a \to b$ and $\Phi \colon B^{\mathrm{op}} \to \mathscr{T}$ is defined by $\Phi(b) = S^0$ and $\Phi(a) = \star$ then Φ is absolute. Since a based space is an ex-fibration over $\star$, Φ is also a parametrized profunctor Φ from $(\mathbb{1}, \star)$ to $(B, \star)$. Further, the coevaluation and evaluation for the profunctor Φ are also a coevaluation and evaluation for the parametrized profunctor Φ.

THEOREM 6.1. *If M_a and M_b are right dualizable ex-fibrations over $\star \times X$, P is an ex-fibration over $X \times X$ and $f_a\colon M_a \to M_a \odot P$, $f_b\colon M_b \to M_b \odot P$, and $i\colon M_a \to M_b$ are fiberwise maps so that the diagram*

$$\begin{array}{ccc} M_a & \xrightarrow{\ i\ } & M_b \\ \downarrow{\scriptstyle f_a} & & \downarrow{\scriptstyle f_b} \\ M_a \odot P & \xrightarrow{\ i\ } & M_b \odot P \end{array}$$

commutes then

$$\operatorname{tr}(\operatorname{colim}^{\Phi}(f)) = \operatorname{tr}(f_b) - \operatorname{tr}(f_a).$$

PROOF. As shown in Example 3.7, Φ is absolute and $\pi_0^s(\langle\!\langle B\rangle\!\rangle) \cong \mathbb{Z}\oplus\mathbb{Z}$. We saw in the proof of Theorem 3.15 that $\operatorname{tr}(\mathrm{id}_{\Phi})$ is determined by morphisms $\phi_a, \phi_b\colon S^n \to S^n$, which we calculated to be $\phi_a = -1$ and $\phi_b = 1$. Thus, for any pointwise dualizable parametrized profunctor M from $(B,\star)$ to $(\mathbb{1}, X)$ and endomorphism $f\colon M \to M \odot_{\mathbb{1}} P$, the desired statement follows from the linearity formula

$$\operatorname{tr}(\operatorname{colim}^{\Phi}(f)) = \phi_a \cdot \operatorname{tr}(f_a) + \phi_b \cdot \operatorname{tr}(f_b). \qquad \square$$

REMARK 6.2. This theorem also holds if M_a and M_b are right dualizable ex-fibrations over $Y \times X$. In the proof we replace Φ with the parametrized profunctor from $(\mathbb{1},\star)$ to (B,Y) whose value on b is S_Y and on a is the identity map of Y regarded as an ex-fibration (no disjoint section is added). In this case, the Φ-weighted colimit of M is the fiberwise homotopy pushout of the diagram on the left below.

$$\begin{array}{ccc} S^0 \wedge M_a & \longrightarrow & S^0 \wedge M_b \\ \downarrow & & \\ 0 \wedge M_a & & \end{array} \qquad\qquad \begin{array}{ccc} M_a & \longrightarrow & M_b \\ \downarrow & & \\ Y \times X & & \end{array}$$

More explicitly, $\operatorname{colim}^{\Phi} M$ is the homotopy pushout of the diagram on the right above; we call this the **fiberwise mapping cone** on i and denote it by $M(i)$.

Now we can state our additivity theorem for the Reidemeister trace.

THEOREM 6.3. *Suppose Y and X are closed smooth manifolds or compact ENRs and $i\colon Y \to X$, $f\colon X \to X$ and $g\colon Y \to Y$ are continuous maps so that $i \circ f = g \circ i$. Then*

$$R(f) - i(R(g)) = R_{X|Y}(f),$$

where $R_{X|Y}(f)$ is the trace of the induced map $\widetilde{M(i)} \to \widetilde{M(i)} \odot X_f$ and $i\colon \Lambda^g Y \to \Lambda^f X$ is induced by i.

We claim that this is a refinement of [**Fer96**, Theorem 3.2.1]. To see this, let $j\colon X \to X/Y$ be the quotient map. If $i\colon Y \to X$ is a cofibration of compact ENR's or closed smooth manifolds, then $j(R_{X|Y}(f)) = R(f/Y)$ [**Pon11**] and $ji(R(f|_Y))$ is equal to the Lefschetz number $L(f|_Y)$. Thus, we recover Ferrario's result

$$j(R(f)) - L(f|_Y) = R(f/Y).$$

PROOF. Regarding $f\colon S_X \to S_X$ as a fiberwise map over $f\colon X \to X$, we have a map $f\colon \widetilde{S_X} \to \widetilde{S_X} \odot X_f$ over X and a similar map $g\colon \widetilde{S_Y} \to \widetilde{S_Y} \odot Y_g$ over Y. The map $i\colon Y \to X$ induces a fiberwise map $i_!(S_Y) \to S_X$, and by Theorem 4.8 the equality $i \circ g = f \circ i$ defines a fiberwise map $k\colon Y_g \odot {}_iX \to {}_iX \odot X_f$ over $Y \times X$. We compose these last two maps to define a twisted endomorphism of $i_!(\widetilde{S_Y})$:

$$\tilde{g}\colon \widetilde{S_Y} \odot {}_iX \xrightarrow{g \odot \mathrm{id}} \widetilde{S_Y} \odot Y_g \odot {}_iX \xrightarrow{\mathrm{id} \odot k} \widetilde{S_Y} \odot {}_iX \odot X_f.$$

This gives a commutative diagram

$$\begin{array}{ccc} i_!(\widetilde{S_Y}) & \longrightarrow & \widetilde{S_X} \\ \Big\downarrow{\scriptstyle \tilde{g}} & & \Big\downarrow{\scriptstyle f} \\ i_!(\widetilde{S_Y}) \odot X_f & \longrightarrow & \widetilde{S_X} \odot X_f \end{array}$$

of ex-fibrations over X. Let M from $(B, \star)$ to $(\mathbb{1}, X)$ be the parametrized profunctor where $M_a = i_!(\widetilde{S_Y})$ and $M_b = \widetilde{S_X}$. The diagram above gives us a natural transformation $f\colon M \to M \odot X_f$. For Φ as defined before Theorem 6.1, the Φ-weighted colimit of M is $\widetilde{M(i)}$, the fiberwise mapping cone on i.

By assumption $\widetilde{S_X}$ and $\widetilde{S_Y}$ are right n-dualizable for some integer n. By Theorem 4.3 the object ${}_iS$ is right dualizable. Then Theorem 4.7 implies $i_!\widetilde{S_Y} \cong \widetilde{S_Y} \odot {}_iX$ is right dualizable. Using Theorem 6.1 we see

$$\mathrm{tr}(h\colon \widetilde{Mi} \to \widetilde{Mi} \odot X_f) = \mathrm{tr}(f\colon \widetilde{S_X} \to \widetilde{S_X} \odot X_f) - \mathrm{tr}(\tilde{g}\colon i_!(\widetilde{S_Y}) \to i_!(\widetilde{S_Y}) \odot X_f).$$

In Theorem 4.6 we saw that the trace of $f\colon \widetilde{S_X} \to \widetilde{S_X} \odot X_f$ is the Reidemeister trace of f. Theorems 4.7 and 4.8 imply

$$\mathrm{tr}(\tilde{g}\colon i_!(\widetilde{S_Y}) \to i_!(\widetilde{S_Y}) \odot X_f) = i(\mathrm{tr}(g\colon \widetilde{S_Y} \to \widetilde{S_Y} \odot Y_g)) = i(R(g)).$$

Finally, $R_{X|Y}(f)$ is defined to be the trace of $\widetilde{Mi} \xrightarrow{M(f)} \widetilde{Mi} \odot X_f$. □

Part 2. Indexed monoidal derivators

It remains to set up the abstract context that encapsulates all the fibrant and cofibrant replacements, so that Theorem 3.8, Theorem 4.7, and Theorem 5.3 are, like [**PS14**, Theorem 5.2], all instances of [**PS13a**, Prop. 7.5]. This is *almost* completed by the companion paper [**PS16**], which studies linearity formulas for trace-like invariants in categorical generality. The examples given therein suffice for the additivity of the Lefschetz number (§3), but for the additivity of Reidemeister trace (§§5–6) we need to combine this theory with that of [**MS06**, **PS12**]. Therefore, in this part of the paper, we will show that the parametrized profunctors used in §§4–6 form a *derivator bicategory*, so that the results of [**PS16**] can then be applied in this context. This provides the foundation for the results in §5.

Relative to Part 1, we now make one change of context: instead of working with ex-fibrations we will consider parametrized spaces. A **parametrized space** over a base space X is an arbitrary space E with maps $s\colon X \to E$ and $p\colon E \to X$ so that $p \circ s = \mathrm{id}$. This is less restrictive than an ex-fibration; we think of an ex-fibration as a very nice parametrized space that is especially well behaved homotopically. This change reflects the shift from choosing specific well behaved examples in the previous sections to the more systematic approach using model categories in the remaining sections.

7. Background

We start by recalling some relevant definitions from [**PS16**]. We will not recall the definition of a *derivator*; all we need to know is that it is a refinement of a homotopy category that also includes information about diagram categories and homotopy limits and colimits. If $\mathscr{D}$ is a derivator and A is a small category, we write $\mathscr{D}(A)$ for the corresponding homotopy category of A-shaped diagrams. We refer to object of $\mathscr{D}(A)$ as (A-shaped) *coherent diagrams* in $\mathscr{D}$. See [**PS16**] or [**Gro13, GPS14**] for more details.

The basic context used in [**PS16**] is a derivator bicategory. By definition, a **derivator bicategory** $\mathscr{W}$ consists of the following data.

- A collection of objects R, S, T, ...
- For each pair of objects R and S a derivator $\mathscr{W}(R,S)$. We think of the category $\mathscr{W}(R,S)(A)$ as the homotopy category of A-shaped diagrams in the "hom-category" $\mathscr{W}(R,S)$.
- For each triple of objects R, S, and T, a morphism of derivators
$$\odot\colon \mathscr{W}(R,S)\times\mathscr{W}(S,T)\to\mathscr{W}(R,T).$$
That is, we have a pseudonatural transformation between 2-functors $\mathcal{C}at^{\mathrm{op}}\to\mathcal{CAT}$, which has components
$$\mathscr{W}(R,S)(A)\times\mathscr{W}(S,T)(A)\to\mathscr{W}(R,T)(A).$$
- We require these morphisms $\odot$ to be cocontinuous in each variable separately [**GPS14**, Definition 3.19].
- For each object R, a morphism of derivators $\mathbb{1}\to\mathscr{W}(R,R)$ (hence an object $\mathbb{I}_{R,A}\in\mathscr{W}(R,R)(A)$, varying pseudonaturally in $A\in\mathcal{C}at$).
- Natural unit and associativity isomorphisms, i.e. invertible modifications

$$\begin{array}{ccc}
\mathscr{W}(R,S)\times\mathscr{W}(S,T)\times\mathscr{W}(T,U) & \xrightarrow{\mathrm{id}\times\odot} & \mathscr{W}(R,S)\times\mathscr{W}(S,U)\\
{\scriptstyle \odot\times\mathrm{id}}\downarrow & \Swarrow\cong & \downarrow{\scriptstyle\odot}\\
\mathscr{W}(R,T)\times\mathscr{W}(T,U) & \xrightarrow[\odot]{} & \mathscr{W}(R,U)
\end{array}$$

$$\begin{array}{ccc}
\mathscr{W}(R,S) & \xrightarrow{(\mathrm{id},\mathbb{I})} & \mathscr{W}(R,S)\times\mathscr{W}(S,S)\\
 & \searrow\ \Swarrow\cong & \downarrow{\scriptstyle\odot}\\
 & & \mathscr{W}(R,S)
\end{array}
\qquad
\begin{array}{ccc}
\mathscr{W}(R,S) & \xrightarrow{(\mathbb{I},\mathrm{id})} & \mathscr{W}(R,R)\times\mathscr{W}(R,S)\\
 & \searrow\ \Swarrow\cong & \downarrow{\scriptstyle\odot}\\
 & & \mathscr{W}(R,S).
\end{array}$$

- The usual pentagon and unit axioms for a bicategory hold.

A derivator bicategory is **closed** if the morphisms $\odot$ participate in a two-variable adjunction of derivators [**GPS14**, Definition 8.1]. A **shadow** on a derivator bicategory $\mathscr{W}$ consists of a derivator $\mathscr{T}$ and cocontinuous morphisms of derivators

$$\langle\!\langle - \rangle\!\rangle\colon \mathscr{W}(R,R)\to\mathscr{T}$$

for each object R, together with invertible modifications

$$\begin{array}{ccccc}
\mathscr{W}(R,S) \times \mathscr{W}(S,R) & & \xrightarrow{\cong} & & \mathscr{W}(S,R) \times \mathscr{W}(R,S) \\
\odot \downarrow & & \Downarrow \cong & & \downarrow \odot \\
\mathscr{W}(R,R) & \xrightarrow[\langle\!\langle - \rangle\!\rangle]{} & \mathscr{T} & \xleftarrow[\langle\!\langle - \rangle\!\rangle]{} & \mathscr{W}(S,S)
\end{array}$$

satisfying the usual compatibility axioms for a shadow ([**PS13a**, Defn. 4.1]). The *underlying bicategory* of a derivator bicategory $\mathscr{W}$ is the bicategory $\mathbf{W}$ with the same objects as $\mathscr{W}$ and with morphisms given by $\mathbf{W}(R,S) = \mathscr{W}(R,S)(\mathbb{1})$.

The remaining sections of this paper are a proof of the following theorem.

THEOREM 10.4. *There is a closed derivator bicategory $\mathcal{E}x$ with a shadow valued in the homotopy derivator of spectra. Its objects are compactly generated spaces, its hom $\mathcal{E}x(R,S)$ is the homotopy derivator of the model category of spectra parametrized over $R \times S$, and its underlying ordinary bicategory is the one constructed in* [**MS06**, Chapter 17].

Before we continue to the proof, we will first recall the generalizations of several of the essential results in the previous sections to derivator bicategories.

THEOREM 7.1 ([**PS16**, Theorem 11.7]). *Given a derivator bicategory $\mathscr{W}$, we can construct a derivator bicategory $\mathcal{P}rof(\mathscr{W})$, with underlying ordinary bicategory denoted* $\mathbf{Prof}(\mathscr{W})$. *The latter is described as follows:*

- *An object is a pair (A,R) where $A \in \mathcal{C}at$ and R is an object of $\mathscr{W}$.*
- *The hom category from (A,R) to (B,S) is $\mathscr{W}(R,S)(A \times B^{\mathrm{op}})$.*
- *The composition functors are constructed by taking homotopy coends.*
- *The unit object of (A,R), denoted $\mathbb{I}_{(A,R)}$, is a diagram whose value at (a,a') is the coproduct of $\hom_A(a',a)$ copies of the unit $\mathbb{I}_R$.*

For the derivator bicategory $\mathcal{P}rof(\mathscr{W})$, the hom-derivators are defined by

$$\mathcal{P}rof(\mathscr{W})((A,R),(B,S))(C) = \mathscr{W}(R,S)(A \times B^{\mathrm{op}} \times C)$$

and the composition and units are defined analogously. If $\mathscr{W}$ is closed, then so is $\mathcal{P}rof(\mathscr{W})$ (and hence also $\mathbf{Prof}(\mathscr{W})$*).*

Finally, if $\mathscr{W}$ has a shadow valued in a derivator $\mathscr{T}$, then so does $\mathcal{P}rof(\mathscr{W})$, defined by taking homotopy coends. It follows that $\mathbf{Prof}(\mathscr{W})$ *also has a shadow valued in $\mathscr{T}(\mathbb{1})$.*

We follow the definitions in the previous section and say a coherent diagram $X \in \mathscr{W}(R,S)(A)$ is **pointwise dualizable** if it is right dualizable when regarded as a 1-cell from (A,R) to $(\mathbb{1},S)$ in $\mathbf{Prof}(\mathscr{W})$. Similarly, a coherent diagram $\Phi \in \mathscr{W}(R,R)(A^{\mathrm{op}})$ is **absolute** if it is right dualizable when regarded as a 1-cell from $(\mathbb{1},R)$ to (A,R) in $\mathbf{Prof}(\mathscr{W})$.

The following statements are the generalizations of Theorem 5.3 and Lemma 5.4 to derivator bicategories; once we have Theorem 10.4 then they will imply Theorem 5.3 and Lemma 5.4.

THEOREM 7.2 ([**PS16**, Theorem 11.13]). *If $X \in \mathscr{W}(R,S)(A)$ is pointwise dualizable and $\Phi \in \mathscr{W}(R,R)(A^{\mathrm{op}})$ is absolute, then for any $f\colon X \to X$ we have*

$$\mathrm{tr}(\mathrm{colim}^{\Phi}(f)) = \mathrm{tr}(f) \circ \mathrm{tr}(\mathrm{id}_{\Phi}).$$

LEMMA 7.3 ([**PS16**, Lemma 11.15]). *If $X \in \mathscr{W}(R,S)(A)$ is pointwise dualizable and $f\colon X \to X$, then for any conjugacy class $[a \xrightarrow{\alpha} a]$ in A, the composite*

$$\langle\!\langle R \rangle\!\rangle \xrightarrow{[\alpha]} \langle\!\langle (A,R) \rangle\!\rangle \xrightarrow{\operatorname{tr}(f)} \langle\!\langle S \rangle\!\rangle$$

is equal to the trace in $\mathscr{W}(R,S)(\mathbb{1})$ of the composite

$$X_a \xrightarrow{X_\alpha} X_a \xrightarrow{f_a} X_a.$$

8. Indexed monoidal derivators

We now introduce an intermediate structure which we will use to construct the derivator bicategory of parametrized spectra. Recall that if $\mathbf{S}$ is a category, an **S-indexed category** is a pseudofunctor $\mathscr{C}$ from $\mathbf{S}^{\mathrm{op}}$ to $\mathcal{CAT}$, and an **S-indexed monoidal category** is a pseudofunctor from $\mathbf{S}^{\mathrm{op}}$ to $\mathcal{MONCAT}$. Thus, for each each object $R \in \mathbf{S}$ we have a (monoidal) category $\mathscr{C}^R$, and for each morphism $f\colon R \to S$ in $\mathbf{S}$ we have a (monoidal) functor $f^*\colon \mathscr{C}^S \to \mathscr{C}^R$.

In an indexed monoidal category, we write the tensor product in each fiber $\mathscr{C}^R$ as $\otimes$, or $\otimes_R$ for emphasis if needed. If $\mathbf{S}$ has finite products, as we will henceforth assume, then there is also an *external* tensor product $\boxtimes\colon \mathscr{C}^R \times \mathscr{C}^S \to \mathscr{C}^{R\times S}$ defined by

$$X \boxtimes Y = \pi_S^* X \otimes_{R\times S} \pi_R^* Y.$$

The external product is associative and unital in a suitable sense, and we can recover the fiberwise products from it as $X \otimes_R Y = (\Delta_R)^*(X \boxtimes Y)$; see [**Shu08**] (where unfortunately the meanings of $\otimes$ and $\boxtimes$ are reversed).

In order to define a bicategory from an indexed monoidal category, we need a little more structure.

DEFINITION 8.1. We say that $\mathscr{C}$ has **S-indexed coproducts** if

(1) Each reindexing functor f^* has a left adjoint $f_!$, and
(2) For any pullback square

$$\begin{array}{ccc} A & \xrightarrow{f} & B \\ {\scriptstyle h}\downarrow & & \downarrow{\scriptstyle g} \\ C & \xrightarrow[k]{} & D \end{array}$$

in $\mathbf{S}$, the composite

$$f_! h^* \longrightarrow f_! h^* k^* k_! \xrightarrow{\cong} f_! f^* g^* k_! \longrightarrow g^* k_!$$

is an isomorphism (the *Beck-Chevalley condition*).

If $\mathscr{C}$ is symmetric monoidal, we say that $\otimes$ **preserves indexed coproducts** (in each variable separately), or that the **projection formula holds**, if

(3) for any $f\colon A \to B$ in $\mathbf{S}$ and any $M \in \mathscr{C}^B$, $N \in \mathscr{C}^A$, the canonical map

$$f_!(f^*M \otimes N) \to f_!(f^*M \otimes f^* f_! N) \cong f_! f^*(M \otimes f_! N) \to M \otimes f_! N$$

is an isomorphism.

LEMMA 8.2 ([**PS12**, Lemma 3.2]). *The projection formula is equivalent to asking that for $f\colon A \to B$, $M \in \mathscr{C}^C$, and $N \in \mathscr{C}^A$, the canonical map*

$$\begin{aligned}(\mathrm{id}_C \times f)_!(M \boxtimes N) &\to (\mathrm{id}_C \times f)_!(M \boxtimes f^* f_! N)\\ &\cong (\mathrm{id}_C \times f)_!(\mathrm{id}_C \times f)^*(M \boxtimes f_! N)\\ &\to M \boxtimes f_! N\end{aligned} \tag{8.3}$$

is an isomorphism.

In [**Shu08**] it is shown that if $\mathscr{C}$ is an indexed monoidal category with indexed coproducts preserved by $\otimes$, then there is a bicategory $\mathscr{C}/_{\mathbf{S}}$, whose 0-cells are the objects of $\mathbf{S}$, and whose hom-categories are $\mathscr{C}/_{\mathbf{S}}(R,S) = \mathscr{C}_{R\times S}$. The compositions and units are defined for $M \in \mathscr{C}/_{\mathbf{S}}(R,S)$ and $N \in \mathscr{C}/_{\mathbf{S}}(S,T)$ by

$$\begin{aligned}M \odot N &= (\mathrm{id}_R \times \pi_S \times \mathrm{id}_T)_!(\mathrm{id}_R \times \Delta_S \times \mathrm{id}_T)^*(M \boxtimes N) \quad \text{and}\\ \mathbb{I}_R &= (\Delta_R)_! {\pi_R}^*(\mathbb{S})\end{aligned}$$

In [**PS12**] it is shown that if $\mathscr{C}$ is symmetric, then $\mathscr{C}/_{\mathbf{S}}$ has a shadow with values in $\mathscr{C}^\star$, where $\star$ is the terminal object of $\mathbf{S}$, defined for $M \in \mathscr{C}/_{\mathbf{S}}(R,R)$ by

$$\langle\!\langle M \rangle\!\rangle = (\pi_R)_!(\Delta_R)^* M.$$

Unfortunately, the examples arising in homotopy theory do not always satisfy the Beck-Chevalley condition for all pullback squares in $\mathbf{S}$. However, they do satisfy it for squares of the following form (which are all automatically pullbacks in any category with products):

$$\begin{array}{ccc} A\times C & \xrightarrow{f\times\mathrm{id}} & B\times C\\ {\scriptstyle \mathrm{id}\times g}\downarrow & & \downarrow{\scriptstyle \mathrm{id}\times g}\\ A\times D & \xrightarrow[f\times\mathrm{id}]{} & B\times D \end{array} \qquad \begin{array}{ccc} A & \xrightarrow{(\mathrm{id},f)} & A\times B\\ {\scriptstyle f}\downarrow & & \downarrow{\scriptstyle f\times\mathrm{id}}\\ B & \xrightarrow[\Delta_B]{} & B\times B \end{array} \qquad \begin{array}{ccc} A & \xrightarrow{\Delta} & A\times A\\ {\scriptstyle \Delta}\downarrow & & \downarrow{\scriptstyle \mathrm{id}\times\Delta}\\ A\times A & \xrightarrow[\Delta\times\mathrm{id}]{} & A\times A\times A \end{array}$$

In [**PS12**] we defined an indexed category to have **indexed homotopy coproducts** if each f^* has a left adjoint $f_!$ satisfying the Beck-Chevalley condition for pullback squares of these three forms, along with any square obtained from these by taking cartesian products with a fixed object, and showed that this suffices for the construction of a bicategory. We will refer to these special pullback squares as **homotopy pullback squares**; this is not an egregious abuse of terminology since all such squares *are* homotopy pullback squares in any reasonable homotopy theory (for instance, in any derivator).

To generalize this construction to derivators, we start by defining an **S-indexed monoidal derivator** to be a pseudofunctor

$$\mathscr{V}\colon \mathbf{S}^{\mathrm{op}} \to \mathcal{MONDER}_{\mathrm{cc}},$$

where $\mathcal{MONDER}_{\mathrm{cc}}$ is the 2-category of monoidal derivators and cocontinuous strong monoidal morphisms [**GPS14**, §3]. Thus, associated to each object R of $\mathbf{S}$ we have a monoidal derivator denoted by $\mathscr{V}^R$ and for each morphism $f\colon R \to S$ in $\mathbf{S}$ we have a cocontinuous strong monoidal morphism of derivators

$$\mathscr{V}^S \to \mathscr{V}^R$$

denoted by f^*. We say that $\mathscr{V}$ is **symmetric** if each $\mathscr{V}^R$ is a symmetric monoidal derivator and each f^* is a symmetric monoidal morphism.

If A is a small category, we denote the value of the derivator $\mathscr{V}^R$ on A by $\mathscr{V}^R(A)$. For each small category A, if we define $(\mathscr{V}^-(A))^R = \mathscr{V}^R(A)$, we have an ordinary **S**-indexed monoidal category $\mathscr{V}^-(A)$. Thus, we can also think of an **S**-indexed monoidal derivator as a derivator consisting of **S**-indexed monoidal categories.

DEFINITION 8.4. An **S**-indexed derivator $\mathscr{V}$ has **S-indexed homotopy coproducts** if

(1) each morphism of derivators $f^*\colon \mathscr{V}^S \to \mathscr{V}^R$ has a left adjoint $f_!$, and
(2) these adjoints satisfy the Beck-Chevalley condition for homotopy pullback squares in **S**.

If $\mathscr{V}$ is additionally monoidal, we say that its indexed coproducts are **preserved by** $\otimes$, or that the **projection formula** holds, if it holds levelwise for each $\mathscr{V}^-(A)$.

Definition 8.4(2) is equivalent to asking that that for each $A \in \mathcal{C}at$, the indexed monoidal category $\mathscr{V}^-(A)$ has **S**-indexed homotopy coproducts.

THEOREM 8.5. *From an* **S**-*indexed monoidal derivator* $\mathscr{V}$ *with indexed homotopy coproducts preserved by* $\otimes$, *we can define a derivator bicategory* $\mathscr{V}/_{\mathbf{S}}$ *whose 0-cells are the objects of* **S**, *with*

$$\mathscr{V}/_{\mathbf{S}}(R,S) = \mathscr{V}^{R\times S},$$

and with composition and units defined by

$$M \odot N = (\mathrm{id}_R \times \pi_S \times \mathrm{id}_T)_!(\mathrm{id}_R \times \Delta_S \times \mathrm{id}_T)^*(M \boxtimes N) \quad \textit{and}$$
$$\mathbb{I}_R = (\Delta_R)_!\pi_R{}^*(U)$$

Moreover, if $\mathscr{V}$ *is symmetric, then* $\mathscr{V}/_{\mathbf{S}}$ *has a shadow with values in* $\mathscr{V}^\star$, *defined by*

$$\langle\!\langle M\rangle\!\rangle = (\pi_R)_!(\Delta_R)^* M.$$

PROOF. Note that for any derivator bicategory $\mathscr{W}$ and any $A \in \mathcal{C}at$, we have an ordinary bicategory $\mathscr{W}(A)$, with hom-categories $\mathscr{W}(A)(R,S) = \mathscr{W}(R,S)(A)$. (The case $A = \mathbb{1}$ yields the underlying ordinary bicategory of $\mathscr{W}$.) Similarly, for any $u\colon A \to B$ in $\mathcal{C}at$, we have an induced pseudofunctor $\mathscr{W}(B) \to \mathscr{W}(A)$ that is the identity on objects, and for any natural transformation in $\mathcal{C}at$ we have an *icon* in the sense of [**Lac10**]. Together these fit together into a pseudofunctor from $\mathcal{C}at^{\mathrm{op}}$ to the 2-category of bicategories, identity-on-objects pseudofunctors, and icons, which encapsulates all the data of $\mathscr{W}$. Thus, a derivator bicategory can equivalently be defined as such a pseudofunctor such that each induced pseudofunctor $\mathscr{W}(R,S)$ is a derivator and the composition morphisms are cocontinuous in each variable.

Now we have observed above that an **S**-indexed monoidal derivator $\mathscr{V}$ can be regarded as a pseudofunctor from $\mathcal{C}at^{\mathrm{op}}$ to **S**-indexed monoidal categories. Thus, the functoriality of the construction in [**Shu08**] automatically yields all the data of a derivator bicategory. Moreover, each $\mathscr{V}^{R\times S}$ is a derivator by assumption, while cocontinuity of composition follows from the cocontinuity of the fiberwise monoidal structures (as assumed in the definition of monoidal derivator), the assumed cocontinuity of the morphisms f^*, and the fact that $f_!$ is cocontinuous as it is a left adjoint [**Gro13**, Prop. 2.9]. Thus, we have a derivator bicategory.

The construction of shadows in [**PS12**] applied at each $A \in \mathcal{C}at$ yields functors $\mathscr{V}/_{\mathbf{S}}(R,S)(A) \to \mathscr{V}^\star(A)$ that satisfy the shadow axioms, so it remains to show that these assemble into cocontinuous morphisms of derivators. However, this is immediate since $(\Delta_R)^*$ and $(\pi_R)_!$ are such. □

9. Closed structures

We now investigate closedness of $\mathscr{V}/_{\mathbf{S}}$, which requires further assumptions on $\mathscr{V}$. We begin with the corresponding structure for ordinary indexed categories.

DEFINITION 9.1. An $\mathbf{S}$-indexed category $\mathscr{C}$ has **$\mathbf{S}$-indexed (homotopy) products** if each reindexing functor f^* has a *right* adjoint f_* that satisfies the Beck-Chevalley condition for all (homotopy) pullback squares in $\mathbf{S}$.

A standard calculation with mates implies that if the functors f^* have both left and right adjoints, then the left adjoints satisfy the Beck-Chevalley condition if and only if the right adjoints do.

DEFINITION 9.2. An $\mathbf{S}$-indexed symmetric monoidal category $\mathscr{C}$ is **closed** if

(1) each monoidal category $\mathscr{C}^R$ is closed, with internal-homs $\blacktriangleright$, and
(2) the reindexing functors f^* are closed monoidal, i.e. for $f\colon R \to S$ and $M, N \in \mathscr{C}^S$ the canonical map

$$\begin{aligned} f^*(M \blacktriangleright N) &\to f^*M \blacktriangleright (f^*(M \blacktriangleright N) \otimes f^*M) \\ &\cong f^*M \blacktriangleright f^*((M \blacktriangleright N) \otimes M) \\ &\to (f^*M \blacktriangleright f^*N) \end{aligned}$$

is an isomorphism.

Recall that we assume $\mathbf{S}$ has finite products.

LEMMA 9.3. *If $\mathscr{C}$ has $\mathbf{S}$-indexed homotopy products, it is closed if and only if*

(1) *each external product functor $\boxtimes\colon \mathscr{C}^R \times \mathscr{C}^S \to \mathscr{C}^{R\times S}$ is part of a two-variable adjunction, with adjoints $\overline{\triangleright}\colon (\mathscr{C}^S)^{\mathrm{op}} \times \mathscr{C}^{R\times S} \to \mathscr{C}^R$, and*
(2) *for each $f\colon R \to T$, $M \in \mathscr{C}^S$, and $N \in \mathscr{C}^{T\times S}$, the canonical map*

$$f^*(M \,\overline{\triangleright}\, N) \to (M \,\overline{\triangleright}\, (f \times \mathrm{id}_S)^* N)$$

is an isomorphism.

PROOF. This is essentially [**Shu08**, Propositions 13.9 and 13.15]; the only difference is the Beck-Chevalley conditions assumed. Proposition 13.15 of *ibid* assumes indexed products ("a strongly BC $*$-fibration") for one direction of the equivalence and a weaker form of indexed homotopy coproducts ("a weakly BC $*$-fibration") for the other, but in fact indexed homotopy coproducts suffice for both directions. □

LEMMA 9.4. *If $\mathscr{C}$ has $\mathbf{S}$-indexed homotopy products and coproducts and satisfies Lemma 9.3(1), then it satisfies Lemma 9.3(2) (hence is closed) if and only if its indexed coproducts are preserved by $\otimes$.*

PROOF. The canonical map in Lemma 8.2 is a mate of that in Lemma 9.3(2), in such a way that one is an isomorphism if and only if the other is. □

It is shown in [**Shu08**] that in the situation of Lemma 9.4, the bicategory $\mathscr{C}/_{\mathbf{S}}$ is closed. We now generalize this to derivators.

DEFINITION 9.5. An $\mathbf{S}$-indexed derivator $\mathscr{V}$ has **$\mathbf{S}$-indexed homotopy products** if each derivator morphism f^* has a right adjoint morphism f_* that satisfies the Beck-Chevalley condition for all homotopy pullback squares in $\mathbf{S}$.

DEFINITION 9.6. An $\mathbf{S}$-indexed symmetric monoidal derivator $\mathscr{V}$ is **closed** if each monoidal derivator $\mathscr{V}^R$ is closed, with internal-homs $\blacktriangleright$, and the reindexing functors f^* are closed monoidal.

Recall that closedness of $\mathscr{V}^R$ means that the derivator morphism $\otimes\colon \mathscr{V}^R \times \mathscr{V}^R \to \mathscr{V}^R$ is part of a two-variable adjunction. In this case, the second condition for closedness of $\mathscr{V}$ is equivalent to closedness of each indexed monoidal category $\mathscr{V}^-(A)$.

LEMMA 9.7. *If an indexed monoidal derivator $\mathscr{V}$ has indexed homotopy products, then it is closed if and only if*

(1) *each derivator morphism $\boxtimes\colon \mathscr{V}^R \times \mathscr{V}^S \to \mathscr{V}^{R\times S}$ is part of a two-variable adjunction, and*
(2) *the condition of Lemma 9.3(2) holds levelwise for each $\mathscr{V}^-(A)$.*

PROOF. The definitions of $\blacktriangleright$ and $\overline{\rhd}$ in terms of each other from [**Shu08**, Proposition 13.9] can be applied verbatim to derivator morphisms. (More precisely, we apply [**GPS14**, Lemma 8.13].) The rest then follows from Lemma 9.3 applied to each $\mathscr{V}^-(A)$. □

LEMMA 9.8. *If an indexed monoidal derivator $\mathscr{V}$ has indexed homotopy products and coproducts and satisfies Lemma 9.7(1), then it is closed if and only if its indexed homotopy coproducts are preserved by $\otimes$.*

PROOF. Apply Lemma 9.4 to each $\mathscr{V}^-(A)$. □

THEOREM 9.9. *If $\mathscr{V}$ is a closed $\mathbf{S}$-indexed monoidal derivator with indexed homotopy products and indexed homotopy coproducts, then the derivator bicategory $\mathscr{V}/_{\mathbf{S}}$ from Theorem 8.5 is closed.*

PROOF. Recall that a derivator bicategory $\mathscr{W}$ is closed if each derivator morphism $\odot\colon \mathscr{W}(R,S) \times \mathscr{W}(S,T) \to \mathscr{W}(R,T)$ is part of a two-variable adjunction. However, the construction of its adjoints in [**Shu08**, Proposition 17.5] can be applied verbatim in to derivator morphisms (again, technically invoking [**GPS14**, Lemma 8.13]). □

10. Indexed monoidal model categories

Finally, we need to be able to construct indexed monoidal derivators, which we do by introducing a notion of *indexed monoidal model category*. However, there is a slightly subtle twist. We could define an $\mathbf{S}$-indexed monoidal model category to be a pseudofunctor from $\mathbf{S}^{\mathrm{op}}$ to the 2-category of monoidal model categories, but this would not include the desired example of parametrized spectra. Instead we ask that the *external* form of the monoidal product is a two-variable Quillen left adjoint.

There is an additional problem in that an isomorphism relating composites of left and right adjoints does not automatically descend to homotopy categories, even if all the functors involved have derived versions. For instance, a Beck-Chevalley isomorphism $f_! h^* \cong g^* k_!$ does not necessarily induce an isomorphism $\mathbf{L}f_! \circ \mathbf{R}h^* \cong \mathbf{R}g^* \circ \mathbf{L}k_!$ of derived functors. However, a transformation of this sort does induce a *not-necessarily-invertible* transformation between composites of derived functors called its *derived transformation*. The main result of [**Shu11**] is that the derived transformation of a mate (such as a Beck-Chevalley transformation) is the corresponding mate at the level of derived functors. Thus, to show that a derived Beck-Chevalley isomorphism (for instance) holds, it suffices to analyze the formula for the derived transformation given in [**Shu11**] and show that it is an isomorphism. This analysis depends on the particular example in question, so in the general definition to follow we simply assume that certain derived transformations are isomorphisms. Fortunately, in our desired example of parametrized spectra, the necessary analysis has already been done by [**MS06**].

DEFINITION 10.1. An **S-indexed (symmetric) monoidal model category** consists of:

(1) A pseudofunctor $\mathscr{C}\colon \mathbf{S}^{\mathrm{op}} \to \mathcal{RMODEL}_{\mathrm{cc}}$, where $\mathcal{RMODEL}_{\mathrm{cc}}$ is the 2-category of model categories and right Quillen functors that preserve homotopy colimits. (By the latter we mean that their induced morphism of derivators is cocontinuous.)
(2) The underlying pseudofunctor $\mathbf{S}^{\mathrm{op}} \to \mathcal{CAT}$ is an indexed (symmetric) monoidal category with indexed homotopy coproducts preserved by $\otimes$.
(3) The external product functors
$$\boxtimes\colon \mathscr{C}^R \times \mathscr{C}^S \to \mathscr{C}^{R\times S}$$
are left Quillen adjoints of two variables.
(4) For any cofibrant $M \in \mathscr{C}^R$ and any cofibrant replacement $Q\mathbb{S}_\star \to \mathbb{S}_\star$ of the unit $\mathbb{S}_\star$ of $\mathscr{C}_\star$, the induced maps
$$Q\mathbb{S}_\star \boxtimes M \to \mathbb{S}_\star \boxtimes M \cong M \qquad \text{and}$$
$$M \boxtimes Q\mathbb{S}_\star \to M \boxtimes \mathbb{S}_\star \cong M$$
are isomorphisms.
(5) The derived transformations of the isomorphisms
$$f^*M \boxtimes g^*N \cong (f\times g)^*(M \boxtimes N)$$
are again isomorphisms.
(6) The derived transformations of the Beck-Chevalley isomorphisms for any homotopy pullback square in $\mathbf{S}$ are again isomorphisms.
(7) The derived functor $(\mathbf{R}f^*)_A\colon \mathscr{H}o(\mathscr{C}^S)(A) \to \mathscr{H}o(\mathscr{C}^R)(A)$ has a right adjoint $(f_*)_A$ for all $A \in \mathcal{C}at$.

Condition (4) may look familiar from the theory of monoidal model categories, but it is actually also a simple example of a derived natural transformation as in [**Shu11**]. Note also that we do not assume $\mathscr{C}$ itself to have indexed products; it may, but all we need for the following theorem is condition (7). The primary example of parametrized spectra does have indexed products on the point-set level, but it is unclear in general whether or how these are related directly to the derived ones, which are constructed using Brown representability.

THEOREM 10.2. *If $\mathscr{C}$ is an* **S***-indexed (symmetric) monoidal model category, then we have an* **S***-indexed closed (symmetric) monoidal derivator $\mathscr{H}o(\mathscr{C})$, defined by $\mathscr{H}o(\mathscr{C})^R = \mathscr{H}o(\mathscr{C}^R)$, which has indexed homotopy products and indexed homotopy coproducts which are preserved by $\otimes$.*

PROOF. Essentially by assumption on the morphisms in $\mathcal{RMODEL}_{\mathrm{cc}}$, passage to homotopy categories and right derived functors induces a pseudofunctor $\mathscr{H}o(\mathscr{C})\colon \mathbf{S}^{\mathrm{op}} \to \mathcal{DER}_{\mathrm{cc}}$. Now it is shown in [**Shu08**] that a (symmetric) monoidal structure on an indexed category can equivalently be described by its external product functors (there represented in terms of a monoidal structure on the corresponding categorical fibration), and this can be extended levelwise to derivators. In our case, Definition 10.1(3) ensures that the external product induces two-variable morphisms of derivators, while (5) ensures that these are appropriately pseudonatural. The associativity isomorphism (and the symmetry isomorphism, if present) descends automatically, since all functors involved are left Quillen, while assumption (4) ensures that the unit isomorphism descends. Coherence for all of these isomorphisms follows from the functoriality of derived transformations. Thus, $\mathscr{H}o(\mathscr{C})$ is an indexed (symmetric) monoidal derivator.

For indexed homotopy coproducts, the Quillen adjunctions $f_! \dashv f^*$ descend to adjunctions of derivators, and assumption (6) ensures that the Beck-Chevalley condition for homotopy pullback squares descends. By Lemma 8.2, for preservation by $\otimes$ it suffices to check that the isomorphism $(\mathrm{id}_C \times f)_!(M \boxtimes N) \to M \boxtimes f_! N$ descends to a derived isomorphism; but this is automatic because all functors involved are left Quillen.

Now since each $\mathbf{R}f^*$ is cocontinuous, by [**Gro13**, Proposition 2.9] the adjoints $(f_*)_A$ assumed by (7) assemble into a morphism of derivators right adjoint to $\mathbf{R}f^*$. Thus, by the remark after Definition 9.1, $\mathscr{H}o(\mathscr{C})$ has indexed homotopy products as well. Finally, closedness follows from Lemma 9.8, since by assumption the external products $\boxtimes$ are two-variable Quillen left adjoints, hence induce two-variable left adjoints of derivators. □

The following is the example we care most about, which we used in §6 to describe the Reidemeister trace.

THEOREM 10.3. *Let* $\mathbf{S}$ *be the category of compactly generated topological spaces. Then there is an* $\mathbf{S}$*-indexed closed symmetric monoidal derivator* $\mathscr{H}o(\mathcal{S}p)$ *with indexed homotopy products and indexed homotopy coproducts preserved by* $\otimes$*, where* $\mathscr{H}o(\mathcal{S}p)^R$ *is the homotopy category of* R*-parametrized spectra.*

PROOF. We apply Theorem 10.2. For $R \in \mathbf{S}$, let $\mathcal{S}p^R$ be the category of parametrized orthogonal spectra over R. By [**MS06**, Theorem 11.4.1 and Proposition 11.4.8], these form an $\mathbf{S}$-indexed closed symmetric monoidal category with indexed products and indexed coproducts preserved by $\otimes$.

Now by [**MS06**, Theorem 12.3.10], each $\mathcal{S}p^R$ has a stable model structure, for which each adjunction $f_! \colon \mathcal{S}p^R \rightleftarrows \mathcal{S}p^S \colon f^*$ is Quillen by [**MS06**, Proposition 12.6.7]. Condition (3) of Definition 10.1 holds by [**MS06**, Prop 12.6.5], while (4) holds because $\mathbb{S}_\star$ is cofibrant (note that $\mathcal{S}p^\star$ coincides with the monoidal model category of ordinary orthogonal spectra from [**MMSS01**]). Conditions (5) and (6) are shown in [**MS06**, Theorem 13.7.2]. and [**MS06**, Theorem 13.7.7] respectively.

It remains to show that each $\mathbf{R}f^*$ is cocontinuous and has a levelwise right adjoint. The proof of [**MS06**, Theorem 13.1.18] shows that $\mathbf{R}f^*$ preserves coproducts. Since it is a right adjoint, it preserves pullbacks; and since its domain and codomain are stable, this implies that it preserves pushouts as well. Thus, by [**PS16**, Theorem 7.13] it is cocontinuous.

Finally, for the existence of a right adjoint to $(\mathbf{R}f^*)_A$, we invoke Brown representability following [**MS06**, Theorem 13.1.18] (which is the case $A = \mathbb{1}$). Since (as we have just shown) $\mathbf{R}f^*$ is a cocontinuous morphism of derivators, $(\mathbf{R}f^*)_A$ is an exact and coproduct-preserving functor of triangulated categories. Thus, by [**MS06**, Theorem 13.1.17] it suffices to show that each triangulated category $\mathscr{H}o(\mathcal{S}p^R)(A)$ is compactly generated. This is shown in the case $A = \mathbb{1}$ by [**MS06**, Lemma 13.1.11]; the generating set was denoted $\mathscr{D}_R$.

For general $A \in \mathcal{C}at$, recall that the "evaluate at a" functors $\mathrm{ev}_a\colon \mathscr{H}o(\mathcal{S}p^R)(A) \to \mathscr{H}o(\mathcal{S}p^R)(\mathbb{1})$ are jointly conservative. Thus, if Lan_a denotes the left adjoint of ev_a (i.e. homotopy left Kan extension), then $\bigcup_a \mathrm{Lan}_a(\mathscr{D}_R)$ is a generating set for $\mathscr{H}o(\mathcal{S}p^R)(A)$. And since ev_a preserves coproducts (being a left adjoint), Lan_a preserves compact objects [**MS06**, Remark 13.1.9]. Thus, $\mathscr{H}o(\mathcal{S}p^R)(A)$ is compactly generated. □

Therefore, by Theorem 8.5 and Theorem 9.9 we have:

THEOREM 10.4. *There is a closed derivator bicategory $\mathcal{E}x$ with a shadow valued in the homotopy derivator of spectra. Its objects are compactly generated spaces, its hom $\mathcal{E}x(R,S)$ is the homotopy derivator of the model category of spectra parametrized over $R \times S$, and its underlying ordinary bicategory is the one constructed in* [**MS06**, Chapter 17].

References

[Ati61] M. F. Atiyah, *Thom complexes*, Proc. London Math. Soc. (3) **11** (1961), 291–310, DOI 10.1112/plms/s3-11.1.291. MR0131880

[DP80] Albrecht Dold and Dieter Puppe, *Duality, trace, and transfer*, Proceedings of the International Conference on Geometric Topology (Warsaw, 1978), PWN, Warsaw, 1980, pp. 81–102. MR656721

[Fer96] Davide Ferrario, *Generalized Lefschetz numbers of pushout maps*, Topology Appl. **68** (1996), no. 1, 67–81, DOI 10.1016/0166-8641(96)00040-5. MR1370927

[GPS14] Moritz Groth, Kate Ponto, and Michael Shulman, *The additivity of traces in monoidal derivators*, J. K-Theory **14** (2014), no. 3, 422–494, DOI 10.1017/is014005011jkt262. MR3349323

[Gro13] Moritz Groth, *Derivators, pointed derivators and stable derivators*, Algebr. Geom. Topol. **13** (2013), no. 1, 313–374, DOI 10.2140/agt.2013.13.313. MR3031644

[Hir03] Philip S. Hirschhorn, *Model categories and their localizations*, Mathematical Surveys and Monographs, vol. 99, American Mathematical Society, Providence, RI, 2003. MR1944041

[Lac10] Stephen Lack, *Icons*, Appl. Categ. Structures **18** (2010), no. 3, 289–307, DOI 10.1007/s10485-008-9136-5. MR2640216

[LMSM86] L. G. Lewis Jr., J. P. May, M. Steinberger, and J. E. McClure, *Equivariant stable homotopy theory*, Lecture Notes in Mathematics, vol. 1213, Springer-Verlag, Berlin, 1986. With contributions by J. E. McClure. MR866482

[May75] J. Peter May, *Classifying spaces and fibrations*, Mem. Amer. Math. Soc. **1** (1975), no. 1, 155, xiii+98. MR0370579

[May01] J. P. May, *The additivity of traces in triangulated categories*, Adv. Math. **163** (2001), no. 1, 34–73, DOI 10.1006/aima.2001.1995. MR1867203

[Mey84] Jean-Pierre Meyer, *Bar and cobar constructions. I*, J. Pure Appl. Algebra **33** (1984), no. 2, 163–207, DOI 10.1016/0022-4049(84)90005-7. MR754954

[MMSS01] M. A. Mandell, J. P. May, S. Schwede, and B. Shipley, *Model categories of diagram spectra*, Proc. London Math. Soc. (3) **82** (2001), no. 2, 441–512, DOI 10.1112/S0024611501012692. MR1806878

[MS06] J. P. May and J. Sigurdsson, *Parametrized homotopy theory*, Mathematical Surveys and Monographs, vol. 132, American Mathematical Society, Providence, RI, 2006. MR2271789

[Pon10] Kate Ponto, *Fixed point theory and trace for bicategories* (English, with English and French summaries), Astérisque **333** (2010), xii+102. MR2741967

[Pon11] Kate Ponto, *Relative fixed point theory*, Algebr. Geom. Topol. **11** (2011), no. 2, 839–886, DOI 10.2140/agt.2011.11.839. MR2782545

[Pon15] Kate Ponto, *Equivariant fixed-point theory*, Homology Homotopy Appl. **17** (2015), no. 2, 161–190, DOI 10.4310/HHA.2015.v17.n2.a9. MR3426379

[Pon16] Kate Ponto, *Coincidence invariants and higher Reidemeister traces*, J. Fixed Point Theory Appl. **18** (2016), no. 1, 147–165, DOI 10.1007/s11784-015-0269-5. MR3463529

[PS12] Kate Ponto and Michael Shulman, *Duality and traces for indexed monoidal categories*, Theory Appl. Categ. **26** (2012), No. 23, 582–659. MR3065938

[PS13a] Kate Ponto and Michael Shulman, *Shadows and traces in bicategories*, J. Homotopy Relat. Struct. **8** (2013), no. 2, 151–200, DOI 10.1007/s40062-012-0017-0. MR3095324

[PS13b] Kate Ponto and Michael Shulman, *Traces in symmetric monoidal categories*, Expo. Math. **32** (2014), no. 3, 248–273, DOI 10.1016/j.exmath.2013.12.003. MR3253568

[PS14] Kate Ponto and Michael Shulman, *The multiplicativity of fixed point invariants*, Algebr. Geom. Topol. **14** (2014), no. 3, 1275–1306, DOI 10.2140/agt.2014.14.1275. MR3190594

[PS16] Kate Ponto and Michael Shulman, *The linearity of traces in monoidal categories and bicategories*, Theory Appl. Categ. **31** (2016), Paper No. 23, 594–689. MR3518981

[Shu06] Michael Shulman. Homotopy limits and colimits and enriched homotopy theory. arXiv:math.CT/0610194, 2006.

[Shu08] Michael Shulman, *Framed bicategories and monoidal fibrations*, Theory Appl. Categ. **20** (2008), No. 18, 650–738. MR2534210

[Shu11] Michael Shulman, *Comparing composites of left and right derived functors*, New York J. Math. **17** (2011), 75–125. MR2781909

University of Kentucky, 719 Patterson Office Tower, Lexington, KY 40508
Email address: kate.ponto@uky.edu

University of San Diego, 5998 Alcala Park, San Diego, CA 92110
Email address: shulman@sandiego.edu

Contemporary Mathematics
Volume **707**, 2018
http://dx.doi.org/10.1090/conm/707/14257

Homotopy coherent centers versus centers of homotopy categories

Markus Szymik

ABSTRACT. Centers of categories capture the natural operations defined on their objects. Homotopy coherent centers are an extension of this notion to categories with an associated homotopy theory. These centers can also be interpreted as Hochschild cohomology type invariants in contexts that are not necessarily linear or stable, and we argue that they are more appropriate to higher categorical contexts than the centers of their homotopy or derived categories. We present an obstruction theory for realizing elements in the centers of homotopy categories, and a Bousfield–Kan type spectral sequence that computes the homotopy groups. Several non-trivial classes of examples are given as illustrations of the general theory throughout.

Introduction

Whenever we are given a small category $\mathbf{C}$, we may ask for a description of all natural operations on its objects. These are the families $\Phi = (\Phi_x \colon x \to x \mid x \in \mathrm{Obj}(\mathbf{C}))$ of endomorphisms that are natural in the objects x, and they form a monoid under composition. In fact, a more conceptual description of this monoid presents it as the endomorphism monoid of the identity functor $\mathbf{C} \to \mathbf{C}$ in the category of all such functors and natural transformations. Naturality implies immediately that this monoid is always abelian. This observation is usually attributed to Eckmann and Hilton. As a classical example, for the category of commutative rings of characteristic p, where p is a prime number, this monoid is the free (abelian) monoid on one generator: Frobenius. For a more recent computation with a similar outcome see [**Szy18**]. On the other hand, if the category in question is a monoid itself–a category with one object, then we have just given a long-winded description of the center of this monoid, the subset of elements that commute with all of its elements. In general, the endomorphism monoid of the identity functor is often referred to as the *center* $\mathrm{Z}(\mathbf{C})$ of $\mathbf{C}$, for instance by Bass [**Bas68**, II, §2] and Mac Lane [**Mac71**, II.5, Exercise 8], and we will follow this terminology. Bernstein, in [**Ber84**, 1.9], defined the center of abelian categories, but his main object of interest was the category of smooth representations of a p-adic group.

In the present paper, we study a refined notion of center, the homotopy coherent center $\mathcal{Z}(\mathbf{C})$, for categories $\mathbf{C}$ that admit a homotopy theory. This center is a Kan

2010 *Mathematics Subject Classification.* 55U40 (18G40, 18G50, 55S35).
Key words and phrases. Centers, homotopy coherence, homotopy limit problems, spectral sequences, obstruction theory.

complex that is defined directly within **C** rather than on the level of the homotopy category. Briefly, its elements will determine families $\Phi = (\Phi_x \colon x \to x \mid x \in \mathrm{Obj}(\mathbf{C}))$ of endomorphisms in **C**, but we no longer require that these are natural in the strict sense. Instead, these elements will also come with continuously chosen homotopies $\Phi_y f \simeq f\Phi_x$ for all arrows $f \colon x \to y$ in **C** and additional higher homotopies that, for instance, show that for any other arrow $g \colon y \to z$ the two evident homotopies $\Phi_z gf \simeq gf\Phi_x$ that we can now obtain from Φ_f, Φ_g, and Φ_{gf}, are also homotopic. These are the coherent natural (rather than strict natural) transformations $\mathrm{Id}_{\mathbf{C}} \to \mathrm{Id}_{\mathbf{C}}$ in the sense of Cordier and Porter [**CP97**], compare Remark 2.4. The actual formulas bear very close resemblance to those used in the definition of Hochschild(–Mitchell) cohomology [**Mit72**]. This suggests that the idea of deriving the notion of the center of a category is not new, in particular in the additive context of homological algebra and differential graded categories. There, it already appeared in an unpublished preprint by Keller [**Kel**]. The ideas presented in the present paper lead to new results in the differential graded context as well. This has been realized in joint work with Neumann [**NS17**], with applications to the characteristic homomorphism from the Hochschild cohomology of a differential graded algebra, or differential graded category, to the graded center of its derived category.

Here we will work in a non-linear and unstable context: categories enriched in spaces. Spaces will be (fibrant) simplicial sets, so that categories enriched in spaces will be categories enriched in (fibrant) simplicial sets or (fibrant) simplicial categories for short. The reader unfamiliar with simplicial categories will be able to replace them by topological categories, but even in that setting, (co)simplicial methods are indispensable for our approach. Simplicial categories are by now a well-established context in which to do homotopy theory, and there is even a homotopy theory of simplicial categories themselves [**Ber07**]. This is relevant here because homotopy coherent centers are invariant under the corresponding notion of weak equivalence for simplicial categories, Dwyer–Kan equivalence (Theorem 4.1). We also note that the present definition automatically extends to other contexts that have an associated homotopy theory, such as Quillen model categories or quasi-categories, by passing to the associated simplicial categories that they define.

The homotopy coherent center of any category enriched in spaces has a canonical E_2 multiplication (Theorem 3.1). This means, in particular, that these centers are A_∞ monoids (coherently associative, i.e. have an action of an $A_\infty = E_1$ operad), and that the multiplication is homotopy commutative (because it extends to an action of an E_2 operad). This is the analogue of Eckmann–Hilton in the present setting, and it can be proved using the methods introduced by McClure and Smith in [**MS02**], [**MS04a**], and [**MS04b**] for their solution of the Deligne conjecture.

In Section 5, we discuss the examples given by simplicial monoids and, in particular, simplicial groups. In the latter case, the homotopy coherent centers are related to other notions of centers studied before in homotopical group theory, see the ICM surveys by Dwyer [**Dwy98**] and Grodal [**Gro10**]. For instance, if G is a simplicial group, then its center consists of the fixed points under the conjugation action, and the homotopy fixed points are equivalent to the homotopy coherent center as defined here (Theorem 5.1). On the other hand, the homotopy center of a p-compact group is defined as the loop space of the space of self-maps of the classifying space based at the identity, see [**MN94**] and [**DW95**], and this is also equivalent to the homotopy coherent center in the case where both of them are defined (Corollary 5.3).

Back in our general context, there is also a very naive and rigid notion $\mathrm{Z}(\mathbf{C})$ of a center for simplicial categories $\mathbf{C}$ based on (strict) equalities. The relation of the homotopy coherent center to this simplicial center takes the form of a morphism

$$\mathrm{Z}(\mathbf{C}) \longrightarrow \mathcal{Z}(\mathbf{C}).$$

The question whether such a map from a strict limit to a homotopy limit is an equivalence (or at least some sort of completion) is called a homotopy limit problem, following Thomason [**Tho83**] and Carlsson [**Car87**]. We will discuss this question for the map displayed above in Section 6 where we also give an example that shows that the situation in the present context is more complicated.

There is a canonical map $\mathcal{Z}(\mathbf{C}) \to \mathrm{Z}(\mathrm{Ho}\,\mathbf{C})$ from the homotopy coherent center to the center (in the ordinary sense) of its homotopy category $\mathrm{Ho}\,\mathbf{C}$. Because the target is discrete, this map factors through the components:

$$\pi_0\mathcal{Z}(\mathbf{C}) \longrightarrow \mathrm{Z}(\mathrm{Ho}\,\mathbf{C}).$$

In general, this latter map will be neither surjective nor injective. Here we will explain the difference between the homotopy coherent centers of simplicial categories $\mathbf{C}$ and the centers of their homotopy categories $\mathrm{Ho}\,\mathbf{C}$. In Section 7, using the methods initiated by Bousfield and Kan, see [**BK72**] and [**Bou89**], we provide means to study the failure of surjectivity by developing an obstruction theory (Theorem 7.5) for the realization of an element in the target by an element in the homotopy coherent center. As for injectivity, we show that there is a spectral sequence (Theorem 7.1) with $\mathrm{E}_2^{0,0}$ related to $\mathrm{Z}(\mathrm{Ho}\,\mathbf{C})$ that not only targets the kernel of that map but also the higher homotopy groups of the homotopy coherent center, information that is entirely invisible from the perspective of the homotopy category. The passage to the center of the homotopy category reappears from this viewpoint as an edge homomorphism.

Fringed Bousfield–Kan type spectral sequences in unstable general contexts may have a less pleasant algebraic behavior than the spectral sequences of abelian groups that we usually meet in homological algebra. In our situation, the homotopy commutative multiplication of the coherent centers leads to some simplification. In an earlier version of this paper, this was illustrated by spelling out the details for the category of groups, where the homotopy theory is induced by conjugation of homomorphisms, so that the homotopy category is the category of representations, and which is related to the theory of bands in the sense of Giraud [**Gir71**]. This has now been developed with Meir [**MS15**] in the general setting of Drinfeld centers of bicategories.

The final Section 8 deals with another class of examples of simplicial categories: simplicial groupoids. We show that the theory presented here is related to very classical and difficult questions in (unstable) homotopy theory such as spaces of homotopy self-equivalences. That section also includes examples of simplicial categories where the homotopy coherent centers have higher homotopy groups, and where the map $\pi_0\mathcal{Z}(\mathbf{C}) \to \mathrm{Z}(\mathrm{Ho}\,\mathbf{C})$ is not injective or not surjective.

All categories considered in this paper are small. The definitions, theorems, and examples exposed here are the basis for the treatment of homotopy coherent centers of large categories such as categories of simplicial diagrams, Bousfield localizations, and categories of universal algebras with Dwyer [**DS17**].

1. Categories enriched in spaces and their centers

We will uses combinatorial models (based on (co)simplicial techniques) throughout this text. If X and Y are simplicial sets, then $\mathrm{Mor}(X,Y)$ will denote the set of simplicial maps $X \to Y$. These are the vertices in the mapping space $\mathrm{Map}(X,Y)$. The general formula for the n-simplices is $\mathrm{Map}(X,Y)_n = \mathrm{Mor}(\Delta^n \times X, Y)$.

A *space* will be a Kan complex, that is a fibrant simplicial set with respect to the Kan model structure. If Y is a space in this sense, so is $\mathrm{Map}(X,Y)$.

We will consider small categories $\mathbf{C}$ that are enriched in spaces, and we will write

$$\mathbf{C}(x,y)$$

for the space of maps from the object x to the object y. Of course, if we are considering the category of spaces, we will continue to write $\mathrm{Map}(X,Y)$. Categories enriched in spaces are the fibrant simplicial categories of [**Ber07**].

Let $\mathbf{C}$ be a category enriched in spaces. For any integer n such that $n \geqslant 0$, we will use $\mathbf{C}_n$ as our notation for the (ordinary) category of n-simplices in $\mathbf{C}$. In particular, we will call $\mathbf{C}_0$ the *underlying category* of $\mathbf{C}$. There is another way to pass from a category $\mathbf{C}$ enriched in spaces to an ordinary category: the *homotopy category* $\mathrm{Ho}\,\mathbf{C}$. It has the same set of objects, but the set of morphisms $x \to y$ in $\mathrm{Ho}\,\mathbf{C}$ is the set $\pi_0\mathbf{C}(x,y)$ of components of the mapping space $\mathbf{C}(x,y)$.

Simplicial centers. We can now define a strict notion of center for categories enriched in spaces.

DEFINITION 1.1. Let $\mathbf{C}$ be a category enriched in spaces. The *simplicial center* $\mathrm{Z}(\mathbf{C})$ of $\mathbf{C}$ is the equalizer (in the category of spaces) of the two maps

$$\prod_x \mathbf{C}(x,x) \longrightarrow \prod_{y,z} \mathrm{Map}(\mathbf{C}(y,z), \mathbf{C}(y,z)) \tag{1.1}$$

that are given by sending a given family $\Phi = (\Phi_x)$ to the map $f \mapsto f\Phi_y$ and the map $f \mapsto \Phi_z f$, respectively, in the (y,z) component.

For discrete categories $\mathbf{C}$, we may replace the target with

$$\prod_{f\colon y \to z} \mathbf{C}(y,z),$$

but the alternative above is adapted to work in the simplicial context as well.

We can determine the set of n-simplices of the simplicial center by direct inspection.

PROPOSITION 1.2. *For every category* $\mathbf{C}$ *enriched in spaces, there are isomorphisms*

$$\mathrm{Z}(\mathbf{C})_n \cong \mathrm{Z}(\mathbf{C}_n)$$

that are natural in n.

In particular, the elements of the center $\mathrm{Z}(\mathbf{C}_0)$ of the underlying category $\mathbf{C}_0$ appear as the vertices of the simplicial center $\mathrm{Z}(\mathbf{C})$ of $\mathbf{C}$.

The simplicial center of a category enriched in spaces is a simplicial monoid: a simplicial submonoid of the product of the endomorphism monoids $\mathbf{C}(x,x)$, by Definition 1.1.

COROLLARY 1.3. *For every category* $\mathbf{C}$ *enriched in spaces, the simplicial center* $\mathrm{Z}(\mathbf{C})$ *is a simplicial abelian monoid.*

PROOF. As has been mentioned in the introduction, by Eckmann and Hilton, the centers $\mathrm{Z}(\mathbf{C}_n)$ of the discrete categories $\mathbf{C}_n$ are abelian. (One way to see it is in [**Kas95**, XI.2.4].) Now the corollary follows from the preceding proposition. □

The following result states that the homotopy types of simplicial centers are as simple as possible.

COROLLARY 1.4. *For every category* $\mathbf{C}$ *enriched in spaces, the simplicial center* $\mathrm{Z}(\mathbf{C})$ *is a product of Eilenberg–Mac Lane spaces.*

PROOF. This follows immediately from the preceding corollary and Moore's theorem about the homotopy types of simplicial abelian monoids, see [**Moo55**]. □

We remark that centers in enriched contexts have been studied before, see Lindner's paper [**Lin80**], for instance.

2. Homotopy coherent centers

The simplicial center of a category enriched in spaces has been defined, in the preceding Section 1, as a certain limit, see also Remark 2.1 below. Now we will define the homotopy coherent center of a category enriched in spaces as the corresponding homotopy limit. We will start with a brief review of the cosimplicial replacement of the homotopy end construction in this particular case, referring to [**CP97**] for more general information. See also [**Wei01**] for the case when the source category is discrete.

Cosimplicial replacements. Let $\mathbf{C}$ be a category enriched in spaces. For any integer $n \geqslant 0$ we can consider the space

$$\Pi^n\mathbf{C} = \prod_{x_0,\dots,x_n} \mathrm{Map}(\mathbf{C}(x_1,x_0)\times\cdots\times\mathbf{C}(x_n,x_{n-1}),\mathbf{C}(x_n,x_0))$$

where the product runs over the $(n+1)$-tuples of objects of $\mathbf{C}$. If Φ is a vertex in $\Pi^n\mathbf{C}$, then it can be evaluated on n-tuples $(f_1,\dots,f_n)$ of composable arrows

$$x_0 \xleftarrow{f_1} x_1 \xleftarrow{f_2} x_2 \longleftarrow \cdots \longleftarrow x_{n-1} \xleftarrow{f_n} x_n$$

to give $\Phi(f_1,\dots,f_n)\in\mathbf{C}(x_n,x_0)$. Together with the evident structure maps, this defines a cosimplicial space $\Pi^\bullet\mathbf{C}$: The coface maps $\Pi^{n-1}\mathbf{C}\to\Pi^n\mathbf{C}$ are given by

$$(d^k\Phi)(f_1,\dots,f_n) = \begin{cases} f_1\Phi(f_2,\dots,f_n) & k=0\\ \Phi(f_1,\dots,f_kf_{k+1},\dots,f_n) & 0<k<n\\ \Phi(f_1,\dots,f_{n-1})f_n & k=n\end{cases}$$

for $k=0,\dots,n$, and the codegeneracy maps $\Pi^{n+1}\mathbf{C}\to\Pi^n\mathbf{C}$ are given by

$$(s^k\Phi)(f_1,\dots,f_n) = \Phi(f_1,\dots,f_k,\mathrm{id},f_{k+1},\dots,f_n)$$

for $k=0,\dots,n$. This cosimplicial space is sometimes referred to as the cobar construction, and it has been extensively used by Bousfield and Kan [**BK72**].

REMARK 2.1. For $n=0$ and $n=1$, we obtain the source and target of (1.1), and we recover the definition of the simplicial center $\mathrm{Z}(\mathbf{C})$ as the equalizer of the two coboundaries $\Pi^0\mathbf{C}\to\Pi^1\mathbf{C}$. This equalizer is the limit $\lim\Pi^\bullet\mathbf{C}$ of the cosimplicial space $\Pi^\bullet\mathbf{C}$.

We note that the cosimplicial space $\Pi^\bullet\mathbf{C}$ is canonically pointed by the families of composition maps (interpreted as identities in low dimensions).

The category of cosimplicial simplicial sets comes with a least three different Quillen model category structures with level-wise equivalences: The most useful one is the Reedy structure that is already implicit in [**BK72**]. The other two are the projective and injective model structure on the diagram category of functors form Δ to simplicial sets. In the projective model structure, the fibrant objects are the cosimplicial spaces, that is the cosimplicial simplicial sets that are level-wise Kan complexes. So $\Pi^\bullet\mathbf{C}$ is a cosimplicial space, as has already been implicit in the terminology used above. It is actually better:

LEMMA 2.2. *If* $\mathbf{C}$ *is a category enriched in spaces, the cosimplicial space* $\Pi^\bullet\mathbf{C}$ *is Reedy fibrant.*

PROOF. This can be checked by hand: It has to be verified that certain maps

$$\Pi^n\mathbf{C} \longrightarrow M^{n-1}\Pi^\bullet\mathbf{C}, \tag{2.1}$$

for all $n \geqslant 0$, are fibrations. Here, the target $M^{n-1}\Pi^\bullet\mathbf{C} \subseteq (\Pi^{n-1}\mathbf{C})^n$ is the matching space, the subspace defined by the n-tuples with consistent codegeneracies, and the map (2.1) is also given by codegeneracies. This already implies the result, because the codegeneracies are given by (projections and) evaluations on identities. The latter, in turn, are induced by insertions of identity factors, which are injective, hence cofibrations. □

Fibrancy in the injective model structure is the strongest condition. It implies the equivalence of the homotopy limit with the actual limit, see [**GJ99**, VIII.2.11]. It will follow from our discussion in Section 6 that $\Pi^\bullet\mathbf{C}$ is not injectively fibrant in general.

Homotopy coherent centers. We now come to the definition that is basic to the rest of this text.

DEFINITION 2.3. Let $\mathbf{C}$ be a category enriched in spaces. The *homotopy coherent center* $\mathcal{Z}(\mathbf{C})$ of $\mathbf{C}$ is defined as the totalization

$$\mathcal{Z}(\mathbf{C}) = \mathrm{Tot}(\Pi^\bullet\mathbf{C}) \simeq \operatorname{holim} \Pi^\bullet\mathbf{C}$$

of the cosimplicial space $\Pi^\bullet\mathbf{C}$.

We note that the indicated equivalence of the totalization with the homotopy limit follows from Lemma 2.2 above: This is a general property of Reedy fibrant cosimplicial spaces, see [**GJ99**, VIII.2.12]. It also follows that $\mathcal{Z}(\mathbf{C})$ is a space (Kan complex) if the category $\mathbf{C}$ is enriched in spaces (locally Kan). See also [**CP97**, Proposition 2.1] for a direct proof of a more general statement.

A vertex in the homotopy coherent center $\mathcal{Z}(\mathbf{C})$ is given by (the adjoint to) a family $\Phi = (\Phi^n \mid n \geqslant 0)$ of maps

$$\Phi^n \colon \Delta^n \times \mathbf{C}(x_1, x_0) \times \cdots \times \mathbf{C}(x_n, x_{n-1}) \longrightarrow \mathbf{C}(x_n, x_0)$$

indexed by all choices of n and $(n+1)$-tuples $x_0, \ldots, x_n$ of objects that are consistent with the cosimplicial structure maps. If we choose $n = 0$, then we obtain a family of morphisms $\Phi^0_x \colon x \to x$, and if we choose $n = 1$, then we obtain homotopies $\Delta^1 \times \mathbf{C}(y, z) \to \mathbf{C}(y, z)$ between the maps $f \mapsto \Phi^0_z f$ and $f \mapsto f\Phi^0_y$. In particular, these give homotopies $\Phi^0_z f \simeq f\Phi^0_y$ that show that the homotopy classes of the Φ^0_x define an element in the center of the homotopy category. For $n \geqslant 2$, the Φ^n contain higher coherence information.

REMARK 2.4. The center of an ordinary category is the monoid of all natural transformations from the identity to itself, and we have here presented a homotopy coherent generalization of this idea that is adapted to the context of categories enriched in spaces. For later purposes, it will be useful to know that there are similar models for spaces of coherent natural transformations between simplicial functors $F, G\colon \mathbf{C} \to \mathbf{D}$ as well: We build a cosimplicial space $\Pi^\bullet(F,G)$ with

$$\Pi^n(F,G) = \prod_{x_0,\dots,x_n} \mathrm{Map}(\mathbf{C}(x_1,x_0) \times \cdots \times \mathbf{C}(x_n,x_{n-1}), \mathbf{D}(Fx_n, Gx_0))$$

in dimension n. Then $\mathcal{N}(F,G) = \mathrm{Tot}\,\Pi^n(F,G)$ may be considered as the space of all coherent natural transformations $F \to G$. Taking $F = \mathrm{id}_{\mathbf{C}} = G$, we recover the homotopy coherent center as $\mathcal{Z}(\mathbf{C}) = \mathcal{N}(\mathrm{id}_{\mathbf{C}}, \mathrm{id}_{\mathbf{C}})$. See [**CP96**] and [**CP97**] for this and generalizations.

3. Multiplicative structure

We will now address the algebraic structure on the homotopy coherent centers that reflects the 'composition' of coherent natural transformations.

An A_∞ multiplication. We have seen in Corollary 1.3 that the simplicial center $\mathrm{Z}(\mathbf{C})$ of a category $\mathbf{C}$ enriched in spaces is a simplicial abelian monoid. I know of no reason why the homotopy coherent center $\mathcal{Z}(\mathbf{C})$ should have a canonical simplicial monoid structure as well. However it does come with a homotopy commutative A_∞ structure, and that is just as good for the purposes of homotopy theory. To see this, we would like to use the criteria for operad actions on totalizations of cosimplicial objects as presented in [**MS02**], [**MS04a**], and [**MS04b**].

These cited papers are written topologically, so as to be able to concatenate paths, for instance. Still their results can be used here: Both geometric realization and singular complex commute with finite products, and the homotopy coherent center will turn out to be invariant under equivalences (Theorem 4.1). Therefore, given a simplicial category $\mathbf{C}$ we can first check that the homotopy coherent center of the topological category $|\mathbf{C}|$ has an action of a topological A_∞ operad using the results from McClure and Smith. Then we infer that the homotopy coherent center of the simplicial category $\mathrm{Sing}(|\mathbf{C}|)$ has an action of a simplicial A_∞ operad. Since $\mathrm{Sing}(|\mathbf{C}|)$ is equivalent to $\mathbf{C}$, this proves our claim.

With these technicalities out of the way, we need to find strictly associative and unital pairings

$$m = m_{p,q}\colon \Pi^p\mathbf{C} \times \Pi^p\mathbf{C} \longrightarrow \Pi^{p+q}\mathbf{C}$$

that satisfy

$$d^k m(\Phi,\Psi) = \begin{cases} m(d^k\Phi,\Psi) & k \leqslant p \\ m(\Phi, d^{k-p}\Psi) & k > p \end{cases}$$

$$m(d^{p+1}\Phi,\Psi) = m(\Phi, d^0\Psi)$$

$$s^k m(\Phi,\Psi) = \begin{cases} m(s^k\Phi,\Psi) & k < p \\ m(\Phi, s^{k-p}\Psi) & k \geqslant p \end{cases}$$

in order to deduce that $\mathcal{Z}(\mathbf{C}) = \mathrm{Tot}(\Pi^\bullet \mathbf{C})$ is an A_∞ monoid [**MS04b**, Section 3]. The reader may check that

$$m_{p,q}(\Phi, \Psi)(f_1, \ldots, f_p, f_{p+1}, \ldots, f_{p+q}) = \Phi(f_1, \ldots, f_p)\Psi(f_{p+1}, \ldots, f_{p+q})$$

satisfies these conditions in our case.

Homotopy commutativity. As suggested by the observation of Eckmann and Hilton, we will now proceed to see that the homotopy coherent center not only has an $A_\infty = E_1$ structure, but in fact is automatically commutative up to homotopy in the sense that this structure can be enhanced to an E_2 structure.

THEOREM 3.1. *For every category* $\mathbf{C}$ *enriched in spaces, the homotopy coherent center* $\mathcal{Z}(\mathbf{C})$ *comes with an* E_2*-multiplication; it is a homotopy commutative* A_∞ *monoid.*

PROOF. Again, we use the work of McClure and Smith cited above, with similar provisos concerning topological versus simplicial categories. Accordingly, we need to turn $\Pi^\bullet \mathbf{C}$ into a (non-symmetric) operad with an associative multiplication with unit [**MS04b**, Section 10].

For the operad structure, it suffices to give the insertion maps

$$\circ_i \colon \Pi^n \mathbf{C} \times \Pi^j \mathbf{C} \longrightarrow \Pi^{n+j-1}\mathbf{C},$$

and we do so by defining

$$(\Phi \circ_i \Psi)(f_1, \ldots, f_{n+j-1}) = \Phi(f_1, \ldots, f_{i-1}, \Psi(f_i, \ldots, f_{i+j-1}), f_{i+j}, \ldots, f_{n+j-1}).$$

The additional structure of an associative multiplication with unit boils down to a pair of vertices $\mu \in \Pi^2\mathbf{C}$ and $\epsilon \in \Pi^0\mathbf{C}$. These have to be chosen such that the two conditions $\mu(\mu, \mathrm{id}) = \mu(\mathrm{id}, \mu)$ and $\mu(\epsilon, \mathrm{id}) = \mathrm{id} = \mu(\mathrm{id}, \epsilon)$ are satisfied. In our case, we can define ϵ to be the family (id_x) of identity morphisms $\mathrm{id}_x \colon x \to x$, and μ to be the family of composition maps $\mathbf{C}(x_1, x_0) \times \mathbf{C}(x_2, x_1) \to \mathbf{C}(x_2, x_0)$ that send (f_1, f_2) to $f_1 f_2$.

Again, it is straightforward to verify that these maps satisfy the required conditions. □

COROLLARY 3.2. *For every category* $\mathbf{C}$ *enriched in spaces, the set* $\pi_0\mathcal{Z}(\mathbf{C})$ *of components of the homotopy coherent center* $\mathcal{Z}(\mathbf{C})$ *is a commutative monoid.*

REMARK 3.3. In the additive context of homological algebra and differential graded categories, the idea of deriving the notion of the center of a category has already appeared in an unpublished preprint by Keller [**Kel**]. In the same context, Tamarkin [**Tam07**] considers a subspace of a more general many objects Hochschild object. Batanin and Markl [**BM12**] obtain the classical Hochschild complex of a differential graded category as a special case of a more general construction. See [**KT05**] for another non-linear version of Deligne's conjecture in a different setting. In the linear context, a version for differential graded categories was proven by Tamarkin [**Tam07**]. Recently, Batanin and Markl [**BM15**] gave a proof of a more general statement which is applicable, in particular, to topologically enriched categories.

4. Functoriality and equivalences

In general, a simplicial functor $F\colon \mathbf{C} \to \mathbf{D}$ between categories $\mathbf{C}$ and $\mathbf{D}$ enriched in spaces does not induce a morphism between the centers, in either direction. This is already clear for discrete monoids. However, centers do allow for some functoriality, and we will discuss this in the present section. The main result will be the invariance of the homotopy coherent center under weak equivalences of simplicial categories as defined by Dwyer and Kan in [**DK80**, 2.4]. Before we do so, let us discuss a different situation that also arises sufficiently often.

Quotients. There is always a morphism

$$\mathcal{Z}(\mathbf{C}) \to \mathrm{Z}(\mathrm{Ho}\,\mathbf{C}),$$

defined by sending a coherent family $\Phi = (\Phi^n)$ to the homotopy class $[\Phi^0]$. (We use the homotopies Φ^1 to show that this is well-defined.) The target of the morphism is discrete, and this map factors to give the second map

$$\pi_0\mathcal{Z}(\mathbf{C}) \longrightarrow \mathrm{Z}(\mathrm{Ho}\,\mathbf{C})$$

displayed in the introduction.

More generally, using functorial Postnikov approximations for spaces, which are simplicial functors $\mathrm{P}_n\colon \mathbf{S} \to \mathbf{S}$ that preserve products, we may also consider the Postnikov approximations $\mathbf{C} \to \mathrm{P}_n\mathbf{C}$ of the categories enriched in spaces $\mathbf{C}$. The homotopy category is the special case $n = 0$, i.e. $\mathrm{P}_0\mathbf{C} = \mathrm{Ho}\,\mathbf{C}$.

We obtain maps $\mathcal{Z}(\mathbf{C}) \to \mathcal{Z}(\mathrm{P}_n\mathbf{C})$ and more generally $\mathcal{Z}(\mathrm{P}_m\mathbf{C}) \to \mathcal{Z}(\mathrm{P}_n\mathbf{C})$ for all $m \geqslant n$. We will later see that $\mathcal{Z}(\mathrm{P}_n\mathbf{C})$ is n-truncated, but it will not be the n-truncation of $\mathcal{Z}(\mathbf{C})$ in general. For instance, in the case $n = 0$, the purpose of the obstruction theory that will be developed in Section 7 is to explain the (potential) failure of the map $\pi_0\mathcal{Z}(\mathbf{C}) \to \mathrm{Z}(\mathrm{Ho}\,\mathbf{C})$ to be a bijection.

Equivalences. Recall from [**DK80**, 2.4] that a simplicial functor $F\colon \mathbf{C} \to \mathbf{D}$ between simplicial categories is called a *weak equivalence* if the following two conditions are satisfied: First, the induced functor $\mathrm{Ho}\,F\colon\ \mathrm{Ho}\,\mathbf{C} \to \mathrm{Ho}\,\mathbf{D}$ between the homotopy categories has to be an equivalence of (ordinary) categories, and second, for each pair of objects x, y of $\mathbf{C}$, the induced map $\mathbf{C}(x, y) \to \mathbf{D}(Fx, Fy)$ has to be a weak equivalence of spaces. Two simplicial categories are *weakly equivalent* if there exists a (finite) zig-zag of weak equivalences between them.

THEOREM 4.1. *Weakly equivalent categories enriched in spaces have weakly equivalent homotopy coherent centers.*

PROOF. Let $\mathbf{C}$ and $\mathbf{D}$ be weakly equivalent categories enriched in spaces. It will be sufficient to deal with the case in which we have a weak equivalence $F\colon \mathbf{C} \to \mathbf{D}$ between them, and to produce a zig-zag of weak equivalences between their homotopy coherent centers.

As an intermediate step, we can use the space $\mathcal{N}(F, F)$ of homotopy coherent natural transformations from the functor F to itself as explained in Remark 2.4. For any simplicial functor $F\colon \mathbf{C} \to \mathbf{D}$, pre- and post-composition with F induces maps

$$\mathcal{Z}(\mathbf{C}) \xrightarrow{F_*} \mathcal{N}(F, F) \xleftarrow{F^*} \mathcal{Z}(\mathbf{D})$$

that are compatible with the multiplications.

If $F\colon \mathbf{C} \to \mathbf{D}$ is a weak equivalence, then so are F_* and F^*, and the result follows. □

5. Monoids and groups

To illustrate the two different notions of centers that we have defined for categories enriched in spaces so far, simplicial centers and homotopy coherent centers, let us study these in the case where the category enriched in spaces has just one object, so that it is a simplicial monoid M whose underlying simplicial set is a Kan complex. This happens, for instance, if M is a simplicial group, or if $M = \mathrm{Map}(X, X)$ for a Kan complex X.

From Section 1, we recall that the simplicial center $\mathrm{Z}(M)$ is a simplicial abelian monoid with n-simplices $\mathrm{Z}(M)_n = \mathrm{Z}(M_n)$, and its homotopy type is determined by its homotopy groups since it is a product of Eilenberg–Mac Lane spaces.

Let us now inspect the homotopy coherent center $\mathfrak{Z}(M)$. The definition from Section 2 gives

$$\mathfrak{Z}(M) = \mathrm{Tot}(\Pi^\bullet M).$$

By inspection, the cosimplicial space $\Pi^\bullet M = \mathrm{Map}(M^\bullet, M)$ is obtained by mapping the bar construction $M^\bullet = \mathrm{B}_\bullet(M)$ into M, but keeping the action of M from the left and the right. Here the bar construction $\mathrm{B}_\bullet(M)$ is the simplicial space with $\mathrm{B}_n(M) = M^n$, faces are given by multiplication in M, except for the first and last one, which omit the corresponding entries, and degeneracies are given by inserting the identity.

The vertices of $\mathfrak{Z}(M)$ are the coherent families of maps

$$\Delta^n \longrightarrow \mathrm{Map}(M^n, M),$$

or equivalently

$$z_n\colon M^n \longrightarrow \mathrm{Map}(\Delta^n, M)$$

by adjunction. (Note the different meanings of the superscript: M^n is the n-th cartesian power, whereas in Δ^n the n indicates the dimension.) A vertex in the homotopy coherent center $\mathfrak{Z}(M)$ therefore gives for $n = 0$ an element $z = z_0$ in M, and for $n = 1$ and each element $m \in M$ a path

$$z_1(m)\colon mz \longrightarrow zm$$

in M, and for $n = 2$ it gives for each pair of elements m, n in M a 2-simplex $z_2(m, n)$ in M that gives a homotopy between the two potentially different paths $mnz \to zmn$ which are $z_1(m \cdot n)$ and $(z_1(m) \cdot n) \circ (m \cdot z_1(n))$.

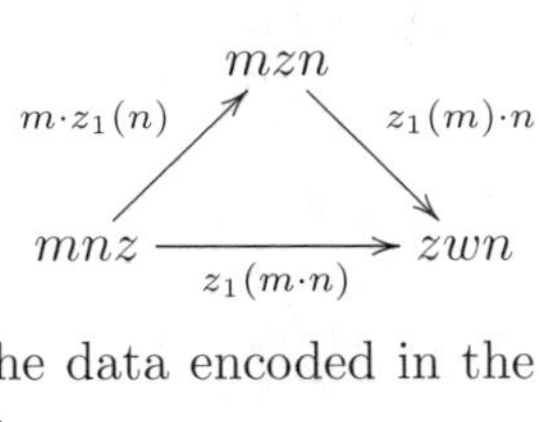

This should give an idea of the data encoded in the homotopy coherent center in the case of simplicial monoids.

We will now specialize a bit more in order to relate our definitions to another branch of contemporary topology: homotopical group theory.

Simplicial groups. In contrast to the general monoid, the elements in a group G all have inverses, so that we have a conjugation action of G on itself. The fixed points of this action form the center $\mathrm{Z}(G)$ of G in the traditional sense.

This suggests that the homotopy fixed point space of the conjugation action of G on itself is another natural candidate for the notion of a homotopy coherent center. It can be defined as the space

$$\mathrm{Map}_G(\mathrm{E}G, \mathrm{Ad}(G))$$

of equivariant maps from a free resolution $\mathrm{E}G \to \star$ of the universal G-fixed point $\star$ to $\mathrm{Ad}(G)$. The following result shows that the homotopy coherent center in our sense agrees with this concept, when the latter is defined.

THEOREM 5.1. *For every simplicial group G, there is an equivalence*

$$\mathfrak{Z}(G) \simeq \mathrm{Map}_G(\mathrm{E}G, \mathrm{Ad}(G))$$

of its homotopy coherent center with the homotopy fixed point space of the conjugation action of G on itself.

PROOF. This follows from the induction-restriction-adjunction associated to the diagonal

$$G \longrightarrow G^{\mathrm{op}} \times G,\ g \longmapsto (g^{-1}, g).$$

Namely, we have

$$G \cong \mathrm{ind}(\star),$$

where G is the G-biset with the action of G by multiplication from the right and from the left, as well as

$$\mathrm{Ad}(G) \cong \mathrm{res}(G).$$

Using this, we can write

$$\begin{aligned} \mathrm{Map}_G(\mathrm{E}G, \mathrm{Ad}(G)) &\cong \mathrm{Hom}(\mathrm{B}(\star, G, G), \mathrm{res}(G)) \\ &\cong \mathrm{Hom}(\mathrm{ind}\,\mathrm{B}(\star, G, G), G) \\ &\cong \mathrm{Hom}(\mathrm{B}(\mathrm{ind}(\star), G, G), G) \\ &\cong \mathrm{Hom}(\mathrm{B}(G, G, G), G). \end{aligned}$$

Since the G-space G, thought of as a functor from G to G-spaces, is full and faithful, this is isomorphic to

$$\mathrm{Hom}(\mathrm{B}(G, G, \mathrm{Id}_G), \mathrm{Id}_G) = \mathfrak{Z}(G),$$

as was to be shown. □

We note that the equivalence in the theorem allows us to model the naturally given map $\mathrm{Z}(G) \to \mathfrak{Z}(G)$ as the inclusion of the fixed points into the homotopy fixed points.

The following (folklore) result gives another presentation of the homotopy fixed point space of the conjugation action of G on itself, and hence also of the homotopy coherent center of G.

PROPOSITION 5.2. *For every simplicial group G, there is an equivalence*

$$\mathrm{Map}_G(\mathrm{E}G, \mathrm{Ad}(G)) \simeq \Omega(\mathrm{Map}(\mathrm{B}G, \mathrm{B}G), \mathrm{id})$$

with the space of loops in the space of self-maps of $\mathrm{B}G$ *which are based at the identity.*

PROOF. The space $\mathrm{Map}_G(\mathrm{E}G, \mathrm{Ad}(G))$ can be identified with the space of sections of the Borel construction bundle

$$\mathrm{E}G \times_G \mathrm{Ad}(G) \longrightarrow \mathrm{B}G,$$

and the space $\Omega(\mathrm{Map}(\mathrm{B}G, \mathrm{B}G), \mathrm{id})$ can be identified with the space of sections of the free loop space evaluation bundle

$$\Lambda \mathrm{B}G \longrightarrow \mathrm{B}G.$$

In remains to note that there is an equivalence

$$\mathrm{E}G \times_G \mathrm{Ad}(G) \simeq \Lambda \mathrm{B}G$$

of spaces over $\mathrm{B}G$. See the original sources [**Goo85**], [**BF86**], or [**BHM93**]. □

COROLLARY 5.3. *For every simplicial group G, there is an equivalence*

$$\mathcal{Z}(G) \simeq \Omega(\mathrm{Map}(\mathrm{B}G, \mathrm{B}G), \mathrm{id})$$

of its homotopy coherent center with the space of loops in the space of self-maps of $\mathrm{B}G$ *which are based at the identity.*

The right hand side of this equivalence is used as a definition for the homotopy center of p-compact groups by Dwyer and Wilkerson [**DW95**].

A still more general formula for the homotopy coherent center of a simplicial groupoid will be given later, see Proposition 8.2.

6. The homotopy limit problem

It is clear from our description of the simplicial center and the homotopy coherent center as a limit and as a homotopy limit, respectively, that there is a natural map

$$\mathrm{Z}(\mathbf{C}) \longrightarrow \mathcal{Z}(\mathbf{C}). \tag{6.1}$$

The question whether this map is an equivalence, or how far off it is, can be compared with the difficult homotopy limit problem, see Thomason [**Tho83**] and Carlsson [**Car87**].

The two notions of center agree for discrete categories.

PROPOSITION 6.1. *For every discrete category* $\mathbf{C}$, *the arrow* (6.1) *is an equivalence; the domain is discrete.*

Of course, this results applies, in particular, to discrete monoids, discrete groups, and discrete groupoids.

We are now in a position to use the different descriptions of the homotopy coherent center of a simplicial group G given in Section 5 in order to discuss the homotopy limit map $\mathrm{Z}(G) \to \mathcal{Z}(G)$ in various other special cases. Here is another case that can be dealt with.

PROPOSITION 6.2. *Let A be a simplicial abelian group. If A is an Eilenberg–Mac Lane space or if A has the homotopy type of an abelian compact Lie group, then the arrow* $\mathrm{Z}(A) \to \mathcal{Z}(A)$ *is an equivalence.*

PROOF. We have to show that the natural map $A = \mathrm{Map}(\star, A) \to \mathrm{Map}(\mathrm{B}A, A)$ is an equivalence. This follows in the first case from Thom's determination [**Tho56**] of the homotopy type of the mapping space into an Eilenberg–Mac Lane space. The second case follows from the same result and the fact that A is a product of a finite group with a torus. □

We will revisit the first case of the preceding result in Example 8.6 from a different perspective.

The consequence of Proposition 6.2 does not hold for all simplicial abelian groups in general, even though they are known to be products of Eilenberg–Mac Lane spaces by Moore's theorem [**Moo55**].

EXAMPLE 6.3. Let us consider the (abelian) group A of all complex line bundles over the circle. Up to homotopy (or isomorphism), there is only one. This immediately gives

$$\pi_0 \mathrm{Z}(A) = \pi_0 A = 0.$$

But as a space (and as a group), this can be modeled as

$$A = \mathrm{Map}(\mathrm{S}^1, \mathbb{C}\mathrm{P}^\infty) \simeq \mathrm{S}^1 \times \mathbb{C}\mathrm{P}^\infty.$$

Therefore

$$\pi_0 \mathcal{Z}(A) \cong \pi_1(\mathrm{Map}(\mathrm{B}A, \mathrm{B}A), \mathrm{id}) \cong \mathbb{Z}.$$

This shows that the behavior of the map $\mathrm{Z}(A) \to \mathcal{Z}(A)$ can be rather unpredictable on components in general.

This example shows that the homotopy limit map (6.1) need not be an equivalence, not even after passage to classifying spaces and on homology with finite coefficients. This may be compared to the following result.

THEOREM 6.4. ([**DW95**, Theorem 1.4]) *Let G be a connected compact Lie group. Then the natural map*

$$\mathrm{BZ}(G) \longrightarrow \mathrm{B}\mathcal{Z}(G)$$

induces an isomorphism on homology with finite coefficients.

Earlier results in this direction had been obtained in [**JMO92**, Theorem 3] in the case when G is simple, and in [**DM87**] when $G = \mathrm{SU}(2)$.

7. Spectral sequences and obstructions

In this section we answer the question about the kernel and image of the map

$$\pi_0 \mathcal{Z}(\mathbf{C}) \longrightarrow \mathrm{Z}(\mathrm{Ho}\,\mathbf{C}),$$

by means of spectral sequences and obstruction theory, respectively.

A ubiquitous problem when discussing realization questions is the potential emptiness of realization spaces. This is irrelevant here, since the (homotopy coherent) center is always nonempty. Moreover, its set of components has the structure of an abelian monoid. This leads to a simplification of another difficulty that we often encounter when computing homotopy groups of spaces by means of such spectral sequences: the lack of (abelian) group structures on π_0 and π_1.

Spectral sequences. If $\mathbf{C}$ is a category enriched in spaces, then its homotopy coherent center $\mathcal{Z}(\mathbf{C})$ has been defined as the totalization of the cosimplicial space $\Pi^\bullet\mathbf{C}$. This is pointed and fibrant, see Lemma 2.2. Bousfield and Kan, see [**BK72**] and [**Bou89**] as well as the textbook [**GJ99**, VIII.1], have shown that there is always a spectral sequence for the totalization of such a cosimplicial space.

THEOREM 7.1. *For every category* $\mathbf{C}$ *enriched in spaces, there exists a fringed spectral sequence with target* $\pi_{t-s}\mathcal{Z}(\mathbf{C})$ *and*

$$\mathrm{E}_2^{s,t} = \pi^s\pi_t\Pi^\bullet\mathbf{C}$$

for $t \geqslant s \geqslant 0$. *In particular,*

$$\mathrm{E}_2^{0,0} \leqslant \mathrm{Z}(\mathrm{Ho}\,\mathbf{C})$$

is a submonoid of the center of its homotopy category.

PROOF. Only the last statement requires proof. The E_1 term is given as

$$\mathrm{E}_1^{s,t} = N^s\pi_t\Pi^\bullet\mathbf{C}.$$

In particular, the 0-line is

$$\mathrm{E}_1^{0,t} = \pi_t\Pi^0\mathbf{C}$$

with differential given by the difference between the two coface operators d_0 and d_1. This gives

$$\mathrm{E}_2^{0,t} = \pi^0\pi_t\Pi^\bullet\mathbf{C}.$$

For $t = 0$, we have

$$\mathrm{E}_1^{0,0} = \prod_x \pi_0\mathbf{C}(x,x),$$

and this consists of the families Φ of homotopy classes $\Phi_x\colon x \to x$. These lie in the equalizer $\pi^0\pi_0\Pi^\bullet\mathbf{C}$ if $f\Phi_x$ and $\Phi_y f$ are homotopic for all $f\colon x \to y$ in $\mathbf{C}$, and these homotopies can be chosen continuously in f. In particular, the family Φ is in the center $\mathrm{Z}(\mathrm{Ho}\,\mathbf{C})$ of the homotopy category $\mathrm{Ho}\,\mathbf{C}$, where continuity in f is not required. □

REMARK 7.2. As the proof has shown, we owe the difference between $\mathrm{E}_2^{0,0}$ and the center of the homotopy category to the fact that the maps

$$\pi_0\,\mathrm{Map}(\mathbf{C}(y,z),\mathbf{C}(y,z)) \longrightarrow \mathrm{Map}(\pi_0\mathbf{C}(y,z),\pi_0\mathbf{C}(y,z))$$

need not be injective.

Compared with the general spectral sequence of Bousfield–Kan type, the spectral sequences in Theorem 7.1 have a relatively well-behaved left lower corner. The monoidal structure on $\Pi^\bullet\mathbf{C}$ leads to abelian group structures from the E_1 term on, with the possible exception of $\mathrm{E}_r^{0,0}$. In that spot, our spectral sequence starts with

$$\mathrm{E}_1^{0,0} = \prod_x \pi_0\mathbf{C}(x,x),$$

which–in the most general case that we consider–is just a monoid. But, the structure $\mathrm{E}_2^{0,0} \leqslant \mathrm{Z}(\mathrm{Ho}\,\mathbf{C})$ on the next term is already an abelian monoid, but not necessarily a group. This is as good as we may hope for in the context of centers. A detailed discussion of the algebraic structure of the terms and differentials in that corner has been given in [**MS15**, Sec. 3] for the analogous situation of bicategories and their Drinfeld centers.

The manner of convergence of Bousfield–Kan type spectral sequences is often a delicate issue, and we do not have to add anything to the original results here. See again [**BK72**], [**Bou89**, §4], and [**GJ99**, VI.2]. We would only like to point out the following result for truncated situations.

PROPOSITION 7.3. *If* $\mathbf{C}$ *is a category enriched in spaces such that all of its mapping spaces* $\mathbf{C}(x,y)$ *are* n*-truncated for an* n *that is independent of the objects* x *and* y*, then the spectral sequence of Theorem* 7.1 *converges completely to its target.*

This proposition applies, in particular, to categories enriched in spaces that arise form categories enriched in groupoids ($n = 1$).

Before we turn towards the obstruction theory that always comes with a fringed Bousfield–Kan type spectral sequence, let us briefly collect the different types of truncations for homotopy coherent centers that we have considered so far. For all $s \geqslant 0$ and $n = t - s \geqslant 0$ there are maps

$$\begin{array}{ccccc} \mathcal{Z}(\mathbf{C}) & \longrightarrow & \mathrm{P}_n\mathcal{Z}(\mathbf{C}) & \longrightarrow & \mathcal{Z}(\mathrm{P}_n\mathbf{C}) \\ \| & & & & \\ \mathrm{Tot}(\Pi^\bullet\mathbf{C}) & \longrightarrow & \mathrm{Tot}_s(\Pi^\bullet\mathbf{C}) & & \end{array}$$

and from the position of the corresponding regions in our spectral sequence it is clear that we cannot expect these to be equivalences in general.

Obstructions. The fringed spectral sequence of Theorem 7.1 comes with an obstruction theory that allows the investigation of the edge homomorphism

$$\pi_0\mathcal{Z}(\mathbf{C}) \longrightarrow \mathrm{Z}(\mathrm{Ho}\,\mathbf{C}),$$

in particular of its image. Note that the range of that map contains the $\mathrm{E}_2^{0,0}$ term of the spectral sequence, whereas the domain belongs to the target of that spectral sequence. Therefore, we can start our discussion with an element of the abelian monoid $\pi_0\,\mathrm{Tot}_1(\Pi^\bullet\mathbf{C}) \leqslant \mathrm{Z}(\mathrm{Ho}\,\mathbf{C})$ and ask whether or not we can lift it to $\pi_0\,\mathrm{Tot}_s(\Pi^\bullet\mathbf{C})$ for a given s. We will be able to do this for every s if and only if the given element lifts to $\pi_0\,\mathrm{Tot}(\Pi^\bullet\mathbf{C}) = \pi_0\mathcal{Z}(\mathbf{C})$.

Since this is by now standard, we will only sketch how the general obstruction theory applies to our specific context and refer to [**Bou89**, §5 and §10] and [**GJ99**, VIII.4] for details.

First of all the representatives of the classes in $\pi_0\,\mathrm{Tot}_s(\Pi^\bullet\mathbf{C})$ can be fairly explicitly described. They are given by $(s+1)$-tuples $(\Phi_0,\dots,\Phi_s)$ of simplices

$$\Phi_p\colon \Delta^p \longrightarrow \Pi^p\mathbf{C}$$

that are compatible with the coface and codegeneracy maps whenever these are defined. The maps $(\Phi_0,\dots,\Phi_s) \mapsto (\Phi_0,\dots,\Phi_{s-1})$ and $(\Phi_0,\dots,\Phi_s) \mapsto \Phi_s$ induce the upper left of the following two pullback squares.

$$
\begin{array}{ccccc}
\mathrm{Tot}_s(\Pi^\bullet \mathbf{C}) & \longrightarrow & \mathrm{Map}(\Delta^s, \Pi^s \mathbf{C}) & & \\
\big\downarrow & & \big\downarrow f & & \\
\mathrm{Tot}_{s-1}(\Pi^\bullet \mathbf{C}) & \longrightarrow & P & \longrightarrow & \mathrm{Map}(\Delta^s, M^{s-1}\Pi^\bullet \mathbf{C}) \\
& & \big\downarrow & & \big\downarrow \\
& & \mathrm{Map}(\partial\Delta^s, \Pi^s \mathbf{C}) & \longrightarrow & \mathrm{Map}(\partial\Delta^s, M^{s-1}\Pi^\bullet \mathbf{C})
\end{array}
$$

The map f is the fibration determined by the fibrant cosimplicial space $\Pi^\bullet \mathbf{C}$. Thus, in order to produce a lift of an element in $\pi_0 \,\mathrm{Tot}_s(\Pi^\bullet \mathbf{C})$, we have to find a lift of its image in the pullback

$$P = \mathrm{Map}(\partial\Delta^s, \Pi^s \mathbf{C}) \times_{\mathrm{Map}(\partial\Delta^s, M^{s-1}\Pi^\bullet \mathbf{C})} \mathrm{Map}(\Delta^s, M^{s-1}\Pi^\bullet \mathbf{C})$$

as illustrated in the following diagram.

$$
\begin{array}{ccc}
\partial\Delta^s & \longrightarrow & \Pi^s \mathbf{C} \\
\big\downarrow & \nearrow & \big\downarrow \\
\Delta^s & \longrightarrow & M^{s-1}\Pi^\bullet \mathbf{C}
\end{array}
$$

The element in $\pi_{s-1}(\Pi^s \mathbf{C})$ represented by $\partial\Delta^s \to \Pi^s \mathbf{C}$ is an obstruction to the existence of such a lift. In fact, this obstruction lives on the E_1 term, more precisely in $\mathrm{E}_1^{s,s-1}$. It turns out that we can pass to the E_2 term to obtain a well-defined obstruction in $\mathrm{E}_2^{s,s-1}$ to the liftability of the restriction in $\pi_0 \,\mathrm{Tot}_{s-1}(\Pi^\bullet \mathbf{C})$ of our element in $\pi_0 \,\mathrm{Tot}_s(\Pi^\bullet \mathbf{C})$ to $\pi_0 \,\mathrm{Tot}_{s+1}(\Pi^\bullet \mathbf{C})$.

REMARK 7.4. A little extra care has to be taken when there are fundamental groups involved. It is common to assume that Whitehead products vanish in the spaces involved in $\Pi^\bullet \mathbf{C}$, or at least that the fundamental groupoids act trivially on all fundamental groups. Both hypotheses are satisfied in the linear case, when the mapping spaces are simplicial abelian groups, or in the stable case, when the mapping spaces are infinite loop spaces.

The following result summarizes the situation.

THEOREM 7.5. *Let* $\mathbf{C}$ *be a category enriched in spaces that satisfies a hypothesis on the fundamental groups as in Remark* 7.4 *above. Then any element in the center*

$$\mathrm{Z}(\mathrm{Ho}\,\mathbf{C}) = \mathrm{Tot}_1(\Pi^\bullet \mathbf{C})$$

of the homotopy category can be lifted to an element in the homotopy coherent center

$$\mathcal{Z}(\mathbf{C}) = \mathrm{Tot}(\Pi^\bullet \mathbf{C})$$

if and only of if it lies in $\mathrm{E}_2^{0,0} \leqslant \mathrm{Z}(\mathrm{Ho}\,\mathbf{C})$ *and the obstruction classes in*

$$\mathrm{E}_2^{s,s-1} = \pi^s \pi_{s-1} \Pi^\bullet \mathbf{C}$$

vanish for all $s \geqslant 2$.

We note that the obstruction classes certainly vanish if the obstruction groups are all trivial. This happens, of course, in the case when $\mathbf{C}$ is homotopically discrete.

8. Groupoids and spaces

In this last section, we will discuss another class of examples of categories enriched in spaces: simplicial groupoids. By definition, a simplicial groupoid $\mathbf{G}$ is a simplicial category such that the categories $\mathbf{G}_n$ of n-simplices are groupoids. For instance, simplicial groups, as discussed already in Section 5, are simplicial groupoids. Here, we are interested mainly in the case where there are many objects. Let us start by reviewing the discrete case first.

Fundamental groupoids. If X is a space, its fundamental groupoid $\Pi_1 X$ is a (discrete) groupoid, and up to equivalence, every discrete groupoid has this form: Take X to be its classifying space.

PROPOSITION 8.1. *For every space X, the center of the fundamental groupoid splits as a product*

$$\mathrm{Z}(\Pi_1 X) \cong \prod_{[x]\in\pi_0 X} \mathrm{Z}(\pi_1(X,x))$$

of centers of its fundamental groups over a set of representatives of its path components.

PROOF. This is certainly true for path connected spaces X, since in that case the fundamental groupoid is equivalent to the fundamental group of any of its points. The general case follows from the compatibility of centers with disjoint unions. □

After this review of the discrete case, let us now move on to simplicial groupoids.

Path groupoids. While the fundamental groupoid $\Pi_1 X$ of a space X is useful for many purposes, the passage to homotopy classes that attends it is a rather drastic simplification of the situation. It is preferable to work with the path groupoid $\mathbf{G}X$ of X that has been introduced by Dwyer and Kan in [**DK84**]. It is a simplicial groupoid that can be thought of as a simplicially enriched refinement of the fundamental groupoid of the space X. In fact, in the cited paper it is shown that the homotopy theory of simplicial groupoids is equivalent to the homotopy theory of spaces. More precisely, inverse equivalences are given by the pair of adjoint functors $\mathbf{G} \mapsto \mathrm{B}\mathbf{G}$, the classifying space, and $X \mapsto \mathbf{G}X$. Consequently, we may expect the centers of simplicial groupoids to be related to the classical homotopy theory of spaces, and as we will see in the examples below, this turns out to be true.

The homotopy category of $\mathbf{G}X$ is the fundamental groupoid $\Pi_1 X$ of X, so that the formula

$$\Pi_1 X = \mathrm{Ho}(\mathbf{G}X),$$

relates the two incarnations of the fundamental groupoid idea.

The center of the fundamental groupoid of X has been described in Proposition 8.1, and we can now ask for the homotopy coherent center $\mathcal{Z}(\mathbf{G}X)$ of the Dwyer–Kan path groupoid of X, and the relation between the two.

We have seen that the homotopy coherent center in general has an E_2 structure by Theorem 3.1. *A fortiori*, it is an $E_1 = A_\infty$ monoid. For simplicial groupoids, it will turn out to be group-like by Corollary 8.3 below. Therefore, by Stasheff's recognition theorem [**Sta63**] (as improved by May [**May74**]), it must be equivalent to the loop space of some (classifying) space. In fact, using the E_2 structure, that space will have its own delooping as well. The reader may wonder what these spaces are, and the following results will answer this question.

PROPOSITION 8.2. *For every simplicial groupoid* $\mathbf{G}$, *we have*

$$\mathcal{Z}(\mathbf{G}) \simeq \Omega(\mathrm{Map}(\mathrm{B}\mathbf{G}, \mathrm{B}\mathbf{G}), \mathrm{id}).$$

PROOF. Similarly to the preceding proposition, this follows from Corollary 5.3 and the compatibility of centers with disjoint unions. □

The center of any groupoid is always an abelian group rather than just an abelian monoid. This is far from obvious in the homotopy coherent setting, but the following result affirms that it still holds true.

COROLLARY 8.3. *For every simplicial groupoid* $\mathbf{G}$, *the abelian monoid* $\pi_0\mathcal{Z}(\mathbf{G})$ *is an abelian group.*

PROOF. Proposition 8.2 allows us to identify $\pi_0\mathcal{Z}(\mathbf{G})$ with a fundamental group of a space. □

If X is a Kan complex, then it is known that the mapping space $\mathrm{Map}(X, X)$ already models the derived mapping space, and it is also a Kan complex. Since there is an equivalence $X \simeq \mathrm{B}\mathbf{G}X$, the proposition above has also the following corollary.

COROLLARY 8.4. *For every Kan complex* X, *we have*

$$\mathcal{Z}(\mathbf{G}X) \simeq \Omega(\mathrm{Map}(X, X), \mathrm{id}).$$

The description of spaces such as $\Omega(\mathrm{Map}(X, X), \mathrm{id})$ is a classical subject of (unstable) homotopy theory. We will now discuss more specific classes of examples in order to demonstrate the complexity of the matter.

Specific classes of spaces. Given a space X, we have the canonical homomorphism

$$\pi_0\mathcal{Z}(\mathbf{G}X) \longrightarrow \mathrm{Z}(\Pi_1 X). \tag{8.1}$$

This is the edge homomorphism of our obstruction theory spectral sequence in Section 7, and we can discuss the problem whether or not this map (8.1) is injective or surjective here. We will see that neither has to be the case.

EXAMPLE 8.5. As the simplest case, let us consider a space that comes to us as a classifying space $X = \mathrm{B}\Gamma = \mathrm{K}(\Gamma, 1)$ for a discrete group Γ. In this case, we should not expect any higher homotopy structure, and indeed, the space $\Omega(\mathrm{Map}(X, X), \mathrm{id})$ is homotopically discrete, with

$$\mathcal{Z}(\mathbf{G}X) \simeq \pi_0\mathcal{Z}(\mathbf{G}X) \cong \mathrm{Z}(\Gamma),$$

the center of the fundamental group.

In order to encounter higher homotopy structure, we may generalize this example in at least two ways: Replace $X = \mathrm{B}\Gamma = \mathrm{K}(\Gamma, 1)$ with an Eilenberg–Mac Lane space $\mathrm{K}(A, n)$ for any abelian group A and any $n \geqslant 2$. Or, replace the aspherical circle $\mathrm{B}\mathbb{Z} = \mathrm{K}(\mathbb{Z}, 1) = \mathrm{S}^1$ by a sphere S^n of dimension $n \geqslant 2$. These will be our next two classes of examples.

EXAMPLE 8.6. Let $X = \mathrm{K}(A,n)$ be an Eilenberg–Mac Lane space of type (A,n) for an abelian group A and an integer $n \geqslant 2$. Then the homotopy type of the space of self-maps is

$$\mathrm{Map}(\mathrm{K}(A,n),\mathrm{K}(A,n)) \simeq \mathrm{Hom}(A,A) \times \mathrm{K}(A,n),$$

see [**Tho56**] for the original argument and [**May67**, 25.2] for the simplicial version. Since the space $\Omega(\mathrm{Map}(X,X),\mathrm{id})$ of loops based at the identity depends only on the component of the identity, we therefore obtain an equivalence

$$\mathcal{Z}(\mathbf{G}\mathrm{K}(A,n)) \simeq \Omega\mathrm{K}(A,n) \simeq \mathrm{K}(A,n-1).$$

This means that the abelian group $\pi_0\mathcal{Z}(\mathbf{G}\mathrm{K}(A,n))$ is trivial. (In particular, it is isomorphic to $\mathrm{Z}(\Pi_1\mathrm{K}(A,n))$ which is also trivial.) But, we see that the center can have arbitrary higher homotopy groups. This information will already be lost when we pass from the center $\mathcal{Z}(\mathbf{G}X)$ to its group $\pi_0\mathcal{Z}(\mathbf{G}X)$ of components, let alone $\mathrm{Z}(\Pi_1 X)$.

EXAMPLE 8.7. Let $X = \mathrm{S}^n$ be an n-dimensional sphere, $n \geqslant 2$. The homotopy groups of the homotopy coherent center $\mathcal{Z}(\mathbf{G}\mathrm{S}^n)$ can be computed from the fibration sequence

$$\Omega^n\mathrm{S}^n \longrightarrow \mathrm{Map}(\mathrm{S}^n,\mathrm{S}^n) \longrightarrow \mathrm{S}^n.$$

Let us first consider the stable case $n \geqslant 3$. In that case, in order to compute π_0, we note that the boundary operator $0 = \pi_2\mathrm{S}^n \to \pi_{n+1}\mathrm{S}^n$ has to be zero, so that $\pi_0\mathcal{Z}(\mathbf{G}\mathrm{S}^n)$ is the stable 1-stem $\mathbb{Z}/2$. The computation of the higher homotopy groups involves the boundary operator

$$\pi_k\mathrm{S}^n \longrightarrow \pi_{k+(n-1)}\mathrm{S}^n.$$

This is the Whitehead product with the identity ι_n of S^n. In particular, the first potentially nonzero contribution is given by the Whitehead square of the identity, and that element figures prominently in the Hopf invariant one problem: the Whitehead square $[\iota_n,\iota_n]$ is zero if and only if there is an element of Hopf invariant one in $\pi_{2n+1}(S^{n+1})$, and this is very rarely the case (by [**Ada60**]: only if $n = 3, 7$ in our range). And, the divisibility of the class $[\iota_n,\iota_n]$ is related to the (strong) Kervaire invariant one problem, see [**BJM83**]. In the meta-stable case $n = 2$, in order to compute $\pi_0\mathcal{Z}(\mathbf{G}\mathrm{S}^2)$, the boundary operator $\mathbb{Z} \cong \pi_2(\mathrm{S}^2) \to \pi_3(\mathrm{S}^2) \cong \mathbb{Z}$ hits $[\iota_2,\iota_2]$ which is divisible by 2 but not by 4, so that $\pi_0\mathcal{Z}(\mathbf{G}\mathrm{S}^2) \cong \mathbb{Z}/2$ also in this case. Indeed, for S^2, the entire homotopy type of the identity component of the mapping space is known, see [**Han83**] and [**Han90**]. The result is

$$\mathrm{Map}(\mathrm{S}^2,\mathrm{S}^2)_{\mathrm{id}} \simeq \mathrm{SO}(3) \times \tilde{\Omega}^2_0\mathrm{S}^2,$$

where $\tilde{\Omega}^2_0\mathrm{S}^2$ is the universal cover of a component of the double loop space of the 2-sphere; it does not matter which component. The existence of such a homotopy equivalence also implies that the abelian group $\pi_0\mathcal{Z}(\mathbf{G}\mathrm{S}^2)$ has order 2, of course.

Example 8.7 shows that the map (8.1) need not be injective, and surjectivity of (8.1) can also fail in this context. In fact, the image has been studied from a different point of view in [**Got65**], and is commonly called the *Gottlieb subgroup* of the fundamental group. It can be as complicated as algebraically possible: If Γ is a discrete group, and G is a subgroup of its center, then there is a connected space X and an isomorphism $\Gamma \cong \pi_1(X)$ such that this isomorphism sends G isomorphically to the Gottlieb subgroup of X, see [**Var74**].

Acknowledgments

This research has been supported by the Danish National Research Foundation through the Centre for Symmetry and Deformation (DNRF92). I am especially grateful to Moritz Groth for many discussions. I thank Tyrone Crisp, Emanuele Dotto, Rosona Eldred, Jesper Grodal, Ehud Meir, Ryszard Nest, and Nathalie Wahl for their interest and comments, and the referees for their helpful reports.

References

[Ada60] J. F. Adams, *On the non-existence of elements of Hopf invariant one*, Ann. of Math. (2) **72** (1960), 20–104, DOI 10.2307/1970147. MR0141119

[BJM83] M. G. Barratt, J. D. S. Jones, and M. E. Mahowald, *The Kervaire invariant problem*, Proceedings of the Northwestern Homotopy Theory Conference (Evanston, Ill., 1982), Contemp. Math., vol. 19, Amer. Math. Soc., Providence, RI, 1983, pp. 9–22, DOI 10.1090/conm/019/711039. MR711039

[Bas68] H. Bass, *Algebraic K-theory*, W. A. Benjamin, Inc., New York-Amsterdam, 1968. MR0249491

[BM12] M. Batanin and M. Markl, *Centers and homotopy centers in enriched monoidal categories*, Adv. Math. **230** (2012), no. 4-6, 1811–1858, DOI 10.1016/j.aim.2012.04.011. MR2927355

[BM15] M. Batanin and M. Markl, *Operadic categories and duoidal Deligne's conjecture*, Adv. Math. **285** (2015), 1630–1687, DOI 10.1016/j.aim.2015.07.008. MR3406537

[Ber07] J. E. Bergner, *A model category structure on the category of simplicial categories*, Trans. Amer. Math. Soc. **359** (2007), no. 5, 2043–2058, DOI 10.1090/S0002-9947-06-03987-0. MR2276611

[Ber84] J. N. Bernstein, *Le "centre" de Bernstein* (French), Representations of reductive groups over a local field, Travaux en Cours, Hermann, Paris, 1984, pp. 1–32. Edited by P. Deligne. MR771671

[BHM93] M. Bökstedt, W. C. Hsiang, and I. Madsen, *The cyclotomic trace and algebraic K-theory of spaces*, Invent. Math. **111** (1993), no. 3, 465–539, DOI 10.1007/BF01231296. MR1202133

[Bou89] A. K. Bousfield, *Homotopy spectral sequences and obstructions*, Israel J. Math. **66** (1989), no. 1-3, 54–104, DOI 10.1007/BF02765886. MR1017155

[BK72] A. K. Bousfield and D. M. Kan, *Homotopy limits, completions and localizations*, Lecture Notes in Mathematics, Vol. 304, Springer-Verlag, Berlin-New York, 1972. MR0365573

[BF86] D. Burghelea and Z. Fiedorowicz, *Cyclic homology and algebraic K-theory of spaces. II*, Topology **25** (1986), no. 3, 303–317, DOI 10.1016/0040-9383(86)90046-7. MR842427

[Car87] G. Carlsson, *Segal's Burnside ring conjecture and the homotopy limit problem*, Homotopy theory (Durham, 1985), London Math. Soc. Lecture Note Ser., vol. 117, Cambridge Univ. Press, Cambridge, 1987, pp. 6–34. MR932258

[CP96] J.-M. Cordier and T. Porter, *Categorical aspects of equivariant homotopy*, Appl. Categ. Structures **4** (1996), no. 2-3, 195–212, DOI 10.1007/BF00122252. The European Colloquium of Category Theory (Tours, 1994). MR1406098

[CP97] J.-M. Cordier and T. Porter, *Homotopy coherent category theory*, Trans. Amer. Math. Soc. **349** (1997), no. 1, 1–54, DOI 10.1090/S0002-9947-97-01752-2. MR1376543

[Dwy98] W. G. Dwyer, *Lie groups and p-compact groups*, Proceedings of the International Congress of Mathematicians, Vol. II (Berlin, 1998), Doc. Math. **Extra Vol. II** (1998), 433–442. MR1648093

[DK80] W. G. Dwyer and D. M. Kan, *Function complexes in homotopical algebra*, Topology **19** (1980), no. 4, 427–440, DOI 10.1016/0040-9383(80)90025-7. MR584566

[DK84] W. G. Dwyer and D. M. Kan, *Homotopy theory and simplicial groupoids*, Nederl. Akad. Wetensch. Indag. Math. **46** (1984), no. 4, 379–385. MR770723

[DM87] W. G. Dwyer and G. Mislin, *On the homotopy type of the components of* $\mathrm{map}_*(BS^3, BS^3)$, Algebraic topology, Barcelona, 1986, Lecture Notes in Math., vol. 1298, Springer, Berlin, 1987, pp. 82–89, DOI 10.1007/BFb0083001. MR928824

[DS17] W. G. Dwyer and M. Szymik, *Frobenius and the derived centers of algebraic theories*, Math. Z. **285** (2017), no. 3-4, 1181–1203, DOI 10.1007/s00209-016-1744-4. MR3623746

[DW95] W. G. Dwyer and C. W. Wilkerson, *The center of a p-compact group*, The Čech centennial (Boston, MA, 1993), Contemp. Math., vol. 181, Amer. Math. Soc., Providence, RI, 1995, pp. 119–157, DOI 10.1090/conm/181/02032. MR1320990

[Gir71] J. Giraud, *Cohomologie non abélienne* (French), Springer-Verlag, Berlin-New York, 1971. Die Grundlehren der mathematischen Wissenschaften, Band 179. MR0344253

[GJ99] P. G. Goerss and J. F. Jardine, *Simplicial homotopy theory*, Progress in Mathematics, vol. 174, Birkhäuser Verlag, Basel, 1999. MR1711612

[Goo85] T. G. Goodwillie, *Cyclic homology, derivations, and the free loopspace*, Topology **24** (1985), no. 2, 187–215, DOI 10.1016/0040-9383(85)90055-2. MR793184

[Got65] D. H. Gottlieb, *A certain subgroup of the fundamental group*, Amer. J. Math. **87** (1965), 840–856, DOI 10.2307/2373248. MR0189027

[Gro10] J. Grodal, *The classification of p-compact groups and homotopical group theory*, Proceedings of the International Congress of Mathematicians. Volume II, Hindustan Book Agency, New Delhi, 2010, pp. 973–1001. MR2827828

[Han83] V. L. Hansen, *The homotopy groups of a space of maps between oriented closed surfaces*, Bull. London Math. Soc. **15** (1983), no. 4, 360–364, DOI 10.1112/blms/15.4.360. MR703761

[Han90] V. L. Hansen, *The space of self-maps on the 2-sphere*, Groups of self-equivalences and related topics (Montreal, PQ, 1988), Lecture Notes in Math., vol. 1425, Springer, Berlin, 1990, pp. 40–47, DOI 10.1007/BFb0083829. MR1070574

[JMO92] S. Jackowski, J. McClure, and B. Oliver, *Homotopy classification of self-maps of BG via G-actions. I*, Ann. of Math. (2) **135** (1992), no. 1, 183–226, DOI 10.2307/2946568. MR1147962

[Kel] B. Keller. Derived invariance of higher structures on the Hochschild complex. Preprint, 2003.

[Kas95] C. Kassel, *Quantum groups*, Graduate Texts in Mathematics, vol. 155, Springer-Verlag, New York, 1995. MR1321145

[KT05] J. Kock and B. Toën, *Simplicial localization of monoidal structures, and a non-linear version of Deligne's conjecture*, Compos. Math. **141** (2005), no. 1, 253–261, DOI 10.1112/S0010437X04001009. MR2099778

[Lin80] H. Lindner, *Center and trace*, Arch. Math. (Basel) **35** (1980), no. 5, 476–496, DOI 10.1007/BF01235372. MR604246

[Mac71] S. Mac Lane, *Categories for the working mathematician*, Springer-Verlag, New York-Berlin, 1971. Graduate Texts in Mathematics, Vol. 5. MR0354798

[May67] J. P. May, *Simplicial objects in algebraic topology*, Van Nostrand Mathematical Studies, No. 11, D. Van Nostrand Co., Inc., Princeton, N.J.-Toronto, Ont.-London, 1967. MR0222892

[May74] J. P. May, *E_∞ spaces, group completions, and permutative categories*, New developments in topology (Proc. Sympos. Algebraic Topology, Oxford, 1972), Cambridge Univ. Press, London, 1974, pp. 61–93. London Math. Soc. Lecture Note Ser., No. 11. MR0339152

[MS02] J. E. McClure and J. H. Smith, *A solution of Deligne's Hochschild cohomology conjecture*, Recent progress in homotopy theory (Baltimore, MD, 2000), Contemp. Math., vol. 293, Amer. Math. Soc., Providence, RI, 2002, pp. 153–193, DOI 10.1090/conm/293/04948. MR1890736

[MS04a] J. E. McClure and J. H. Smith, *Operads and cosimplicial objects: an introduction*, Axiomatic, enriched and motivic homotopy theory, NATO Sci. Ser. II Math. Phys. Chem., vol. 131, Kluwer Acad. Publ., Dordrecht, 2004, pp. 133–171, DOI 10.1007/978-94-007-0948-5_5. MR2061854

[MS04b] J. E. McClure and J. H. Smith, *Cosimplicial objects and little n-cubes. I*, Amer. J. Math. **126** (2004), no. 5, 1109–1153. MR2089084

[MS15] E. Meir and M. Szymik, *Drinfeld centers for bicategories*, Doc. Math. **20** (2015), 707–735. MR3398725

[Mit72] B. Mitchell, *Rings with several objects*, Advances in Math. **8** (1972), 1–161, DOI 10.1016/0001-8708(72)90002-3. MR0294454

[Moo55] J.C. Moore. Homotopie des complexes monoïdaux, II. Séminaire Henri Cartan, 1955, Exp. 19, 1–7.

[MN94] J. M. Møller and D. Notbohm, *Centers and finite coverings of finite loop spaces*, J. Reine Angew. Math. **456** (1994), 99–133. MR1301453

[NS17] F. Neumann and M. Szymik, *Spectral sequences for Hochschild cohomology and graded centers of derived categories*, Selecta Math. (N.S.) **23** (2017), no. 3, 1997–2018, DOI 10.1007/s00029-017-0331-9. MR3663600

[Sta63] J. D. Stasheff, *Homotopy associativity of H-spaces. I, II*, Trans. Amer. Math. Soc. 108 (1963), 275-292; ibid. **108** (1963), 293–312. MR0158400

[Szy18] M. Szymik, *Permutations, power operations, and the center of the category of racks*, Comm. Algebra **46** (2018), no. 1, 230–240, DOI 10.1080/00927872.2017.1316857. MR3764859

[Tam07] D. Tamarkin, *What do dg-categories form?*, Compos. Math. **143** (2007), no. 5, 1335–1358, DOI 10.1112/S0010437X07002771. MR2360318

[Tho56] R. Thom, *L'homologie des espaces fonctionnels* (French), Colloque de topologie algébrique, Louvain, 1956, Georges Thone, Liège; Masson & Cie, Paris, 1957, pp. 29–39. MR0089408

[Tho83] R. W. Thomason, *The homotopy limit problem*, Proceedings of the Northwestern Homotopy Theory Conference (Evanston, Ill., 1982), Contemp. Math., vol. 19, Amer. Math. Soc., Providence, R.I., 1983, pp. 407–419. MR711065

[Var74] K. Varadarajan, *On a certain problem of realization in homotopy theory*, Pacific J. Math. **54** (1974), 277–291. MR0358771

[Wei01] C. Weibel, *Homotopy ends and Thomason model categories*, Selecta Math. (N.S.) **7** (2001), no. 4, 533–564, DOI 10.1007/s00029-001-8098-3. MR1872081

Department of Mathematical Sciences, NTNU Norwegian University of Science and Technology, 7491 Trondheim, NORWAY

Email address: `markus.szymik@ntnu.no`

URL: `https://folk.ntnu.no/markussz`

Contemporary Mathematics
Volume **707**, 2018
http://dx.doi.org/10.1090/conm/707/14258

Recent developments on noncommutative motives

Gonçalo Tabuada

To Lily, for being by my side.

ABSTRACT. This survey covers some of the recent developments on noncommutative motives and their applications. Among other topics, we compute the additive invariants of relative cellular spaces and orbifolds; prove Kontsevich's semi-simplicity conjecture; prove a far-reaching noncommutative generalization of the Weil conjectures; prove Grothendieck's standard conjectures of type C^+ and D, Voevodsky's nilpotence conjecture, and Tate's conjecture, in several new cases; embed the (cohomological) Brauer group into secondary K-theory; construct a noncommutative motivic Gysin triangle; compute the localizing $\mathbb{A}^1$-invariants of corner skew Laurent polynomial algebras and of noncommutative projective schemes; relate Kontsevich's category of noncommutative mixed motives to Morel-Voevodsky's stable $\mathbb{A}^1$-homotopy category, to Voevodsky's triangulated category of mixed motives, and to Levine's triangulated category of mixed motives; prove the Schur-finiteness conjecture for quadric fibrations over low-dimensional bases; and finally extend Grothendieck's theory of periods to the broad noncommutative setting of dg categories.

Introduction

After the release of the monograph [Noncommutative Motives. With a preface by Yuri I. Manin. University Lecture Series **63**, American Mathematical Society, 2015], several important results on the theory of noncommutative motives have been established. The purpose of this survey, written for a broad mathematical audience, is to give a rigorous overview of some of these recent results. We will follow closely the notations, as well as the writing style, of the monograph [**81**]. Therefore, we suggest the reader to have it at his/her desk while reading this survey. The monograph [**81**] is divided into the following chapters:

Chapter 1. Differential graded categories.
Chapter 2. Additive invariants.
Chapter 3. Background on pure motives.
Chapter 4. Noncommutative pure motives.
Chapter 5. Noncommutative (standard) conjectures.
Chapter 6. Noncommutative motivic Galois groups.
Chapter 7. Jacobians of noncommutative Chow motives.

The author was partially supported by the National Science Foundation CAREER Award #1350472 and by the Portuguese Foundation for Science and Technology grant PEst-OE/MAT/UI0297/2014.

In this survey we cover some of the recent developments concerning the Chapters 2, 4, 5, 6, 8, and 9. These developments are described in Sections 1, 2, 3, 4, 5, and 6, respectively. The final Section 7, entitled "Noncommutative realizations and periods", discusses a recent research subject which was not addressed in [**81**].

Preliminaries. Throughout the survey, k will denote a base field. We will assume the reader is familiar with the language of differential graded (=dg) categories; for a survey on dg categories, we invite the reader to consult Keller's ICM address [**45**]. In particular, we will freely use the notions of *Morita equivalence* of dg categories (see [**81**, §1.6]) and *smooth/proper dg category* in the sense of Kontsevich (see [**81**, §1.7]). We will write $\mathrm{dgcat}(k)$ for the category of (small) dg categories and $\mathrm{dgcat}_{\mathrm{sp}}(k)$ for the full subcategory of smooth proper dg categories. Given a k-scheme X (or more generally an algebraic stack $\mathcal{X}$), we will denote by $\mathrm{perf}_{\mathrm{dg}}(X)$ the canonical dg enhancement of the category of perfect complexes $\mathrm{perf}(X)$; see [**81**, Example 1.27].

1. Additive invariants

Recall from [**81**, §2.3] the construction of the *universal additive invariant* of dg categories $U\colon \mathrm{dgcat}(k) \to \mathrm{Hmo}_0(k)$. In [**81**, §2.4] we described the behavior of U with respect to semi-orthogonal decompositions, full exceptional collections, purely inseparable field extensions, central simple algebras, sheaves of Azumaya algebras, twisted flag varieties, nilpotent ideals, finite-dimensional algebras of finite global dimension, etc. In §1.1-1.2 we describe the behavior of U with respect to relative cellular spaces and orbifolds. As explained in [**81**, Thm. 2.9], all the results in §1.1-1.2 are *motivic* in the sense that they hold similarly for every additive invariant such as algebraic K-theory, mod-n algebraic K-theory, Karoubi-Villamayor K-theory, nonconnective algebraic K-theory, homotopy K-theory, étale K-theory, Hochschild homology, cyclic homology, negative cyclic homology, periodic cyclic homology, topological Hochschild homology, topological cyclic homology, topological periodic cyclic homology, etc. Consult §7.1 for further examples of additive invariants.

NOTATION 1.1. Given a k-scheme X (or more generally an algebraic stack $\mathcal{X}$), we will write $U(X)$ instead of $U(\mathrm{perf}_{\mathrm{dg}}(X))$.

1.1. Relative cellular spaces. A flat morphism of k-schemes $p\colon X \to Y$ is called an *affine fibration* of relative dimension d if for every point $y \in Y$ there exists a Zariski open neighborhood $y \in V$ such that $X_V := p^{-1}(V) \simeq Y \times \mathbb{A}^d$ with $p_V : X_V \to Y$ isomorphic to the projection onto the first factor. Following Karpenko [**43**, Def. 6.1], a smooth projective k-scheme X is called a *relative cellular space* if it admits a filtration by closed subschemes

$$\varnothing = X_{-1} \hookrightarrow X_0 \hookrightarrow \cdots \hookrightarrow X_i \hookrightarrow \cdots \hookrightarrow X_{n-1} \hookrightarrow X_n = X$$

and affine fibrations $p_i\colon X_i \backslash X_{i-1} \to Y_i, 0 \leq i \leq n$, of relative dimension d_i with Y_i a smooth projective k-scheme. The smooth k-schemes $X_i \backslash X_{i-1}$ are called the *cells* and the smooth projective k-schemes Y_i the *bases* of the cells.

EXAMPLE 1.1 ($\mathbb{G}_m$-schemes). The celebrated Bialynicki-Birula decomposition [**9**] provides a relative cellular space structure on smooth projective k-schemes equipped with a $\mathbb{G}_m$-action. In this case, the bases of the cells are given by the connected components of the fixed point locus. This class of relative cellular spaces includes the isotropic flag varieties considered by Karpenko in [**43**] as well as the isotropic homogeneous spaces considered by Chernousov-Gille-Merkurjev in [**16**].

THEOREM 1.2 ([**95**, Thm. 2.7]). *Given a relative cellular space X, we have an isomorphism $U(X) \simeq \bigoplus_{i=0}^n U(Y_i)$.*

Theorem 1.2 shows that the additive invariants of relative cellular spaces X are completely determined by the basis Y_i of the cells $X_i \backslash X_{i-1}$. Among other ingredients, its proof makes use of Theorem 5.3; consult [**95**, §9] for details.

EXAMPLE 1.3 (Knörrer periodicity). Let $q = fg + q'$, where f, g, and q', are forms of degrees $a > 0$, $b > 0$, and $a + b$, in disjoint sets of variables $(x_i)_{i=1,\ldots,m}$, $(y_j)_{j=1,\ldots,n}$, and $(z_l)_{l=1,\ldots,p}$, respectively. Such a decomposition holds, for example, in the case of isotropic quadratic forms q. Let us write Q and Q' for the projective hypersurfaces defined by q and q', respectively. Assume that Q is smooth. Under this assumption, we have a $\mathbb{G}_m$-action on Q given by $\lambda \cdot (\underline{x}, \underline{y}, \underline{z}) := (\lambda^b \underline{x}, \lambda^{-a} \underline{y}, \underline{z})$ with fixed point locus $\mathbb{P}^{m-1} \amalg \mathbb{P}^{n-1} \amalg Q'$; this implies that Q' is also smooth. By combining Theorem 1.2 and Example 1.1 with the fact that $U(\mathbb{P}^n) \simeq U(k)^{\oplus(n+1)}$ (see [**81**, §2.4.2]), we obtain an induced isomorphism $U(Q) \simeq U(k)^{\oplus(m+n)} \oplus U(Q')$. Morally speaking, this shows that the additive invariants of Q and Q' are the same.

1.2. Orbifolds. Let G be a finite group of order n (we assume that $1/n \in k$), φ the set of all cyclic subgroups of G, $\varphi/\!\sim$ a set of representatives of the conjugacy classes in φ, X a smooth k-scheme equipped with a G-action, and $[X/G]$ the associated orbifold. As explained in [**96**, §3], the assignment $[V] \mapsto V \otimes_k -$, where V stands for a G-representation, gives rise to an action of the representation ring $R(G)$ on $U([X/G])$. Given $\sigma \in \varphi$, let e_σ be the unique idempotent of $R(\sigma)_{1/n}$ whose image under all the restrictions $R(\sigma)_{1/n} \to R(\sigma')_{1/n}$, with $\sigma' \subsetneq \sigma$, is zero. The normalizer $N(\sigma)$ of σ acts naturally on $[X^\sigma/\sigma]$ and hence on $U([X^\sigma/\sigma])$. By functoriality, this action restricts to the direct summand $e_\sigma U([X^\sigma/\sigma])_{1/n}$.

THEOREM 1.4 ([**96**, Thm. 1.1 and Cor. 1.6]). *The following computations hold:*

(i) *We have an induced isomorphism*

$$U([X/G])_{1/n} \simeq \bigoplus_{\sigma \in \varphi/\sim} (e_\sigma U([X^\sigma/\sigma])_{1/n})^{N(\sigma)} \tag{1.2}$$

in the $\mathbb{Z}[1/n]$-linearized (and idempotent completed) category $\mathrm{Hmo}_0(k)_{1/n}$.

(ii) *If k contains the nth roots of unity, then (1.2) reduces to an isomorphism*

$$U([X/G])_{1/n} \simeq \bigoplus_{\sigma \in \varphi/\sim} (U(X^\sigma)_{1/n} \otimes_{\mathbb{Z}[1/n]} e_\sigma R(\sigma)_{1/n})^{N(\sigma)}, \tag{1.3}$$

where $- \otimes_{\mathbb{Z}[1/n]} -$ stands for the canonical action of the category of finitely generated projective $\mathbb{Z}[1/n]$-modules on $\mathrm{Hmo}_0(k)_{1/n}$.

(iii) *If k contains the nth roots of unity and F is a field which contains the nth roots of unity and $1/n \in F$, then we have induced isomorphisms*

$$U([X/G])_F \simeq \bigoplus_{g \in G/\sim} U(X^g)_F^{C(g)} \simeq (\bigoplus_{g \in G} U(X^g)_F)^G \tag{1.4}$$

in the category $\mathrm{Hmo}_0(k)_F$*, where* $C(g)$ *stands for the centralizer of* g*. Moreover,* (1.3)-(1.4) *are isomorphisms of (commutative) monoids.*

Roughly speaking, Theorem 1.4 shows that the additive invariants of orbifolds can be computed using solely "ordinary" schemes.

EXAMPLE 1.5 (McKay correspondence). In many cases, $\mathrm{perf}_{\mathrm{dg}}([X/G])$ is known to be Morita equivalent to $\mathrm{perf}_{\mathrm{dg}}(Y)$ for a crepant resolution Y of the (singular) geometric quotient $X/\!\!/G$; see [**8**, **15**, **42**, **44**]. This is generally referred to as the "McKay correspondence". Whenever it holds, we can replace $[X/G]$ by Y in the formulas (1.2)-(1.4). Here is an illustrative example (with k algebraically closed): the cyclic group $G = C_2$ acts on any abelian surface S by the involution $a \mapsto -a$ and the Kummer surface $\mathrm{Km}(S)$ is defined as the blow-up of $S/\!\!/C_2$ at its 16 singular points. In this case, the dg category $\mathrm{perf}_{\mathrm{dg}}([S/C_2])$ is Morita equivalent to $\mathrm{perf}_{\mathrm{dg}}(\mathrm{Km}(S))$. Consequently, Theorem 1.4(ii) leads to an isomorphism:

$$U(\mathrm{Km}(S))_{1/2} \simeq U(S)^{C_2}_{1/2} \oplus U(k)^{\oplus 16}_{1/2} \,. \tag{1.5}$$

Note that since the Kummer surface is Calabi-Yau, the category $\mathrm{perf}(\mathrm{Km}(S))$ does *not* admit non-trivial semi-orthogonal decompositions. This shows that the isomorphism (1.5) is *not* induced from a semi-orthogonal decomposition.

COROLLARY 1.6 (Algebraic K-theory). If k contains the nth roots of unity, then we have the following isomorphism of $\mathbb{Z}$-graded commutative $\mathbb{Z}[1/n]$-algebras:

$$K_*([X/G])_{1/n} \simeq \bigoplus_{\sigma \in \varphi/\sim} (K_*(X^\sigma)_{1/n} \otimes_{\mathbb{Z}[1/n]} e_\sigma R(\sigma)_{1/n})^{N(\sigma)} \,. \tag{1.7}$$

The formula (1.7) was originally established by Vistoli in [**103**, Thm. 1]. Among other ingredients, Vistoli's proof makes essential use of dévissage. The proof of Theorem 1.4, and hence of (1.7), is not only different but moreover avoids the use of dévissage; consult [**96**, §6] for details.

COROLLARY 1.8 (Cyclic homology). If k contains the nth roots of unity, then we have the following isomorphisms of $\mathbb{Z}$-graded commutative k-algebras:

$$HC_*([X/G]) \simeq \bigoplus_{g \in G/\sim} HC_*(X^g)^{C(g)} \simeq (\bigoplus_{g \in G} HC_*(X^g))^G \,. \tag{1.9}$$

The formula (1.9) was originally established by Baranovsky in [**4**, Thm. 1.1]. Baranovsky's proof is very specific to cyclic homology. In constrast, the proof of Theorem 1.4, and hence of (1.9), avoids all the specificities of cyclic homology and is moreover quite conceptual; consult [**96**, §6] for details.

COROLLARY 1.10 (Topological periodic cyclic homology). Let k be a perfect field of characteristic $p > 0$, $W(k)$ the associated ring of p-typical Witt vectors, and $K := W(k)[1/p]$ the fraction field of $W(k)$. If k contains the nth roots of unity, then we have the following isomorphisms of $\mathbb{Z}/2$-graded commutative K-algebras:

$$TP_*([X/G])_{1/p} \simeq \bigoplus_{g \in G/\sim} TP_*(X^g)^{C(g)}_{1/p} \simeq (\bigoplus_{g \in G} TP_*(X^g)_{1/p})^G \,. \tag{1.11}$$

To the best of the author's knowledge, the formula (1.11) is new in the literature; consult [**96**, §1] for further corollaries of Theorem 1.4.

1.2.1. *Twisted analogues.* Given a sheaf of Azumaya algebras $\mathcal{F}$ over $[X/G]$, *i.e.* a G-equivariant sheaf of Azumaya algebras over X, all the computations of Theorem 1.4 admit $\mathcal{F}$-twisted analogues; consult [**96**, Thm. 1.27 and Cor. 1.29].

2. Noncommutative pure motives

In §2.1 we recall the definition of the different categories of noncommutative pure motives. Subsections §2.2-2.4 are devoted to three structural properties of these categories (relation with the Brauer group, semi-simplicity and rigidity). In 2.5 we prove to a far-reaching noncommutative generalization of the Weil conjectures; see Theorem 2.12. Finally, in §2.6 we describe some of the equivariant analogues of the theory of noncommutative pure motives.

2.1. Recollections. Recall from [**81**, §4.1] that the category of *noncommutative Chow motives* $\mathrm{NChow}(k)$ is defined as the idempotent completion of the full subcategory of $\mathrm{Hmo}_0(k)$ consisting of the objects $U(\mathcal{A})$, with $\mathcal{A}$ a smooth proper dg category. By construction, this category is additive, rigid symmetric monoidal, and comes equipped with a symmetric monoidal functor $U\colon \mathrm{dgcat}_{\mathrm{sp}}(k) \to \mathrm{NChow}(k)$. Moreover, given smooth proper dg categories $\mathcal{A}$ and $\mathcal{B}$, we have isomorphisms:

$$\mathrm{Hom}_{\mathrm{NChow}(k)}(U(\mathcal{A}), U(\mathcal{B})) \simeq K_0(\mathcal{D}_c(\mathcal{A}^{\mathrm{op}} \otimes \mathcal{B})) =: K_0(\mathcal{A}^{\mathrm{op}} \otimes \mathcal{B})\,. \tag{2.1}$$

Given a rigid symmetric monoidal category $(\mathcal{C}, \otimes, \mathbf{1})$, consider the $\otimes$-ideal

$$\otimes_{\mathrm{nil}}(a,b) := \{f \in \mathrm{Hom}_{\mathcal{C}}(a,b) \mid f^{\otimes n} = 0 \text{ for some } n \gg 0\}\,.$$

Recall from [**81**, §4.4] that the category of *noncommutative $\otimes$-nilpotent motives* $\mathrm{NVoev}(k)$ is defined as the idempotent completion of the quotient $\mathrm{NChow}(k)/\otimes_{\mathrm{nil}}$.

As explained in [**81**, §4.5], periodic cyclic homology gives rise to an additive symmetric monoidal functor $HP^{\pm}\colon \mathrm{NChow}(k) \to \mathrm{Vect}_{\mathbb{Z}/2}(k)$, with values in the category of finite-dimensional $\mathbb{Z}/2$-graded k-vector spaces. Recall from *loc. cit.* that the category of *noncommutative homological motives* $\mathrm{NHom}(k)$ is defined as the idempotent completion of the quotient $\mathrm{NChow}(k)/\mathrm{Ker}(HP^{\pm})$.

Given a rigid symmetric monoidal category $(\mathcal{C}, \otimes, \mathbf{1})$, consider the $\otimes$-ideal

$$\mathcal{N}(a,b) := \{f \in \mathrm{Hom}_{\mathcal{C}}(a,b) \mid \forall g \in \mathrm{Hom}_{\mathcal{C}}(b,a) \text{ we have } \mathrm{tr}(g\circ f) = 0\}\,,$$

where $\mathrm{tr}(g \circ f)$ stands for the categorical trace of the endomorphism $g \circ f$. Recall from [**81**, §4.6] that the category of *noncommutative numerical motives* $\mathrm{NNum}(k)$ is defined as the idempotent completion of the quotient $\mathrm{NChow}(k)/\mathcal{N}$.

2.2. Relation with the Brauer group. Let $\mathrm{Br}(k)$ be the Brauer group of the base field k. Given a central simple k-algebra A, let $[A]$ be its Brauer class.

EXAMPLE 2.1 (Local fields). Recall that a local field k is isomorphic to $\mathbb{R}$, to $\mathbb{C}$, to a finite field extension of $\mathbb{Q}_p$, or to a finite field extension of $\mathbb{F}_p((t))$. Thanks to local class field theory, we have $\mathrm{Br}(\mathbb{R}) \simeq \mathbb{Z}/2$, $\mathrm{Br}(\mathbb{C}) = 0$, and $\mathrm{Br}(k) \simeq \mathbb{Q}/\mathbb{Z}$ in all the remaining cases. Moreover, every element of the Brauer group $\mathrm{Br}(k)$ can be represented by a cyclic k-algebra.

Recall from [**81**, §2.4.4] that we have the following equivalence

$$[A] = [B] \Leftrightarrow U(A) \simeq U(B) \tag{2.2}$$

for any two central simple k-algebras A and B. Intuitively speaking, (2.2) shows that the Brauer class $[A]$ and the noncommutative Chow motive $U(A)$ contain exactly the same information. Let $K_0(\mathrm{NChow}(k))$ be the Grothendieck ring of the additive symmetric monoidal category $\mathrm{NChow}(k)$. Given a central simple k-algebra A, we write $[U(A)]$ for the Grothendieck class of $U(A)$. The (proof of the) next result is contained in [**80**, Thm. 6.12][**88**, Thm. 1.3]:

THEOREM 2.2. *Given central simple k-algebras A and B, we have:*

$$U(A) \simeq U(B) \Leftrightarrow [U(A)] \simeq [U(B)]\,. \tag{2.3}$$

Roughly speaking, Theorem 2.2 shows that the noncommutative Chow motives of central simple k-algebras are insensitive to the Grothendieck group relations. By combining the equivalences (2.2) and (2.3), we obtain the following result:

COROLLARY 2.4. The following map is injective:

$$\mathrm{Br}(k) \longrightarrow K_0(\mathrm{NChow}(k)) \qquad [A] \mapsto [U(A)]\,.$$

Consult §5.1.1 for some applications of Corollary 2.4 to secondary K-theory.

REMARK 2.3 (Generalizations). Theorem 2.2 and Corollary 2.4 hold more generally with k replaced by a base k-scheme X. Furthermore, instead of the Brauer group $\mathrm{Br}(X)$, we can consider the second étale cohomology group[1] $H^2_{\mathrm{et}}(X, \mathbb{G}_m)$; consult [**80, 88**] for details. In the case of an affine cone over a smooth irreducible plane complex curve of degree ≥ 4, the latter étale cohomology group contains non-torsion classes. The same phenomenon occurs, for example, in the case of Mumford's (celebrated) singular surface [**67**, page 75]; see [**88**, Example 1.32].

REMARK 2.4 (Jacques Tits' motivic measure). The Grothendieck ring of varieties $K_0\mathrm{Var}(k)$, introduced in a letter from Grothendieck to Serre in the sixties, is defined as the quotient of the free abelian group on the set of isomorphism classes of k-schemes by the "cut-and-paste" relations. Although very important, the structure of this ring still remains poorly understood. Among other ingredients, Theorem 2.2 was used in the construction of a new motivic measure μ_T entitled *Tits motivic measure*; consult [**87**] for details. This new motivic measure led to the proof of several new structural properties of $K_0\mathrm{Var}(k)$. For example, making use of μ_T, it was proved in *loc. cit.* that two quadric hypersurfaces (or more generally involution varieties), associated to quadratic forms of degree 6, have the same Grothendieck class if and only if they are isomorphic. In the same vein, it was proved in *loc. cit.* that two products of conics have the same Grothendieck class if and only if they are isomorphic; this refines a previous result of Kollár [**49**].

2.3. Semi-simplicity. Let F be a field of characteristic zero. The following result is obtained by combining [**81**, Thm. 4.27] with [**93**, Thm. 1.1]:

THEOREM 2.5. *The category* $\mathrm{NNum}(k)_F$ *is abelian semi-simple.*

Assuming certain (polarization) conjectures, Kontsevich conjectured in his seminal talk [**53**] that the category $\mathrm{NNum}(k)_F$ was abelian semi-simple. Theorem 2.5 not only proves this conjecture but moreover shows that Kontsevich's insight holds unconditionally. Let $\mathrm{Num}(k)_F$ be the (classical) category of numerical motives; see [**1**, §4]. The next result is obtained by combining [**81**, Rk. 4.32] with [**93**, Cor. 1.2]:

COROLLARY 2.5. The category $\mathrm{Num}(k)_F$ is abelian semi-simple.

---

[1] As proved by Gabber [**23**] and de Jong [**38**], in the case where X admits an ample line bundle (e.g. X affine), the Brauer group $\mathrm{Br}(X)$ may be identified with the torsion subgroup of $H^2_{\mathrm{et}}(X, \mathbb{G}_m)$.

Assuming certain (standard) conjectures, Grothendieck conjectured in the sixties that the category $\mathrm{Num}(k)_F$ was abelian semi-simple. This conjecture was proved unconditionally by Jannsen [**36**] in the nineties using étale cohomology. Corollary 2.5 provides us with an alternative proof of Grothendieck's conjecture.

2.3.1. *Numerical Grothendieck group.* The Grothendieck group $K_0(\mathcal{A})$ of a proper dg category $\mathcal{A}$ comes equipped with the following Euler bilinear pairing:

$$\chi\colon K_0(\mathcal{A}) \times K_0(\mathcal{A}) \longrightarrow \mathbb{Z} \qquad ([M],[N]) \mapsto \sum_n (-1)^n \mathrm{dim}_k \mathrm{Hom}_{\mathcal{D}_c(\mathcal{A})}(M, N[n])\,.$$

This bilinear pairing is, in general, not symmetric neither skew-symmetric. Nevertheless, when $\mathcal{A}$ is moreover smooth the associated left and right kernels of χ agree; see [**81**, Prop. 4.24]. Consequently, under these assumptions on $\mathcal{A}$, we have a well-defined *numerical Grothendieck group* $K_0(\mathcal{A})/_{\sim \mathrm{num}} := K_0(\mathcal{A})/\mathrm{Ker}(\chi)$. Following [**81**, Thm. 4.26], given smooth proper dg categories $\mathcal{A}$ and $\mathcal{B}$, we have:

$$\mathrm{Hom}_{\mathrm{NNum}(k)}(U(\mathcal{A}), U(\mathcal{B})) \simeq K_0(\mathcal{A}^{\mathrm{op}} \otimes \mathcal{B})/\mathrm{Ker}(\chi)\,. \tag{2.6}$$

The next result, whose proof makes use of Theorem 2.5, is obtained by combining [**94**, Thm. 1.2] with [**93**, Thm. 6.2]:

THEOREM 2.6. *$K_0(\mathcal{A})/_{\sim \mathrm{num}}$ is a finitely generated free abelian group.*

Given a smooth proper k-scheme X, let us write $\mathcal{Z}^*(X)/_{\sim \mathrm{num}}$ for the (graded) group of algebraic cycles on X up to numerical equivalence. By combining Theorem 2.6 with the Hirzebruch-Riemann-Roch theorem, we obtain the following result:

COROLLARY 2.7. $\mathcal{Z}^*(X)/_{\sim \mathrm{num}}$ is a finitely generated free abelian group.

2.4. Rigidity. Recall that a field extension l/k is called *primary* if the algebraic closure of k in l is purely inseparable over k. When k is algebraically closed, every field extension l/k is primary.

THEOREM 2.7 ([**90**, Thm. 2.1(i)]). *Given a primary field extension l/k and a field F of characteristic zero, the functor $-\otimes_k l : \mathrm{NNum}(k)_F \to \mathrm{NNum}(l)_F$ is fully-faithful. The same holds integrally when k is algebraically closed.*

Intuitively speaking, Theorem 2.7 shows that the theory of noncommutative numerical motives is "rigid" under base-change along primary field extensions. Alternatively, thanks to the isomorphisms (2.6), Theorem 2.7 shows that the numerical Grothendieck group is "rigid" under primary field extensions. The commutative counterpart, resp. mixed analogue, of Theorem 2.7 was established by Kahn in [**40**, Prop. 5.5], resp. is provided by Theorem 6.8.

REMARK 2.8 (Extra functoriality). Let l/k be a primary field extension. As proved in [**90**, Thm. 2.3], Theorems 2.5 and 2.7 imply that the base-change functor $-\otimes_k l : \mathrm{NNum}(k)_F \to \mathrm{NNum}(l)_F$ admits a left=right adjoint. Without the assumption that the field extension l/k is primary, such an adjoint functor does *not* exists in general; consult [**90**, Rk. 2.4] for details.

2.5. Zeta functions of endomorphisms. Let $N\!M \in \mathrm{NChow}(k)_{\mathbb{Q}}$ be a noncommutative Chow motive and f an endomorphism of $N\!M$. Following Kahn [**39**, Def. 3.1], the *zeta function of f* is defined as the following formal power series

$$Z(f;t) := \exp\left(\sum_{n\geq 1} \mathrm{tr}(f^{\circ n})\frac{t^n}{n}\right) \in \mathbb{Q}[\![t]\!]\,, \tag{2.8}$$

where $f^{\circ n}$ stands for the composition of f with itself n-times, $\mathrm{tr}(f^{\circ n}) \in \mathbb{Q}$ stands for the categorical trace of $f^{\circ n}$, and $\exp(t) := \sum_{m\geq 0} \frac{t^m}{m!} \in \mathbb{Q}[\![t]\!]$.

REMARK 2.9. When $N\!M = U(\mathcal{A})_{\mathbb{Q}}$ and $f = [\mathrm{B}]_{\mathbb{Q}}$, with $\mathrm{B} \in \mathcal{D}_c(\mathcal{A}^{\mathrm{op}} \otimes \mathcal{A})$ a dg $\mathcal{A}$-$\mathcal{A}$-bimodule (see §2.1), we have the following computation

$$\mathrm{tr}(f^{\circ n}) = [HH(\mathcal{A}; \underbrace{\mathrm{B} \otimes^{\mathbf{L}}_{\mathcal{A}} \cdots \otimes^{\mathbf{L}}_{\mathcal{A}} \mathrm{B}}_{n\text{-times}})] \in K_0(k) \simeq \mathbb{Z}\,, \tag{2.9}$$

where $HH(\mathcal{A}; \mathrm{B} \otimes^{\mathbf{L}}_{\mathcal{A}} \cdots \otimes^{\mathbf{L}}_{\mathcal{A}} \mathrm{B})$ stands for the Hochschild homology of $\mathcal{A}$ with coefficients in $\mathrm{B} \otimes^{\mathbf{L}}_{\mathcal{A}} \cdots \otimes^{\mathbf{L}}_{\mathcal{A}} \mathrm{B}$; see [**81**, Prop. 2.26]. Intuitively speaking, the integer (2.9) is the "number of fixed points" of the dg $\mathcal{A}$-$\mathcal{A}$-bimodule $\mathrm{B} \otimes^{\mathbf{L}}_{\mathcal{A}} \cdots \otimes^{\mathbf{L}}_{\mathcal{A}} \mathrm{B}$.

EXAMPLE 2.10 (Zeta function). Let $k = \mathbb{F}_q$ be a finite field, X a smooth proper k-scheme, and Fr the geometric Frobenius. When $\mathcal{A} = \mathrm{perf}_{\mathrm{dg}}(X)$ and B is the dg bimodule associated to the pull-back dg functor $\mathrm{Fr}^* \colon \mathrm{perf}_{\mathrm{dg}}(X) \to \mathrm{perf}_{\mathrm{dg}}(X)$, (2.9) reduces to $[HH(X; \Gamma_{\mathrm{Fr}^{\circ n}})] = \langle \Delta \cdot \Gamma_{\mathrm{Fr}^{\circ n}} \rangle = |X(\mathbb{F}_{q^n})|$. Consequently, (2.8) reduces to the (classical) zeta function $Z_X(t) := \exp(\sum_{n\geq 1} |X(\mathbb{F}_{q^n})| \frac{t^n}{n})$ of X.

REMARK 2.11 (Witt vectors). Recall from [**32**] the definition of the ring of (big) Witt vectors $\mathrm{W}(\mathbb{Q}) = (1 + t\mathbb{Q}[\![t]\!], \times, *)$. Since the leading term of (2.8) is equal to 1, the zeta function $Z(f;t)$ of f belongs to $\mathrm{W}(\mathbb{Q})$. Moreover, given endomorphisms f and f' of noncommutative Chow motives $N\!M$ and $N\!M'$, we have $Z(f \oplus f';t) = Z(f;t) \times Z(f';t)$ and $Z(f \otimes f';t) = Z(f;t) * Z(f';t)$ in $\mathrm{W}(\mathbb{Q})$.

Let $B = \prod_i B_i$ be a finite-dimensional semi-simple $\mathbb{Q}$-algebra, Z_i the center of B_i, δ_i for the degree $[Z_i : \mathbb{Q}]$, and d_i the index $[B_i : Z_i]^{1/2}$. Given a unit $b \in B^\times$, its ith *reduced norm* $\mathrm{Nrd}_i(b) \in \mathbb{Q}$ is defined as the composition $(\mathrm{N}_{Z_i/\mathbb{Q}} \circ \mathrm{Nrd}_{B_i/Z_i})(b_i)$.

Let $N\!M \in \mathrm{NChow}(k)_{\mathbb{Q}}$ be a noncommutative Chow motive. Thanks to Theorem 2.5, $B := \mathrm{End}_{\mathrm{NNum}(k)_{\mathbb{Q}}}(N\!M)$ is a finite-dimensional semi-simple $\mathbb{Q}$-algebra; let us write $e_i \in B$ for the central idempotent corresponding to the summand B_i. Given an invertible endomorphism f of $N\!M$, its *determinant* $\det(f) \in \mathbb{Q}$ is defined as the following product $\prod_i \mathrm{Nrd}_i(f)^{\mu_i}$, where $\mu_i := \frac{\mathrm{tr}(e_i)}{\delta_i d_i}$.

THEOREM 2.12 ([**93**, Thm. 5.8]).

(i) *The series* $Z(f;t)$ *is* rational, *i.e.* $Z(f;t) = \frac{p(t)}{q(t)}$ *with* $p(t), q(t) \in \mathbb{Q}[t]$, *and* $\deg(q(t)) - \deg(p(t)) = \mathrm{tr}(\mathrm{id}_{N\!M})$.

(ii) *When f is invertible, we have the following functional equation:*

$$Z(f^{-1};t^{-1}) = (-t)^{\mathrm{tr}(\mathrm{id}_{N\!M})}\det(f)Z(f;t)\,.$$

COROLLARY 2.10 (Weil conjectures). Let $k = \mathbb{F}_q$ be a finite field, X a smooth proper k-scheme X of dimension d, and $\mathrm{E} := \langle \Delta \cdot \Delta \rangle \in \mathbb{Z}$ the self-intersection number of the diagonal Δ of $X \times X$.

(i) The zeta function $Z_X(t)$ of X is rational, and $\deg(q(t)) - \deg(p(t)) = \mathrm{E}$.

(ii) We have the following functional equation $Z_X(\frac{1}{q^d t}) = \pm t^{\mathrm{E}} q^{\frac{d}{2}\mathrm{E}} Z_X(t)$.

Weil conjectured[2] in [**110**] that the zeta function $Z_X(t)$ of X was rational and that it satisfied a functional equation. These conjectures were proved independently by Dwork [**20**] and Grothendieck [**26**] using p-adic analysis and étale cohomology, respectively. Corollary 2.10 provides us with an alternative proof of the

[2]Weil conjectured also that the zeta function $Z_X(t)$ of X satisfied an analogue of the Riemann hypothesis. This conjecture was proved by Deligne [**18**] using Lefschetz pencils.

Weil conjectures; see [**93**, Cor. 5.12]. Moreover, Theorem 2.12 proves a far-reaching noncommutative generalization of the Weil conjectures.

2.6. Equivariant noncommutative motives. Let G be a finite group of order n (we assume that $1/n \in k$). Recall from [**82**, Def. 4.1] the definition of a *G-action* on a dg category $\mathcal{A}$. Given a G-action $G \circlearrowright \mathcal{A}$, we have an associated dg category $\mathcal{A}^G$ of G-equivariant objects. From a topological viewpoint, $\mathcal{A}^G$ may be understood as the "homotopy fixed points" of the G-action on $\mathcal{A}$.

EXAMPLE 2.13 (G-schemes). Given a G-scheme X, the dg category $\mathrm{perf}_{\mathrm{dg}}(X)$ inherits a G-action. In this case, the dg category $\mathrm{perf}_{\mathrm{dg}}(X)^G$ is Morita equivalent to the dg category of G-equivariant perfect complexes $\mathrm{perf}^G_{\mathrm{dg}}(X) = \mathrm{perf}_{\mathrm{dg}}([X/G])$.

EXAMPLE 2.14 (Cohomology classes). Given a cohomology class $[\alpha] \in H^2(G, k^\times)$, the dg category k inherits a G-action $G \circlearrowright_\alpha k$. In this case, the dg category of G-equivariant objects is Morita equivalent to the twisted group algebra $k_\alpha[G]$.

Let $\mathrm{dgcat}^G(k)$ be the category of (small) dg categories equipped with a G-action, and $\mathrm{dgcat}^G_{\mathrm{sp}}(k)$ the full subcategory of smooth proper dg categories. As explained in [**82**, §5], the category $\mathrm{NChow}(k)$ admits a G-equivariant counterpart $\mathrm{NChow}^G(k)$. Recall from *loc. cit.* that the latter category is additive, rigid symmetric monoidal, and comes equipped with a symmetric monoidal functor $U^G \colon \mathrm{dgcat}^G_{\mathrm{sp}}(k) \to \mathrm{NChow}^G(k)$. Moreover, we have isomorphisms

$$\mathrm{Hom}_{\mathrm{NChow}^G(k)}(U^G(G \circlearrowright \mathcal{A}), U^G(G \circlearrowright \mathcal{B})) \simeq K_0^G(\mathcal{A}^{\mathrm{op}} \otimes \mathcal{B}),$$

where the right-hand side stands for the G-equivariant Grothendieck group. In particular, the ring of endomorphisms of the $\otimes$-unit $U^G(G \circlearrowright_1 k)$ agrees with the representation ring[3] $R(G)$. Let us write I for the augmentation ideal associated to the rank homomorphism $R(G) \twoheadrightarrow \mathbb{Z}$.

2.6.1. *Relation with equivariant Chow motives.* Making use of Edidin-Graham's work [**21**] on equivariant intersection theory, Laterveer [**59**], and Iyer and Müller-Stach [**35**], extended the theory of Chow motives to the G-equivariant setting. In particular, they constructed a category of G-equivariant Chow motives $\mathrm{Chow}^G(k)$ and a (contravariant) symmetric monoidal functor $\mathfrak{h}^G \colon \mathrm{SmProj}^G(k) \to \mathrm{Chow}^G(k)$, defined on smooth projective G-schemes.

THEOREM 2.15 ([**82**, Thm. 8.4]). *There exists a $\mathbb{Q}$-linear, fully-faithful, symmetric monoidal $\Phi^G_{\mathbb{Q}}$ making the following diagram commute*

$$\begin{array}{ccc}
\mathrm{SmProj}^G(k) & \xrightarrow{X \mapsto G \circlearrowright \mathrm{perf}_{\mathrm{dg}}(X)} & \mathrm{dgcat}^G_{\mathrm{sp}}(k) \\
\downarrow{\scriptstyle \mathfrak{h}^G(-)_{\mathbb{Q}}} & & \downarrow{\scriptstyle U^G(-)_{\mathbb{Q}}} \\
\mathrm{Chow}^G(k)_{\mathbb{Q}} & & \mathrm{NChow}^G(k)_{\mathbb{Q}} \\
\downarrow & & \downarrow{\scriptstyle (-)_{I_{\mathbb{Q}}}} \\
\mathrm{Chow}^G(k)_{\mathbb{Q}}/_{-\otimes \mathbb{Q}(1)} & \xrightarrow[\Phi^G_{\mathbb{Q}}]{} & \mathrm{NChow}^G(k)_{\mathbb{Q}, I_{\mathbb{Q}}},
\end{array}$$

[3]Recall that when $k = \mathbb{C}$ and G is abelian, we have an isomorphism $R(G) \simeq \mathbb{Z}[\widehat{G}]$.

where $\mathrm{Chow}^G(k)_{\mathbb{Q}}/_{-\otimes\mathbb{Q}(1)}$ *stands for the orbit category of* $\mathrm{Chow}^G(k)_{\mathbb{Q}}$ *with respect to the G-equivariant Tate motive* $\mathbb{Q}(1)$ *(see* [**81**, §4.2]*), and* $(-)_{I_{\mathbb{Q}}}$ *for the localization functor associated to the augmentation ideal* $I_{\mathbb{Q}}$.

Roughly speaking, Theorem 2.15 shows that in order to compare the equivariant commutative world with the equivariant noncommutative world, we need to "$\otimes$-trivialize" the G-equivariant Tate motive $\mathbb{Q}(1)$ on one side and to localize at the augmentation ideal $I_{\mathbb{Q}}$ on the other side. Only after these two reductions, the equivariant commutative world embeds fully-faithfully into the equivariant noncommutative world. As illustrated in §2.6.2, this shows that the G-equivariant Chow motive $\mathfrak{h}^G(X)_{\mathbb{Q}}$ and the G-equivariant noncommutative Chow motive $U^G(G \circlearrowright \mathrm{perf}_{\mathrm{dg}}(X))$ contain (important) independent information about X.

2.6.2. *Full exceptional collections.* Let X be a smooth projective G-scheme. In order to study it, we can proceed into two distinct directions. On one direction, we can associate to X its G-equivariant Chow motive $\mathfrak{h}^G(X)_{\mathbb{Q}}$. On another direction, we can associate to X the G-action $G \circlearrowright \mathrm{perf}_{\mathrm{dg}}(X)$. The following result, whose proof makes use of Theorem 2.15, relates these two distinct directions of study:

THEOREM 2.16 ([**82**, Thm. 1.2]). *If the category* $\mathrm{perf}(X)$ *admits a full exceptional collection* $(\mathcal{E}_1, \ldots, \mathcal{E}_n)$ *of G-invariant objects* ($\neq$ *G-equivariant objects), then there exists a choice of integers* $r_1, \ldots, r_n \in \{0, \ldots, \dim(X)\}$ *such that*

$$\mathfrak{h}^G(X)_{\mathbb{Q}} \simeq \mathbb{L}^{\otimes r_1} \oplus \cdots \oplus \mathbb{L}^{\otimes r_n}\,, \tag{2.11}$$

where $\mathbb{L} \in \mathrm{Chow}^G(k)_{\mathbb{Q}}$ *stands for the G-equivariant Lefschetz motive.*

Theorem 2.16 can be applied, for example, to any G-action on projective spaces, quadrics, Grassmannians, etc; consult [**82**, Examples 9.9-9.11]. Morally speaking, Theorem 2.16 shows that the existence of a full exceptional collection of G-invariant objects completely determines the G-equivariant Chow motive $\mathfrak{h}^G(X)_{\mathbb{Q}}$. In particular, $\mathfrak{h}^G(X)_{\mathbb{Q}}$ loses all the information about the G-action on X. In contrast, as explained in [**82**, Rmk. 9.4 and Prop. 9.8], the G-invariant objects $\mathcal{E}_1, \ldots, \mathcal{E}_n$ yield (non-trivial) cohomology classes $[\alpha_1], \ldots, [\alpha_n] \in H^2(G, k^\times)$ such that

$$U^G(G \circlearrowright \mathrm{perf}_{\mathrm{dg}}(X)) \simeq U^G(G \circlearrowright_{\alpha_1} k) \oplus \cdots \oplus U^G(G \circlearrowright_{\alpha_n} k)\,. \tag{2.12}$$

Taking into account (2.11)-(2.12), the G-equivariant Chow motive $\mathfrak{h}^G(X)_{\mathbb{Q}}$ and the G-equivariant noncommutative Chow motive $U^G(G \circlearrowright \mathrm{perf}_{\mathrm{dg}}(X))$ should be considered as complementary. While the former keeps track of the Tate twists but not of the G-action, the latter keeps track of the G-action but not of the Tate twists.

3. Noncommutative (standard) conjectures

In §3.1 we recall some important conjectures of Grothendieck, Voevodsky, and Tate. Subsection §3.2 is devoted to their noncommutative counterparts. As a first application of the noncommutative viewpoint, we prove that the original conjectures of Grothendieck, Voevodsky, and Tate, are invariant under homological projective duality. This leads to a proof of these original conjectures in several new cases. As a second application, we extend the original conjectures from schemes to algebraic stacks and prove them in the case of "low-dimensional" orbifolds.

3.1. Recollections. Let k be a base field of characteristic zero. Given a smooth proper k-scheme X and a Weil cohomology theory H^*, let us write π^n_X for the nth Künneth projector of $H^*(X)$, $Z^*(X)_\mathbb{Q}$ for the $\mathbb{Q}$-vector space of algebraic cycles on X, and $Z^*(X)_\mathbb{Q}/_{\sim \mathrm{nil}}$, $Z^*(X)_\mathbb{Q}/_{\sim \mathrm{hom}}$, and $Z^*(X)_\mathbb{Q}/_{\sim \mathrm{num}}$, for the quotient of $Z^*(X)_\mathbb{Q}$ with respect to the smash-nilpotence, homological, and numerical equivalence relation, respectively. Recall from [**81**, §3.0.8-3.0.11] that:

(i) The Grothendieck's standard conjecture[4] of type C^+, denoted by $C^+(X)$, asserts that the even Künneth projector $\pi^+_X := \sum_n \pi^{2n}_X$ is algebraic.
(ii) The Grothendieck's standard conjecture of type D, denoted by $D(X)$, asserts that $Z^*(X)_\mathbb{Q}/_{\sim \mathrm{hom}} = Z^*(X)_\mathbb{Q}/_{\sim \mathrm{num}}$.
(iii) The Voevodsky's nilpotence conjecture $V(X)$ (which implies conjecture $D(X)$) asserts that $Z^*(X)_\mathbb{Q}/_{\sim \mathrm{nil}} = Z^*(X)_\mathbb{Q}/_{\sim \mathrm{num}}$.
(iv) The Schur-finiteness conjecture[5], denoted by $S(X)$, asserts that the Chow motive $\mathfrak{h}(X)_\mathbb{Q}$ is Schur-finite in the sense of Deligne [**17**, §1].

REMARK 3.1 (Status).

(i) Thanks to the work of Grothendieck and the work of Kleiman (see [**25**,**47**, **48**]), the conjecture $C^+(X)$ holds when $\dim(X) \leq 2$, and also for abelian varieties. Moreover, the standard conjecture of type C^+ is stable under products.
(ii) Thanks to the work of Lieberman [**62**], the conjecture $D(X)$ holds when $\dim(X) \leq 4$, and also for abelian varieties.
(iii) Thanks to the work Voevodsky [**106**] and Voisin [**108**], the conjecture $V(X)$ holds when $\dim(X) \leq 2$. Thanks to the work of Kahn-Sebastian [**41**], the conjecture $V(X)$ holds moreover when X is an abelian 3-fold.
(iv) Thanks to the work of Kimura [**46**] and Shermenev [**74**], the conjecture $S(X)$ holds when $\dim(X) \leq 1$, and also for abelian varieties.

Let $k = \mathbb{F}_q$ be a finite base field of characteristic $p > 0$. Given a smooth proper k-scheme X and a prime number $l \neq p$, recall from [**97**, **98**] that the Tate conjecture, denoted by $T^l(X)$, asserts that the cycle class map is surjective:

$$\mathcal{Z}^*(X)_{\mathbb{Q}_l} \longrightarrow H^{2*}_{l\text{-adic}}(X_{\overline{k}}, \mathbb{Q}_l(*))^{\mathrm{Gal}(\overline{k}/k)}.$$

REMARK 3.2 (Status). Thanks to the work of Tate [**98**], the conjecture $T^l(X)$ holds when $\dim(X) \leq 1$, and also for abelian varieties. Thanks to the work of several other people (consult Totaro's survey [**102**]), the conjecture $T^l(X)$ holds moreover when X is a $K3$-surface (and $p \neq 2$).

3.2. Noncommutative counterparts. Let k be a base field of characteristic zero. Recall from §2 that periodic cyclic homology descends to the category of noncommutative Chow motives yielding a functor $HP^\pm \colon \mathrm{NChow}(k)_\mathbb{Q} \to \mathrm{Vect}_{\mathbb{Z}/2}(k)$. Given a smooth proper dg category $\mathcal{A}$, consider the following $\mathbb{Q}$-vector spaces

$$K_0(\mathcal{A})_\mathbb{Q}/_{\sim ?} := \mathrm{Hom}_?(U(k)_\mathbb{Q}, U(\mathcal{A})_\mathbb{Q}),$$

[4]The standard conjecture of type C^+ is also known as the *sign conjecture*. If the even Künneth projector π^+_X is algebraic, then the odd Künneth projector $\pi^-_X := \sum_n \pi^{2n+1}_X$ is also algebraic.

[5]Consult §6.5 for the mixed analogue of the Schur-finiteness conjecture.

where ? belongs to the sets {nil, hom, num} and $\{\mathrm{NVoev}(k), \mathrm{NHom}(k)_\mathbb{Q}, \mathrm{NNum}(k)_\mathbb{Q}\}$, respectively. Under these notations, the important conjectures in §3.1 admit the following noncommutative counterparts:

Conjecture $C^+_{\mathrm{nc}}(\mathcal{A})$: The even Künneth projector $\pi^+_\mathcal{A}$ of $HP^\pm(\mathcal{A})$ is *algebraic*, *i.e.* there exists an endomorphism $\underline{\pi}^+_\mathcal{A}$ of $U(\mathcal{A})_\mathbb{Q}$ such that $HP^\pm(\underline{\pi}^+_\mathcal{A}) = \pi^+_\mathcal{A}$.

Conjecture $D_{\mathrm{nc}}(\mathcal{A})$: The equality $K_0(\mathcal{A})_\mathbb{Q}/_{\sim \mathrm{hom}} = K_0(\mathcal{A})_\mathbb{Q}/_{\sim \mathrm{num}}$ holds.

Conjecture $V_{\mathrm{nc}}(\mathcal{A})$: The equality $K_0(\mathcal{A})_\mathbb{Q}/_{\sim \mathrm{nil}} = K_0(\mathcal{A})_\mathbb{Q}/_{\sim \mathrm{num}}$ holds.

Conjecture $S_{\mathrm{nc}}(\mathcal{A})$: The noncommutative Chow motive $U(\mathcal{A})_\mathbb{Q}$ is Schur-finite.

Let $k = \mathbb{F}_q$ be a finite base field of characteristic $p > 0$. Given a smooth proper dg category $\mathcal{A}$ and a prime number $l \neq p$, consider the following abelian groups

$$\mathrm{Hom}\left(\mathbb{Z}(l^\infty), \pi_{-1} L_{KU} K(\mathcal{A} \otimes_{\mathbb{F}_q} \mathbb{F}_{q^n})\right) \qquad n \geq 1, \tag{3.1}$$

where $\mathbb{Z}(l^\infty)$ stands for the Prüfer l-group and $L_{KU}K(\mathcal{A} \otimes_k k_n)$ for the Bousfield localization of the algebraic K-theory spectrum $K(\mathcal{A} \otimes_{\mathbb{F}_q} \mathbb{F}_{q^n})$ with respect to topological complex K-theory KU. Under these notations, Tate's conjecture admits the following noncommutative counterpart:

Conjecture $T^l_{\mathrm{nc}}(\mathcal{A})$: The abelian groups (3.1) are zero.

We now relate the conjectures in §3.1 with their noncommutative counterparts:

THEOREM 3.3. *Given a smooth proper k-scheme X, we have the equivalences:*

$$C^+(X) \Leftrightarrow C^+_{\mathrm{nc}}(\mathrm{perf}_{\mathrm{dg}}(X)) \tag{3.2}$$

$$D(X) \Leftrightarrow D_{\mathrm{nc}}(\mathrm{perf}_{\mathrm{dg}}(X)) \tag{3.3}$$

$$V(X) \Leftrightarrow V_{\mathrm{nc}}(\mathrm{perf}_{\mathrm{dg}}(X)) \tag{3.4}$$

$$S(X) \Leftrightarrow S_{\mathrm{nc}}(\mathrm{perf}_{\mathrm{dg}}(X)) \tag{3.5}$$

$$T^l(X) \Leftrightarrow T^l_{\mathrm{nc}}(\mathrm{perf}_{\mathrm{dg}}(X)). \tag{3.6}$$

Morally speaking, Theorem 3.3 shows that the important conjectures in §3.1 belong not only to the realm of algebraic geometry but also to the broad noncommutative setting of smooth proper dg categories. Consult [**81**, §5], and the references therein, for the implications $\Rightarrow$ in (3.2)-(3.3) and also for the equivalences (3.4)-(3.5). The converse implications $\Leftarrow$ in (3.2)-(3.3) were established in [**84**, Thm. 1.1]. Finally, the equivalence (3.6) was proved in [**91**, Thm. 1.2].

3.2.1. *Homological projective duality.* For a survey on homological projective duality (=HPD), we invite the reader to consult Kuznetsov's ICM address [**55**]. Let X be a smooth projective k-scheme equipped with a line bundle $\mathcal{L}_X(1)$; we write $X \to \mathbb{P}(W)$ for the associated morphism where $W := H^0(X, \mathcal{L}_X(1))^*$. Assume that the category $\mathrm{perf}(X)$ admits a Lefschetz decomposition $\langle \mathbb{A}_0, \mathbb{A}_1(1), \ldots, \mathbb{A}_{i-1}(i-1)\rangle$ with respect to $\mathcal{L}_X(1)$ in the sense of [**56**, Def. 4.1]. Following [**56**, Def. 6.1], let Y be the HP-dual of X, $\mathcal{L}_Y(1)$ the HP-dual line bundle, and $Y \to \mathbb{P}(W^*)$ the morphism associated to $\mathcal{L}_Y(1)$. Given a linear subspace $L \subset W^*$, consider the linear sections $X_L := X \times_{\mathbb{P}(W^*)} \mathbb{P}(L)$ and $Y_L := Y \times_{\mathbb{P}(W)} \mathbb{P}(L^\perp)$. The next result, whose proof makes use of Theorem 3.3, is obtained by concatenating [**81**, §5.3-5.4] with [**84**, Thm. 1.4][**85**, Thm. 1.1][**91**, Thm. 1.3]:

THEOREM 3.4 (HPD-invariance[6]). *Let X and Y be as above. Assume that X_L and Y_L are smooth, that* $\dim(X_L) = \dim(X) - \dim(L)$, *that* $\dim(Y_L) = \dim(Y) -$

[6]Consult Theorem 7.7 for another HPD-invariance type result.

$\dim(L^{\perp})$, *and that the following conjectures hold*

$$(3.7)\quad C^+_{\mathrm{nc}}(\mathbb{A}_{0,\mathrm{dg}}) \qquad D_{\mathrm{nc}}(\mathbb{A}_{0,\mathrm{dg}}) \qquad V_{\mathrm{nc}}(\mathbb{A}_{0,\mathrm{dg}}) \qquad S_{\mathrm{nc}}(\mathbb{A}_{0,\mathrm{dg}}) \qquad T^l_{\mathrm{nc}}(\mathbb{A}_{0,\mathrm{dg}})\,,$$

where $\mathbb{A}_{0,\mathrm{dg}}$ *stands for the dg enhancement of* $\mathbb{A}_0$ *induced by* $\mathrm{perf}_{\mathrm{dg}}(X)$. *Under these assumptions, we have the following equivalences of conjectures:*

$$?(X_L) \Leftrightarrow ?(Y_L) \quad \textit{with} \quad ? \in \{C^+, D, V, S, T^l\}\,.$$

REMARK 3.5. The conjectures (3.7) hold, for example, whenever the triangulated category $\mathbb{A}_0$ admits a full exceptional collection (this is the case in all the examples in the literature). Furthermore, Theorem 3.4 holds more generally when Y (or X) is singular. In this case, we need to replace Y by a noncommutative resolution of singularities in the sense of [**55**, §2.4].

Theorem 3.4 shows that the conjectures in §3.1 are invariant under homological projective duality. As a consequence, we obtain the following practical result:

COROLLARY 3.8. Let X_L and Y_L be smooth linear sections as in Theorem 3.4.
(a) If $\dim(Y_L) \leq 2$, then the conjectures $C^+(X_L)$ and $V(X_L)$ hold.
(b) If $\dim(Y_L) \leq 4$, then the conjecture $D(X_L)$ holds.
(c) If $\dim(Y_L) \leq 1$, then the conjectures $S(X_L)$ and $T^l(X_L)$ hold.

By applying Corollary 3.8 to the Veronese-Clifford duality, to the spinor duality, to the Grassmannian-Pfaffian duality, to the determinantal duality, and to other (incomplete) HP-dualities (see [**55**, §4]), we obtain a proof of the conjectures in §3.1 in several new cases; consult [**7**, **84**, **85**, **91**] for details. In the particular case of the Veronese-Clifford duality, Corollary 3.8 leads furthermore to an alternative proof of the Tate conjecture for smooth complete intersections of two quadrics (the original (geometric) proof, based on the notion of variety of maximal planes, is due to Reid [**72**]); consult [**91**, Thm. 1.7] for details.

3.2.2. *Algebraic stacks.* Theorem 3.3 allows us to easily extend the important conjectures in §3.1 from smooth proper schemes to smooth proper algebraic stacks $\mathcal{X}$ by setting $?(\mathcal{X}) :=?_{\mathrm{nc}}(\mathrm{perf}_{\mathrm{dg}}(\mathcal{X}))$, where $? \in \{C^+, D, V, S, T^l\}$. The next result, obtained by combining [**96**, Thm. 9.2] with [**91**, Thm. 1.9], proves these conjectures in the case of "low-dimensional" orbifolds; consult [**84**] for further examples of algebraic stacks satisfying these conjectures.

THEOREM 3.6. *Let G be a finite group, X a smooth projective k-scheme equipped with a G-action, and $\mathcal{X} := [X/G]$ the associated orbifold.*
(a) *The conjectures $C^+(\mathcal{X})$ and $V(\mathcal{X})$ hold when* $\dim(X) \leq 2$. *The conjecture $C^+(\mathcal{X})$ also holds when X is an abelian variety and G acts by group homomorphisms.*
(b) *The conjecture $D(\mathcal{X})$ holds when* $\dim(X) \leq 4$.
(c) *The conjectures $S(\mathcal{X})$ and $T^l(\mathcal{X})$ hold when* $\dim(X) \leq 1$.

Roughly speaking, Theorem 3.6 shows that the above conjectures are "insensitive" to the G-action. Among other ingredients, its proof makes use of Theorem 1.4.

REMARK 3.7 (Generalizations). Theorem 3.6 holds more generally under the assumption that the conjectures in §3.1 are satisfied by the fixed point locus $\{X^\sigma\}_\sigma$, with $\sigma \in \varphi/\!\sim$. For example, the conjecture $T^l(\mathcal{X})$ also holds when X is an abelian surface and the group $G = C_2$ acts by the involution $a \mapsto -a$.

4. Noncommutative motivic Galois groups

Let F be a field of characteristic zero and $\mathrm{NNum}(k)_F$ the abelian category of numerical motives. The next result was proved in [**81**, Thm. 6.4] and [**93**, Thm. 7.1]:

THEOREM 4.1. *The category* $\mathrm{NNum}(k)_F$ *is super-Tannakian in the sense of Deligne* [**17**]. *When F is algebraically closed,* $\mathrm{NNum}(k)_F$ *is neutral super-Tannakian.*

By combining Theorem 4.1 with Deligne's super-Tannakian formalism [**17**], we obtain an affine super-group F-scheme $\mathrm{sGal}(\mathrm{NNum}(k)_F)$ called the *noncommutative motivic Galois super-group.* The following result relates this super-group with the (classical) motivic Galois super-group $\mathrm{sGal}(\mathrm{Num}(k)_F)$:

THEOREM 4.2 ([**93**, Thm. 7.4]). *Assume that F is algebraically closed. Then, there exists a faithfully flat morphism of affine super-group F-schemes*

$$\mathrm{sGal}(\mathrm{NNum}(k)_F) \twoheadrightarrow \mathrm{Ker}(\mathrm{sGal}(\mathrm{Num}(k)_F) \overset{t^*}{\twoheadrightarrow} \mathbb{G}_m)\,,$$

where $\mathbb{G}_m$ stands for the multiplicative (super-)group scheme and t for the inclusion of the category of Tate motives into numerical motives.

Theorem 4.2 was envisioned by Kontsevich; see his seminal talk [**53**]. Intuitively speaking, it shows that the "$\otimes$-symmetries" of the commutative world which can be lifted to the noncommutative world are precisely those which become trivial when restricted to Tate motives. Theorem 4.2 also holds when F is *not* algebraically closed. However, in this case the super-group schemes are only defined over a (very big) commutative F-algebra.

REMARK 4.3 (Simplification). The analogue of Theorem 4.2, with k of characteristic zero, was proved in [**81**, Thm. 6.7(ii)]. However, therein we assumed the noncommutative counterparts of the standard conjectures of type C^+ and D and moreover used Deligne-Milne's theory of Tate-triples. In contrast, Theorem 4.2 is unconditional and its proof avoids the use of Tate-triples; consult [**93**, §7].

Base-change. Recall from [**81**, §6] the definition of the (conditional[7]) noncommutative motivic Galois group $\mathrm{Gal}(\mathrm{NNum}^\dagger(k)_F)$.

THEOREM 4.4 ([**90**, Thm. 2.2]). *Given a primary field extension l/k, the induced base-change functor $-\otimes_k l\colon \mathrm{NNum}^\dagger(k)_F \to \mathrm{NNum}^\dagger(l)_F$ gives rise to a faithfully flat morphism of affine group F-schemes* $\mathrm{Gal}(\mathrm{NNum}^\dagger(l)_F) \to \mathrm{Gal}(\mathrm{NNum}^\dagger(k)_F)$.

Roughly speaking, Theorem 4.2 shows that every "$\otimes$-symmetry" of the category of noncommutative numerical k-linear motives can be extended to a "$\otimes$-symmetry" of the category of noncommutative l-linear motives. Among other ingredients, its proof makes use of Theorem 2.7. In the particular case of an extension of algebraically closed fields l/k, the commutative counterpart of Theorem 4.4 was established by Deligne-Milne in [**19**, Prop. 6.22(b)].

5. Localizing invariants

Recall from [**81**, §8.1] the notion of a *short exact sequence* of dg categories in the sense of Drinfeld/Keller. In §5.1 we describe a key structural property of these short exact sequences and explain its implications to secondary K-theory.

---

[7]We assume the noncommutative counterparts of the standard conjectures of type C^+ and D.

Recall from [**81**, §8.5.1] the construction of the *universal localizing $\mathbb{A}^1$-invariant* of dg categories $\mathrm{U}\colon \mathrm{dgcat}(k) \to \mathrm{NMot}(k)$; in *loc. cit.* we used the explicit notation $\mathrm{U}^{\mathbb{A}^1}_{\mathrm{loc}}\colon \mathrm{dgcat}(k) \to \mathrm{NMot}^{\mathbb{A}^1}_{\mathrm{loc}}(k)$. In [**81**, §8.5.3] we described the behavior of U with respect to dg orbit categories and dg cluster categories. In §5.2-5.4 we describe the behavior of U with respect to open/closed scheme decompositions, corner skew Laurent polynomial algebras, and noncommutative projective schemes. As explained in [**81**, Thm. 8.25], all the results in §5.2-5.4 are *motivic* in the sense that they hold similarly for every localizing $\mathbb{A}^1$-invariant such as mod-n algebraic K-theory (when $1/n \in k$), homotopy K-theory, étale K-theory, periodic cyclic homology[8] (when $\mathrm{char}(k) = 0$), etc. The results of §5.4 do *not* require $\mathbb{A}^1$-homotopy invariance. Hence, they hold similarly for every localizing invariant; see Remark 5.18.

NOTATION 5.1. Given a k-scheme X (or more generally an algebraic stack $\mathcal{X}$), we will write $\mathrm{U}(X)$ instead of $\mathrm{U}(\mathrm{perf}_{\mathrm{dg}}(X))$.

5.1. Short exact sequences. Recall from [**81**, §8.4] the notion of a *split short exact sequence* of dg categories $0 \to \mathcal{A} \to \mathcal{B} \to \mathcal{C} \to 0$. Up to Morita equivalence, this data is equivalent to inclusions of dg categories $\mathcal{A}, \mathcal{C} \subseteq \mathcal{B}$ yielding a semi-orthogonal decomposition of triangulated categories $\mathrm{H}^0(\mathcal{B}) = \langle \mathrm{H}^0(\mathcal{A}), \mathrm{H}^0(\mathcal{C})\rangle$ in the sense of Bondal-Orlov [**13**]; by definition, the category $\mathrm{H}^0(\mathcal{A})$ has the same objects as $\mathcal{A}$ and morphisms $\mathrm{H}^0(\mathcal{A})(x,y) := H^0(\mathcal{A}(x,y))$.

THEOREM 5.1 ([**80**, Thm. 4.4]). *Let $0 \to \mathcal{A} \to \mathcal{B} \to \mathcal{C} \to 0$ be a short exact sequence of dg categories in the sense of Drinfeld/Keller. If $\mathcal{A}$ is smooth and proper and $\mathcal{B}$ is proper, then the short exact sequence is split.*

Morally speaking, Theorem 5.1 shows that the smooth proper dg categories behave as "injective" objects. In the setting of triangulated categories, this conceptual idea goes back to the pioneering work of Bondal-Kapranov [**11**].

5.1.1. *Secondary K-theory.* Two decades ago, Bondal-Larsen-Lunts introduced in [**12**] the *Grothendieck ring of smooth proper dg categories $\mathcal{PT}(k)$*. This ring is defined by generators and relations. The generators are the Morita equivalence classes of smooth proper dg categories[9] and the relations $[\mathcal{B}] = [\mathcal{A}] + [\mathcal{C}]$ arise from semi-orthogonal decompositions $\mathrm{H}^0(\mathcal{B}) = \langle \mathrm{H}^0(\mathcal{A}), \mathrm{H}^0(\mathcal{C})\rangle$. The multiplication law is induced by the tensor product of dg categories. One decade ago, Toën introduced in [**101**] a "categorified" version of the Grothendieck ring named *secondary Grothendieck ring $K_0^{(2)}(k)$*. By definition, $K_0^{(2)}(k)$ is the quotient of the free abelian group on the Morita equivalence classes of smooth proper dg categories by the relations $[\mathcal{B}] = [\mathcal{A}] + [\mathcal{C}]$ arising from short exact sequences $0 \to \mathcal{A} \to \mathcal{B} \to \mathcal{C} \to 0$. The multiplication law is also induced by the tensor product of dg categories.

Theorem 5.1 directly leads to the following result:

COROLLARY 5.2. The rings $\mathcal{PT}(k)$ and $K_0^{(2)}(k)$ are isomorphic.

Morally speaking, Corollary 5.2 shows that the secondary Grothendieck ring is *not* a new mathematical notion.

---

[8]Periodic cyclic homology is not a localizing $\mathbb{A}^1$-invariant in the sense of [**81**, §8.5] because it does not preserves filtered (homotopy) colimits. Nevertheless, all the results of §5.2-5.4 hold similarly for periodic cyclic homology.

[9]Bondal-Larsen-Lunts worked originally with (pretriangulated) dg categories. In this generality, the classical Eilenberg's swindle argument implies that the Grothendieck ring is trivial.

By construction, the universal additive invariant U (see §1) sends semi-orthogonal decompositions to direct sums. Therefore, it gives rise to a ring homomorphism $\mathcal{PT}(k) \to K_0(\mathrm{NChow}(k))$. Making use of Corollary 2.4, we then obtain the result:

COROLLARY 5.3. The following map is injective:

$$\mathrm{Br}(k) \longrightarrow \mathcal{PT}(k) \simeq K_0^{(2)}(k) \qquad [A] \mapsto [A]\,. \tag{5.4}$$

The map (5.4) may be understood as the "categorification" of the canonical map from the Picard group $\mathrm{Pic}(k)$ to the Grothendieck ring $K_0(k)$. In contrast with $\mathrm{Pic}(k) \to K_0(k)$, the map (5.4) does not seems to admit a "determinant" map in the converse direction. Nevertheless, Corollary 5.3 shows that this map is injective.

REMARK 5.2 (Generalizations). Similarly to Remark 2.3, Corollaries 5.2-5.3 hold more generally with k replaced by a base k-scheme X.

5.2. Gysin triangle. Let X be a smooth k-scheme, $i\colon Z \hookrightarrow X$ a smooth closed subscheme, and $j\colon V \hookrightarrow X$ the open complement of Z.

THEOREM 5.3 ([**95**, Thm. 1.9]). *We have a distinguished "Gysin" triangle*

$$\mathrm{U}(Z) \xrightarrow{\mathrm{U}(i_*)} \mathrm{U}(X) \xrightarrow{\mathrm{U}(j^*)} \mathrm{U}(V) \xrightarrow{\partial} \mathrm{U}(Z)[1]\,, \tag{5.5}$$

where i_*, resp. j^*, stands for the push-forward, resp. pull-back, dg functor.

REMARK 5.4 (Generalizations). As explained in [**95**, §7], Theorem 5.3 holds not only for smooth schemes but also for smooth algebraic Artin spaces.

Roughly speaking, Theorem 5.3 shows that the difference between the localizing $\mathbb{A}^1$-invariants of X and the localizing $\mathbb{A}^1$-invariants of V is completely determined by the subscheme Z. Consult Remark 6.3, resp. 6.5, for the relation between (5.5) and the motivic Gysin triangles constructed by Morel-Voevodsky, resp. Voevodsky.

5.2.1. *Quillen's localization theorem.* Homotopy K-theory is a localizing $\mathbb{A}^1$-invariant which agrees with Quillen's algebraic K-theory when restricted to smooth k-schemes. Therefore, Theorem 5.3 leads to the K-theoretical localization theorem

$$K(Z) \xrightarrow{K(i_*)} K(X) \xrightarrow{K(j^*)} K(V) \xrightarrow{\partial} K(Z)[1] \tag{5.6}$$

originally established by Quillen in [**70**, Chapter 7 §3]. Among other ingredients, Quillen's proof makes essential use of dévissage. The proof of Theorem 5.3, and hence of (5.6), is quite different and avoids the use of dévissage.

5.2.2. *Six-term exact sequence in de Rham cohomology.* Periodic cyclic homology is a localizing $\mathbb{A}^1$-invariant (when $\mathrm{char}(k) = 0$). Moreover, thanks to the Hochschild-Kostant-Rosenberg theorem, we have an isomorphism of $\mathbb{Z}/2$-graded k-vector spaces $HP^\pm(X) \simeq (\bigoplus_{n\,\mathrm{even}} H^n_{dR}(X), \bigoplus_{n\,\mathrm{odd}} H^n_{dR}(X))$, where H^*_{dR} stands for de Rham cohomology. Furthermore, the maps i and j give rise to homomorphisms $H^n_{dR}(i_*)\colon H^n_{dR}(Z) \to H^{n+2c}_{dR}(X)$ and $H^n_{dR}(j^*)\colon H^n_{dR}(X) \to H^n_{dR}(V)$, where c stands for the codimension of i. Therefore, Theorem 5.3 leads to the following six-term exact sequence in de Rham cohomology:

$$\begin{array}{ccccc}
\bigoplus_{n\,\mathrm{even}} H^n_{dR}(Z) & \xrightarrow{\bigoplus_n H^n_{dR}(i_*)} & \bigoplus_{n\,\mathrm{even}} H^n_{dR}(X) & \xrightarrow{\bigoplus_n H^n_{dR}(j^*)} & \bigoplus_{n\,\mathrm{even}} H^n_{dR}(V) \\
{\scriptstyle\partial}\uparrow & & & & \downarrow{\scriptstyle\partial} \\
\bigoplus_{n\,\mathrm{odd}} H^n_{dR}(V) & \xleftarrow[\bigoplus_n H^n_{dR}(j^*)]{} & \bigoplus_{n\,\mathrm{odd}} H^n_{dR}(X) & \xleftarrow[\bigoplus_n H^n_{dR}(i_*)]{} & \bigoplus_{n\,\mathrm{odd}} H^n_{dR}(Z)\,.
\end{array}$$

This exact sequence is the "2-periodization" of the Gysin long exact sequence on de Rham cohomology originally constructed by Hartshorne in [**31**, Chapter II §3].

5.2.3. *Reduction to projective schemes.* As a byproduct of Theorem 5.3, the study of the localizing $\mathbb{A}^1$-invariants of smooth k-schemes can be reduced to the study of the localizing $\mathbb{A}^1$-invariants of smooth *projective* k-schemes:

THEOREM 5.5 ([**95**, Thm. 2.1]). *Let X a smooth k-scheme.*

(i) *If k is of characteristic zero, then $\mathrm{U}(X)$ belongs to the smallest triangulated subcategory of $\mathrm{NMot}(k)$ containing the objects $\mathrm{U}(Y)$, with Y a smooth projective k-scheme.*
(ii) *If k is a perfect field of characteristic $p > 0$, then $\mathrm{U}(X)_{1/p}$ belongs to the smallest thick triangulated subcategory of $\mathrm{NMot}(k)_{1/p}$ containing the objects $\mathrm{U}(Y)_{1/p}$, with Y a smooth projective k-scheme.*

Among other ingredients, the proof of item (i), resp. item (ii), of Theorem 5.5 makes use of resolution of singularities, resp. of Gabber's refined version of de Jong's theory of alterations; consult [**95**, §8] for details.

REMARK 5.6 (Dualizable objects). Given a smooth projective k-scheme Y, the associated dg category $\mathrm{perf}_{\mathrm{dg}}(Y)$ is smooth and proper; see [**81**, Example 1.42]. Therefore, since the universal localizing $\mathbb{A}^1$-invariant U is symmetric monoidal, it follows from [**81**, Thm. 1.43] that $\mathrm{U}(Y)$ is a dualizable object of the symmetric monoidal category $\mathrm{NMot}(k)$. Given a smooth k-scheme X, the associated dg category $\mathrm{perf}_{\mathrm{dg}}(X)$ is smooth but *not* necessarily proper. Nevertheless, Theorem 5.5 implies that $\mathrm{U}(X)$, resp. $\mathrm{U}(X)_{1/n}$, is still a dualizable object of the symmetric monoidal category $\mathrm{NMot}(k)$, resp. $\mathrm{NMot}(k)_{1/p}$.

5.3. Corner skew Laurent polynomial algebras. Let A be a unital k-algebra, e an idempotent of A, and $\phi\colon A \xrightarrow{\sim} eAe$ a "corner" isomorphism. The associated *corner skew Laurent polynomial algebra* $A[t_+, t_-; \phi]$ is defined as follows: the elements are formal expressions $t_-^m a_{-m} + \cdots + t_- a_{-1} + a_0 + a_1 t_+ \cdots + a_n t_+^n$ with $a_{-i} \in \phi^i(1)A$ and $a_i \in A\phi^i(1)$ for every $i \geq 0$; the addition is defined componentwise; the multiplication is determined by the distributive law and by the relations $t_- t_+ = 1$, $t_+ t_- = e$, $at_- = t_-\phi(a)$ for every $a \in A$, and $t_+ a = \phi(a)t_+$ for every $a \in A$. Note that $A[t_+, t_-; \phi]$ admits a canonical $\mathbb{Z}$-grading with $\deg(t_\pm) = \pm 1$. As proved in [**2**, Lem. 2.4], the corner skew Laurent polynomial algebras can be characterized as those $\mathbb{Z}$-graded algebras $C = \bigoplus_{n\in\mathbb{Z}} C_n$ containing elements $t_+ \in C_1$ and $t_- \in C_{-1}$ such that $t_- t_+ = 1$. Concretely, we have $C = A[t_+, t_-; \phi]$ with $A := C_0$, $e := t_+ t_-$, and $\phi\colon C_0 \to t_+ t_- C_0 t_+ t_-$ given by $c_0 \mapsto t_+ c_0 t_-$.

EXAMPLE 5.7 (Skew Laurent polynomial algebras). When $e = 1$, $A[t_+, t_-; \phi]$ reduces to the classical skew Laurent polynomial algebra $A \rtimes_\phi \mathbb{Z}$. In the particular case where ϕ is the identity, $A \rtimes_\phi \mathbb{Z}$ reduces furthermore to $A[t, t^{-1}]$.

EXAMPLE 5.8 (Leavitt algebras). Following [**60**], the *Leavitt algebra L_n, $n \geq 0$*, is the k-algebra generated by elements $x_0, \ldots, x_n, y_0, \ldots, y_n$ subject to the relations $y_i x_j = \delta_{ij}$ and $\sum_{i=0}^n x_i y_i = 1$. Note that the canonical $\mathbb{Z}$-grading, with $\deg(x_i) = 1$ and $\deg(y_i) = -1$, makes L_n into a corner skew Laurent polynomial algebra. Note also that $L_0 \simeq k[t, t^{-1}]$. In the remaining cases $n \geq 1$, L_n is the universal example of a k-algebra of *module type* $(1, n+1)$, *i.e.* $L_n \simeq L_n^{\oplus(n+1)}$ as right L_n-modules.

EXAMPLE 5.9 (Leavitt path algebras). Let $Q = (Q_0, Q_1, s, r)$ be a finite quiver with no sources; Q_0 and Q_1 stand for the sets of vertices and arrows, respectively, and s and r for the source and target maps, respectively. Consider the double quiver $\overline{Q} = (Q_0, Q_1 \cup Q_1^*, s, r)$ obtained from Q by adding an arrow α^*, in the converse direction, for each arrow $\alpha \in Q_1$. The *Leavitt path algebra L_Q of Q* is the quotient of the quiver algebra $k\overline{Q}$ (which is generated by elements $\alpha \in Q_1 \cup Q_1^*$ and e_i with $i \in Q_0$) by the Cuntz-Krieger's relations: $\alpha^*\beta = \delta_{\alpha\beta}e_{r(\alpha)}$ for every $\alpha, \beta \in Q_1$ and $\sum_{\{\alpha \in Q_1 | s(\alpha)=i\}} \alpha\alpha^* = e_i$ for every non-sink $i \in Q_0$. Note that L_Q admits a canonical $\mathbb{Z}$-grading with $\deg(\alpha) = 1$ and $\deg(\alpha^*) = -1$. For every vertex $i \in Q_0$ choose an arrow α_i such that $r(\alpha_i) = i$ and consider the associated elements $t_+ := \sum_{i \in Q_0} \alpha_i$ and $t_- := t_+^*$. Since $\deg(t_\pm) = \pm 1$ and $t_- t_+ = 1$, L_Q is an example of a corner skew Laurent polynomial algebra. In the particular case where Q is the quiver with one vertex and $n+1$ arrows, L_Q reduces to the Leavitt algebra L_n.

THEOREM 5.10 ([**86**, Thm. 3.1]). *We have an induced distinguished triangle*

$$\mathrm{U}(A) \stackrel{\mathrm{id}-\mathrm{U}({}_\phi A)}{\longrightarrow} \mathrm{U}(A) \longrightarrow \mathrm{U}(A[t_+, t_-; \phi]) \stackrel{\partial}{\longrightarrow} \mathrm{U}(A)[1],$$

where ${}_\phi A$ stands for the A-A-bimodule associated to ϕ.

Roughly speaking, Theorem 5.10 shows that $A[t_+, t_-; \phi]$ may be considered as a model for the orbits of the $\mathbb{N}$-action on $\mathrm{U}(A)$ induced by the endomorphism $\mathrm{U}({}_\phi A)$.

5.3.1. *Leavitt path algebras.* Let $Q = (Q_0, Q_1, s, r)$ be a quiver as in Example 5.9, with v vertices and v' sinks. Assume that the set Q_0 is ordered with the first v' elements corresponding to the sinks. Let I'_Q be the incidence matrix of Q, I_Q the matrix obtained from I'_Q by removing the first v' rows (which are zero), and I_Q^t the transpose of I_Q. Under these notations, Theorem 5.10 (concerning the Leavitt path algebra L_Q) admits the following refinement:

THEOREM 5.11 ([**86**, Thm. 3.7]). *We have an induced distinguished triangle:*

$$\mathrm{U}(k)^{(v-v')\oplus} \stackrel{\binom{0}{\mathrm{id}} - I_Q^t}{\longrightarrow} \mathrm{U}(k)^{v\oplus} \longrightarrow \mathrm{U}(L_Q) \stackrel{\partial}{\longrightarrow} \mathrm{U}(k)[1]^{(v-v')\oplus}. \tag{5.7}$$

Roughly speaking, Theorem 5.11 shows that all the information about the localizing $\mathbb{A}^1$-homotopy invariants of Leavitt path algebras L_Q is encoded in the incidence matrix of the quiver Q. As an application, Theorem 5.11 directly leads to the following explicit model for the mod-n Moore construction[10]:

EXAMPLE 5.12 (mod-n Moore construction). Let Q be the quiver with one vertex and $n+1$ arrows. In this particular case, (5.7) reduces to the triangle

$$\mathrm{U}(k) \stackrel{n \cdot \mathrm{id}}{\longrightarrow} \mathrm{U}(k) \longrightarrow \mathrm{U}(L_n) \stackrel{\partial}{\longrightarrow} \mathrm{U}(k)[1].$$

This shows that the Leavitt algebra $L_n, n \geq 2$, is a model for the mod-n Moore object of $\mathrm{U}(k)$. Therefore, since the universal localizing $\mathbb{A}^1$-invariant U is symmetric monoidal, given a small dg category $\mathcal{A}$, we conclude that the tensor product $\mathcal{A} \otimes L_n$ is a model for the mod-n Moore object of $\mathrm{U}(\mathcal{A})$.

[10]Explicit models for the suspension construction, namely the Waldhausen's $S_\bullet$-construction and the Calkin algebra, are described in [**81**, §8.3.2 and §8.4.4].

5.4. Noncommutative projective schemes. Let $A = \bigoplus_{n\geq 0} A_n$ be a $\mathbb{N}$-graded Noetherian k-algebra. In what follows, we assume that A is *connected*, *i.e.* $A_0 = k$, and *locally finite-dimensional*, i.e. $\dim_k(A_n) < \infty$ for every n. Following Manin [**63**], Gabriel [**24**], Artin-Zhang [**3**], and others, the *noncommutative projective scheme* $\mathrm{qgr}(A)$ associated to A is defined as the quotient abelian category $\mathrm{gr}(A)/\mathrm{tors}(A)$, where $\mathrm{gr}(A)$ stands for the abelian category of finitely generated $\mathbb{Z}$-graded (right) A-modules and $\mathrm{tors}(A)$ for the Serre subcategory of torsion A-modules. This definition was motivated by Serre's celebrated result [**73**, Prop. 7.8], which asserts that in the particular case where A is commutative and generated by elements of degree 1 the quotient category $\mathrm{qgr}(A)$ is equivalent to the abelian category of coherent $\mathcal{O}_{\mathrm{Proj}(A)}$-modules $\mathrm{coh}(\mathrm{Proj}(A))$. For example, when A is the polynomial k-algebra $k[x_1, \ldots, x_d]$, with $\deg(x_i) = 1$, we have the equivalence of categories $\mathrm{qgr}(k[x_1, \ldots, x_d]) \simeq \mathrm{coh}(\mathbb{P}^{d-1})$. For a survey on noncommutative projective geometry, we invite the reader to consult Stafford's ICM address [**76**].

Assume that A is Koszul and has finite global dimension d. Under these assumptions, the Hilbert series $h_A(t) := \sum_{n\geq 0} \dim_k(A_n)t^n \in \mathbb{Z}[\![t]\!]$ is invertible and its inverse $h_A(t)^{-1}$ is a polynomial $1 - \beta_1 t + \beta_2 t^2 - \cdots + (-1)^d \beta_d t^d$ of degree d, with β_i the dimension of the k-vector space $\mathrm{Tor}_i^A(k,k)$ (or $\mathrm{Ext}_A^i(k,k)$).

EXAMPLE 5.13 (Quantum polynomial algebras). Choose constants $q_{ij} \in k^\times$ with $1 \leq i < j \leq d$. Following Manin [**64**, §1], the $\mathbb{N}$-graded Noetherian k-algebra

$$A := k\langle x_1, \ldots, x_d\rangle / \langle x_j x_i - q_{ij} x_i x_j \,|\, 1 \leq i < j \leq d\rangle\,,$$

with $\deg(x_i) = 1$, is called the *quantum polynomial algebra* associated to q_{ij}. This algebra is Koszul, has global dimension d, and $h_A(t)^{-1} = (1-t)^d$.

EXAMPLE 5.14 (Quantum matrix algebras). Choose a constant $q \in k^\times$. Following Manin [**64**, §1], the $\mathbb{N}$-graded Noetherian k-algebra A defined as the quotient of $k\langle x_1, x_2, x_3, x_4\rangle$ by the following relations

$$\begin{array}{lll} x_1x_2 = qx_2x_1 & x_1x_3 = qx_3x_1 & x_1x_4 - x_4x_1 = (q-q^{-1})x_2x_3 \\ x_2x_3 = x_3x_2 & x_2x_4 = qx_4x_2 & x_1x_4 = qx_4x_3\,, \end{array}$$

with $\deg(x_i) = 1$, is called the *quantum matrix algebra* associated to q. This algebra is Koszul, has global dimension 4, and $h_A(t)^{-1} = (1-t)^4$.

EXAMPLE 5.15 (Sklyanin algebras). Let C be a smooth elliptic curve, σ an automorphism of C given by translation under the group law, and $\mathcal{L}$ a line bundle on C of degree $d \geq 3$. We write $\Gamma_\sigma \subset C \times C$ for the graph of σ and W for the d-dimensional k-vector space $H^0(C, \mathcal{L})$. Following Feigin-Odesskii [**22**] and Tate-Van den Bergh [**99**, §1], the $\mathbb{N}$-graded Noetherian k-algebra $A := T(W)/R$, where

$$R := H^0(C \times C, (\mathcal{L} \boxtimes \mathcal{L})(-\Gamma_\sigma)) \subset H^0(C \times C, \mathcal{L} \boxtimes \mathcal{L}) = W \otimes W\,,$$

is called the *Sklyanin algebra* associated to the triple $(C, \sigma, \mathcal{L})$. This algebra is Koszul, has global dimension d, and $h_A(t)^{-1} = (1-t)^d$.

EXAMPLE 5.16 (Homogenized enveloping algebras). Let $\mathfrak{g}$ be a finite-dimensional Lie algebra. Following Smith [**75**, §12], the $\mathbb{N}$-graded Noetherian k-algebra

$$A := T(\mathfrak{g} \oplus kz)/\langle \{z \otimes x - x \otimes z \,|\, x \in \mathfrak{g}\} \cup \{x \otimes y - y \otimes x - [x,y] \otimes z \,|\, x, y \in \mathfrak{g}\}\rangle\,,$$

is called the *homogenized enveloping algebra* of $\mathfrak{g}$. This algebra is Koszul, has global dimension $d := \dim(\mathfrak{g}) + 1$, and $h_A(t)^{-1} = (1-t)^d$.

Given a $\mathbb{N}$-graded k-algebra A as above, let us write $\mathcal{D}^b_{\mathrm{dg}}(\mathrm{qgr}(A))$ for the canonical dg enhancement of the bounded derived category of $\mathrm{qgr}(A)$. This dg category is, in general, *not* proper; see [**89**, §1]. The following result is contained [**89**, Thm. 1.2]:

THEOREM 5.17. *We have an induced distinguished triangle*

$$\bigoplus_{-\infty}^{+\infty} \mathrm{U}(k) \xrightarrow{\mathrm{M}} \bigoplus_{-\infty}^{+\infty} \mathrm{U}(k) \longrightarrow \mathrm{U}(\mathcal{D}^b_{\mathrm{dg}}(\mathrm{qgr}(A))) \xrightarrow{\partial} \bigoplus_{-\infty}^{+\infty} \mathrm{U}(k)[1]\,, \tag{5.8}$$

where M *stands for the (infinite) matrix* $\mathrm{M}_{ij} := (-1)^j(-1)^{(i-j)}\beta_{i-j}$. *Moreover, when* $\beta_d = 1$, *the triangle* (5.8) *induces an isomorphism* $\mathrm{U}(\mathcal{D}^b_{\mathrm{dg}}(\mathrm{qgr}(A))) \simeq \mathrm{U}(k)^{\oplus d}$.

As proved in [**78**, Cor. 0.2], we have $h_A(t)^{-1} = (1-t)^3$ whenever $d = 3$.

REMARK 5.18 (Localizing invariants). The proof of Theorem 5.17 does *not* makes use of $\mathbb{A}^1$-homotopy invariance. Consequently, as explained in [**81**, Thm. 8.5], Theorem 5.17 holds similarly for every localizing invariant in the sense of [**81**, Def. 8.3]. Examples of localizing invariants which are *not* $\mathbb{A}^1$-homotopy invariant include nonconnective algebraic K-theory, Hochschild homology, cyclic homology, negative cyclic homology, periodic cyclic homology (when $\mathrm{char}(k) = p > 0$), topological Hochschild homology, topological cyclic homology, topological periodic cyclic homology, etc.

Roughly speaking, Theorem 5.17 (and Remark 5.18) shows that the localizing invariants of a noncommutative projective scheme $\mathrm{qgr}(A)$ are completely determined by the Hilbert series $h_A(t)$.

6. Noncommutative mixed motives

In this section we assume that the base field k is perfect. Kontsevich introduced in [**50, 51, 53**] a certain rigid symmetric monoidal triangulated category of noncommutative mixed motives $\mathrm{NMix}(k)$. As explained in [**81**, §9.1.1], this category can be (conceptually) described as the smallest thick triangulated subcategory of $\mathrm{NMot}(k)$ (see §5) containing the objects $\mathrm{U}(\mathcal{A})$, with $\mathcal{A}$ smooth and proper.

In §6.1 we compute the Picard group of the thick triangulated subcategory of $\mathrm{NMix}(k)$ generated by the noncommutative mixed motives of central simple k-algebras. Subsections §6.2-6.4 are devoted to the precise relation between the category $\mathrm{NMix}(k)$ and Morel-Voevodsky's stable $\mathbb{A}^1$-homotopy category, Voevodsky's triangulated category of geometric mixed motives, and Levine's triangulated category of mixed motives, respectively. In §6.5 we address the Schur-finiteness conjecture in the case of quadric fibrations. Finally, subsection §6.6 is devoted to the rigidity property of the category of mod-n noncommutative mixed motives.

6.1. Picard group. The computation of the Picard group of the category of noncommutative mixed motives is a major challenge which seems completely out of reach at the present time. However, this major challenge can be met if we restrict ourselves to central simple k-algebras. Let $\mathrm{NMix}_{\mathrm{csa}}(k)$ be the thick triangulated subcategory of $\mathrm{NMix}(k)$ generated by the noncommutative mixed motives $\mathrm{U}(A)$ of central simple k-algebras A. Similarly to §2.2, $[A] = [B] \Leftrightarrow \mathrm{U}(A) \simeq \mathrm{U}(B)$ for any two central simple k-algebras A and B. Moreover, as proved in [**81**, Thm. 8.28], we have the following non-trivial Ext-groups:

$$\mathrm{Hom}_{\mathrm{NMix}(k)}(\mathrm{U}(A), \mathrm{U}(B)[-n]) \simeq K_n(A^{\mathrm{op}} \otimes B) \qquad n \in \mathbb{Z}\,. \tag{6.1}$$

This shows that $\mathrm{NMix}_{\mathrm{csa}}(k)$ contains information not only about the Brauer group $\mathrm{Br}(k)$ but also about all the higher algebraic K-theory of central simple k-algebras.

THEOREM 6.1 ([**14**, Thm. 2.22]). *We have the following isomorphism:*

$$\mathrm{Br}(k) \times \mathbb{Z} \xrightarrow{\sim} \mathrm{Pic}(\mathrm{NMix}_{\mathrm{csa}}(k)) \qquad ([A], n) \mapsto \mathrm{U}(A)[n] .$$

Roughly speaking, Theorem 6.1 shows that, although the category $\mathrm{NMix}_{\mathrm{csa}}(k)$ contains information about all the higher algebraic K-theory of central simple k-algebras, none of the noncommutative mixed motives which are built using the non-trivial Ext-groups (6.1) is $\otimes$-invertible.

6.2. Morel-Voevodsky's motivic category. Morel-Voevodsky introduced in [**66**, **105**] the stable $\mathbb{A}^1$-homotopy category of $(\mathbb{P}^1, \infty)$-spectra $\mathrm{SH}(k)$. By construction, we have a symmetric monoidal functor $\Sigma^\infty(-_+)\colon \mathrm{Sm}(k) \to \mathrm{SH}(k)$ defined on smooth k-schemes. Let $\mathrm{KGL} \in \mathrm{SH}(k)$ be the ring $(\mathbb{P}^1, \infty)$-spectrum representing homotopy K-theory and $\mathrm{Mod}(\mathrm{KGL})$ the homotopy category of KGL-modules.

THEOREM 6.2. (i) *If* $\mathrm{char}(k) = 0$, *then there exists a fully-faithful, symmetric monoidal, triangulated functor* Ψ *making the diagram commute*

$$\begin{array}{ccccc}
\mathrm{Sm}(k) & & \xrightarrow{X \mapsto \mathrm{perf}_{\mathrm{dg}}(X)} & & \mathrm{dgcat}(k) \\
{\scriptstyle \Sigma^\infty(-_+)}\downarrow & \searrow & & & \downarrow{\scriptstyle \mathrm{U}} \\
\mathrm{SH}(k) & & \mathrm{NMix}(k) & \longrightarrow & \mathrm{NMot}(k) \\
{\scriptstyle -\wedge \mathrm{KGL}}\downarrow & & {\scriptstyle (-)^\vee}\downarrow & & \downarrow{\scriptstyle \underline{\mathrm{Hom}}(-,\mathrm{U}(k))} \\
\mathrm{Mod}(\mathrm{KGL}) & \xrightarrow[\Psi]{} & \mathrm{NMix}(k)^\oplus & \longrightarrow & \mathrm{NMot}(k) ,
\end{array} \tag{6.2}$$

where $\underline{\mathrm{Hom}}(-,-)$ *stands for the internal-Hom of the closed symmetric monoidal category* $\mathrm{NMot}(k)$, $(-)^\vee$ *for the (contravariant) duality functor, and* $\mathrm{NMix}(k)^\oplus$ *for the smallest triangulated subcategory of* $\mathrm{NMot}(k)$ *which contains* $\mathrm{NMix}(k)$ *and is stable under arbitrary direct sums.*

(ii) *If* $\mathrm{char}(k) = p > 0$, *then there exists a* $\mathbb{Z}[1/p]$*-linear, fully-faithful, symmetric monoidal, triangulated functor* $\Psi_{1/p}$ *making the diagram commute:*

$$\begin{array}{ccccc}
\mathrm{Sm}(k) & & \xrightarrow{X \mapsto \mathrm{perf}_{\mathrm{dg}}(X)} & & \mathrm{dgcat}(k) \\
{\scriptstyle \Sigma^\infty(-_+)_{1/p}}\downarrow & \searrow & & & \downarrow{\scriptstyle \mathrm{U}(-)_{1/p}} \\
\mathrm{SH}(k)_{1/p} & & \mathrm{NMix}(k)_{1/p} & \longrightarrow & \mathrm{NMot}(k)_{1/p} \\
{\scriptstyle -\wedge \mathrm{KGL}_{1/p}}\downarrow & & {\scriptstyle (-)^\vee}\downarrow & & \downarrow{\scriptstyle \underline{\mathrm{Hom}}(-,\mathrm{U}(k)_{1/p})} \\
\mathrm{Mod}(\mathrm{KGL}_{1/p}) & \xrightarrow[\Psi_{1/p}]{} & \mathrm{NMix}(k)^\oplus_{1/p} & \longrightarrow & \mathrm{NMot}(k)_{1/p} .
\end{array}$$

Intuitively speaking, Theorem 6.2 shows that as soon as we pass to KGL-modules, the commutative world embeds fully-faithfully into the noncommutative world. Consult [**81**, §9.4], and the references therein, for the construction of the two outer commutative diagrams. The inner commutative squares follow from the combination of Theorem 5.5 with Remark 5.6; consult [**95**, Thm. 3.1] for details.

REMARK 6.3 (Morel-Voevodsky's motivic Gysin triangle). Let X be a smooth k-scheme, $Z \hookrightarrow X$ a smooth closed subscheme with normal vector bundle N, and $j\colon V \hookrightarrow X$ the open complement of Z. Making use of homotopy purity, Morel-Voevodsky constructed in [**66**, §3.2][**105**, §4] a motivic Gysin triangle

$$\Sigma^\infty(V_+) \xrightarrow{\Sigma^\infty(j_+)} \Sigma^\infty(X_+) \longrightarrow \Sigma^\infty(\mathrm{Th}(N)) \xrightarrow{\partial} \Sigma^\infty(V_+)[1] \tag{6.3}$$

in $\mathrm{SH}(k)$, where $\mathrm{Th}(N)$ stands for the Thom space of N. Since homotopy K-theory is an orientable and periodic cohomology theory, $\Sigma^\infty(\mathrm{Th}(N))\wedge \mathrm{KGL}$ is isomorphic to $\Sigma^\infty(Z_+)\wedge\mathrm{KGL}$. Using the commutative diagram (6.2), we hence observe that the image of (6.3) under the composed functor $\Psi\circ(-\wedge\mathrm{KGL})\colon \mathrm{SH}(k)\to \mathrm{NMix}(k)^\oplus$ agrees with the dual of the Gysin triangle (5.5). In other words, the Gysin triangle (5.5) is the dual of the "KGL-linearization" of (6.3).

6.3. Voevodsky's motivic category. Voevodsky introduced in [**104**, §2] the triangulated category of geometric mixed motives $\mathrm{DM}_{\mathrm{gm}}(k)$. By construction, it comes equipped with a symmetric monoidal functor $M\colon \mathrm{Sm}(k)\to \mathrm{DM}_{\mathrm{gm}}(k)$ and is the natural setting for the study of algebraic cycle (co)homology theories such as higher Chow groups, Suslin homology, motivic cohomology, etc.

THEOREM 6.4. *There exists a $\mathbb{Q}$-linear, fully-faithful, symmetric monoidal functor $\Phi_{\mathbb{Q}}$ making the following diagram commute:*

$$\begin{array}{ccccc}
\mathrm{Sm}(k) & & \xrightarrow{X\mapsto \mathrm{perf}_{\mathrm{dg}}(X)} & & \mathrm{dgcat}(k) \\
{\scriptstyle M(-)_{\mathbb{Q}}}\downarrow & \searrow & & & \downarrow{\scriptstyle \mathrm{U}(-)_{\mathbb{Q}}} \\
\mathrm{DM}_{\mathrm{gm}}(k)_{\mathbb{Q}} & & \mathrm{NMix}(k)_{\mathbb{Q}} & \longrightarrow & \mathrm{NMot}(k)_{\mathbb{Q}} \\
\downarrow & & {\scriptstyle (-)^\vee}\downarrow & & \downarrow{\scriptstyle \underline{\mathrm{Hom}}(-,\mathrm{U}(k)_{\mathbb{Q}})} \\
\mathrm{DM}_{\mathrm{gm}}(k)_{\mathbb{Q}}/_{-\otimes\mathbb{Q}(1)[2]} & \xrightarrow[\Phi_{\mathbb{Q}}]{} & \mathrm{NMix}(k)_{\mathbb{Q}} & \longrightarrow & \mathrm{NMot}(k)_{\mathbb{Q}}\,.
\end{array} \tag{6.4}$$

Intuitively speaking, Theorem 6.4 shows that as soon as we "$\otimes$-trivialize" the Tate motive $\mathbb{Q}(1)[2]$, the commutative world embedds fully-faithfully into the noncommutative world. Consult [**81**, §9.5], and the references therein, for the construction of the outer commutative diagram. The inner commutative square follows from the combination of Theorem 5.5 with Remark 5.6; consult [**95**, Thm. 3.7].

REMARK 6.5 (Voevodsky's motivic Gysin triangle). Let X be a smooth k-scheme, $Z \hookrightarrow X$ a smooth closed subscheme of codimension c, and $j\colon V\hookrightarrow X$ the open complement of Z. Making use of deformation to the normal cone, Voevodsky constructed in [**104**, §2] a motivic Gysin triangle

$$M(V)_{\mathbb{Q}} \xrightarrow{M(j)_{\mathbb{Q}}} M(X)_{\mathbb{Q}} \longrightarrow M(Z)_{\mathbb{Q}}(c)[2c] \xrightarrow{\partial} M(V)_{\mathbb{Q}}[1] \tag{6.5}$$

in $\mathrm{DM}_{\mathrm{gm}}(k)_{\mathbb{Q}}$. Using the commutative diagram (6.4), we observe that the image of (6.5) under the (composed) functor $\Phi_{\mathbb{Q}}\colon \mathrm{DM}_{\mathrm{gm}}(k)_{\mathbb{Q}}\to\mathrm{NMix}(k)_{\mathbb{Q}}$ agrees with the dual of the rationalized Gysin triangle (5.5). In other words, the rationalized Gysin triangle (5.5) is the dual of the "Tate $\otimes$-trivialization" of (6.5).

Let $\mathrm{DM}^{\mathrm{et}}_{\mathrm{gm}}(k)$ be the étale variant of $\mathrm{DM}_{\mathrm{gm}}(k)$ introduced by Voevodsky in [**104**, §3.3]. As proved in *loc. cit.*, $\mathrm{DM}_{\mathrm{gm}}(k)_{\mathbb{Q}}$ is equivalent to $\mathrm{DM}^{\mathrm{et}}_{\mathrm{gm}}(k)_{\mathbb{Q}}$. Consequently, Theorem 6.4 leads to the following result (see [**95**, Thm. 3.13]):

COROLLARY 6.6 (Étale descent). The presheaf of noncommutative mixed motives $\mathrm{Sm}(k)^{\mathrm{op}} \to \mathrm{NMot}(k)_{\mathbb{Q}}, X \mapsto \mathrm{U}(X)_{\mathbb{Q}}$, satisfies *étale descent*, i.e. for every étale cover $\mathcal{V} = \{V_i \to X\}_{i\in I}$ of X, we have an isomorphism $\mathrm{U}(X)_{\mathbb{Q}} \simeq \mathrm{holim}_{n\geq 0}\mathrm{U}(\check{\mathrm{C}}_n\mathcal{V})_{\mathbb{Q}}$, where $\check{\mathrm{C}}_\bullet\mathcal{V}$ stands for the Čech simplicial k-scheme associated to the cover $\mathcal{V}$.

6.4. Levine's motivic category. Levine introduced in [**61**, Part I] a certain triangulated category of mixed motives $\mathcal{DM}(k)$ and a (contravariant) symmetric monoidal functor $h\colon \mathrm{Sm}(k) \to \mathcal{DM}(k)$. As proved in [**34**, Thm. 4.2], the assignment $h(X)_{\mathbb{Q}}(n) \mapsto \underline{\mathrm{Hom}}(M(X), \mathbb{Q}(n))$ gives rise to an equivalence of categories $\mathcal{DM}(k)_{\mathbb{Q}} \to \mathrm{DM}_{\mathrm{gm}}(k)_{\mathbb{Q}}$ whose precomposition with $h(-)_{\mathbb{Q}}$ agrees with $X \mapsto M(X)^{\vee}_{\mathbb{Q}}$. Consequently, thanks to Theorem 6.4, there exists a $\mathbb{Q}$-linear, fully-faithful, symmetric monoidal functor $\Phi_{\mathbb{Q}}$ making the following diagram commute:

$$
\begin{array}{ccccc}
\mathrm{Sm}(k) & & \xrightarrow{X\mapsto \mathrm{perf}_{\mathrm{dg}}(X)} & & \mathrm{dgcat}(k) \\
\downarrow{\scriptstyle h(-)_{\mathbb{Q}}} & & & & \\
\mathcal{DM}(k)_{\mathbb{Q}} & & & & \downarrow{\scriptstyle \mathrm{U}(-)_{\mathbb{Q}}} \\
\downarrow & & & & \\
\mathcal{DM}(k)_{\mathbb{Q}}/_{-\otimes\mathbb{Q}(1)[2]} & \xrightarrow[\Phi_{\mathbb{Q}}]{} & \mathrm{NMix}(k)_{\mathbb{Q}} & \longrightarrow & \mathrm{NMot}(k)_{\mathbb{Q}}\,.
\end{array}
\tag{6.7}
$$

Note that in contrast with the diagrams of Theorems 6.2 and 6.4, the commutative diagram (6.7) does *not* uses any kind of duality functor.

6.5. Schur-finiteness conjecture. Given a smooth k-scheme X, the Schur-finiteness conjecture, denoted by $S(X)$, asserts that the mixed motive $M(X)_{\mathbb{Q}}$ is Schur-finite in the sense of Deligne [**17**, §1]. Thanks to the (independent) work of Guletskii [**27**] and Mazza [**65**], the conjecture $\mathrm{S}(X)$ holds when $\dim(X) \leq 1$, and also for abelian varieties. In addition to these cases, it remains wide open.

THEOREM 6.6 ([**85**, Thm. 1.1]). *Let $q\colon Q \to B$ a flat quadric fibration of relative dimension $d-2$. Assume that B and Q are k-smooth and that q has only* simple degenerations, i.e. *that all the fibers of q have corank ≤ 1 and that the locus $D \subset B$ of the critical values of q is k-smooth. Under these assumptions, the following holds:*

(i) *If d is even, then we have the following equivalence $S(B)+S(\widetilde{B}) \Leftrightarrow S(Q)$, where $\widetilde{B}$ stands for the discriminant 2-fold cover of B (ramified over D).*
(ii) *If d is odd and* $\mathrm{char}(k) \neq 2$, *then we have the following implication $\{S(V_i)\} + \{S(\widetilde{D}_i)\} \Rightarrow S(Q)$, where V_i is any affine open of B and $\widetilde{D}_i$ is any Galois 2-fold cover of $D_i := D \cap V_i$.*

Roughly speaking, Theorem 6.6 relates the Schur-finiteness conjecture for the total space Q with the Schur-finiteness conjecture for certain coverings/subschemes of the base B. Among other ingredients, its proof makes use of Theorem 6.4 and of the twisted analogue of Theorem 1.4 (see §1.2.1). Theorem 6.6 enables the proof of the Schur-finiteness conjecture in the following new cases:

COROLLARY 6.8 ([**92**, Cor. 1.3 and 1.5]). Let $q\colon Q \to B$ be as in Theorem 6.6.

(i) Assume that B is a curve, and that $\mathrm{char}(k) \neq 2$ when d is odd. Under these assumptions, the conjecture $S(Q)$ holds.

(ii) Assume that B is a surface, that d is odd, that $\mathrm{char}(k) \neq 2$, and that the conjecture $S(B)$ holds. Under these assumptions, $S(Q)$ also holds.

Corollary 6.8(ii) can be applied, for example, to the case where B is an open subscheme of an abelian surface or smooth projective surface with $p_g = 0$ satisfying Bloch's conjecture (see Guletskii-Pedrini [**28**, §4 Thm. 7]). Recall that Bloch's conjecture holds for surfaces not of general type (see Bloch-Kas-Leiberman [**10**]), for surfaces which are rationally dominated by a product of curves (see Kimura [**46**]), for Godeaux, Catanese and Barlow surfaces (see Voisin [**107**, **109**]), etc.

REMARK 6.7 (Bass-finiteness conjecture). Let $k = \mathbb{F}_q$ be a finite field and X a smooth k-scheme. The Bass-finiteness conjecture (see [**5**, §9]) asserts that the algebraic K-theory groups $K_n(X), n \geq 0$, are finitely generated. Thanks to the work of Quillen [**29**, **69**, **71**], the Bass-finiteness conjecture holds when $\dim(X) \leq 1$.

In the same vein, we can consider the *mod 2-torsion* Bass-finiteness conjecture, where $K_n(X)$ is replaced by $K_n(X)_{1/2}$. As proved in [**92**], Theorem 6.6 and Corollary 6.8 hold similarly with the Schur-finiteness conjecture replaced by the mod 2-torsion Bass-finiteness conjecture. As a consequence, we obtain a proof of the (mod 2-torsion) Bass-finiteness conjecture in new cases.

6.6. Rigidity. Given an integer $n \geq 2$, recall from [**81**, §9.9] the definition of the category of mod-n noncommutative mixed motives $\mathrm{NMix}(k; \mathbb{Z}/n)$. By construction, given smooth proper dg categories $\mathcal{A}$ and $\mathcal{B}$, we have isomorphisms

$$\mathrm{Hom}_{\mathrm{NMix}(k;\mathbb{Z}/n)}(\mathrm{U}(\mathcal{A}), \mathrm{U}(\mathcal{B})[-n]) \simeq K_n(\mathcal{A}^{\mathrm{op}} \otimes \mathcal{B}; \mathbb{Z}/n) \quad n \in \mathbb{Z}\,, \tag{6.9}$$

where the right-hand side stands for mod-n algebraic K-theory.

THEOREM 6.8 ([**90**, Thm. 2.1(ii)]). *Given an extension of separably closed fields l/k, the base-change functor $-\otimes_k l\colon \mathrm{NMix}(k; \mathbb{Z}/n) \to \mathrm{NMix}(l; \mathbb{Z}/n)$ is fully-faithful whenever n is coprime to the characteristic of k.*

Theorem 6.8 is the mixed analogue of Theorem 2.7. Intuitively speaking, it shows that the theory of mod-n noncommutative mixed motives is "rigid" under extensions of separably closed fields. Alternatively, thanks to the isomorphisms (6.9), Theorem 6.8 shows that mod-n algebraic K-theory is "rigid" under extensions of separably closed fields. This is a far-reaching noncommutative generalization of Suslin's celebrated rigidity theorem [**79**]; consult [**90**, §2] for details and also for applications to equivariant and twisted algebraic K-theory. In the particular case of an extension of algebraically closed fields, the commutative counterpart of Theorem 6.8 was established by Haesemeyer-Hornbostel in [**30**, Thm. 30].

7. Noncommutative realizations and periods

In this section we assume that the base field k is perfect. Subsection §7.1 is devoted to the noncommutative realizations associated to the (classical) cohomology theories of schemes. In §7.2, making use of the noncommutative realization associated to de Rham-Betti cohomology theory, we extend Grothendieck's theory of periods to the broad noncommutative setting of dg categories. As an application, we prove that (modulo $2\pi i$) Grothendieck's theory of periods is invariant under homological projective duality.

7.1. Noncommutative realizations. Let F be a field of characteristic zero and $(\mathcal{C}, \otimes, \mathbf{1})$ an F-linear neutral Tannakian category equipped with a $\otimes$-invertible "Tate" object $\mathbf{1}(1)$. In what follows, we write $\mathrm{Gal}(\mathcal{C})$ for the Tannakian Galois group of $\mathcal{C}$ and $\mathrm{Gal}_0(\mathcal{C})$ for the kernel of the homomorphism $\mathrm{Gal}(\mathcal{C}) \twoheadrightarrow \mathbb{G}_m$, where $\mathbb{G}_m$ agrees with the Tannakian Galois group of the smallest Tannakian subcategory of $\mathcal{C}$ containing $\mathbf{1}(1)$. As explained in [**83**, §1-2], given a cohomology theory $H^*\colon \mathrm{Sm}(k) \to \mathrm{Gr}^b_{\mathbb{Z}}(\mathcal{C})$, we can consider the associated *modified* cohomology theory:

$$H^{\frac{*}{2}}\colon \mathrm{Sm}(k) \longrightarrow \mathrm{Rep}_{\mathbb{Z}/2}(\mathrm{Gal}_0(\mathcal{C})) \qquad X \mapsto \Big(\bigoplus_{n \text{ even}} H^n(X), \bigoplus_{n \text{ odd}} H^n(X)\Big)$$

with values in the category of finite-dimensional $\mathbb{Z}/2$-graded continuous representations of $\mathrm{Gal}_0(\mathcal{C})$. Examples of cohomology theories include Nori's cohomology theory H^*_N (with values in Nori's Tannakian category of mixed motives [**33**, §8]), Jannsen's cohomology theory H^*_J (with values in Jannsen's Tannakian category of mixed motives [**37**, Part I]), de Rham cohomology theory H^*_{dR} (with values in the Tannakian category of finite-dimensional k-vector spaces), Betti cohomology theory H^*_B (with values in the Tannakian category finite-dimensional $\mathbb{Q}$-vector spaces), de Rham-Betti cohomology theory H^*_{dRB} (with values in the Tannakian category $\mathrm{Vect}(k, \mathbb{Q})$ of triples (V, W, ω), where V is a finite-dimensional k-vector space, W is a finite-dimensional $\mathbb{Q}$-vector space, and ω is an isomorphism $V \otimes_k \mathbb{C} \simeq W \otimes_{\mathbb{Q}} \mathbb{C}$), étale l-adic cohomology theory $H^*_{l\text{-adic}}$ (with values in the Tannakian category of finite-dimensional l-adic representations of the absolute Galois group of k), Hodge cohomology theory H^*_{Hod} (with values in the Tannakian category of mixed $\mathbb{Q}$-Hodge structures [**77**, §1]), etc; consult [**83**, §2] for details and for further examples. Each one of these cohomology theories gives rise to a modified cohomology theory.

The (proof of the) next result is contained in [**83**, Thms. 1.2, 2.2 and Prop. 3.1]:

THEOREM 7.1. *Given H^* as above, there exists an additive invariant*

$$H^{\frac{*}{2}}_{\mathrm{nc}}\colon \mathrm{dgcat}(k) \longrightarrow \mathrm{Ind}(\mathrm{Rep}_{\mathbb{Z}/2}(\mathrm{Gal}_0(\mathcal{C}))), \tag{7.1}$$

with values in the category of ind-objects, such that $H^{\frac{*}{2}}_{\mathrm{nc}}(\mathrm{perf}_{\mathrm{dg}}(X)) \simeq H^{\frac{*}{2}}(X)$ for every smooth k-scheme X.

The additive invariant (7.1) is called the *noncommutative realization* associated to the cohomology theory H^*. Morally speaking, Theorem 7.1 shows that the modified cohomology theories belong not only to the realm of algebraic geometry but also to the broad noncommutative setting of dg categories. This insight goes back to Kontsevich's definition of noncommutative étale cohomology theory; see [**52**]. Among other ingredients, the proof of Theorem 7.1 makes use of Theorem 6.4.

REMARK 7.2 (Generalizations).
(i) In the case where k is of characteristic zero, $\mathrm{Sm}(k)$ can be replaced by the category of k-schemes.
(ii) By construction, (7.1) can be promoted to a localizing invariant.

The following result describes the behavior of the noncommutative realizations with respect to sheaves of differential operators in characteristic zero[11]:

[11] Consult [**81**, Example 2.20] for the description of the behavior of *all* additive invariants with respect to sheaves of differential operators in positive characteristic.

THEOREM 7.3 ([**83**, Thm. 3.4]). *Let k be a field of characteristic zero, X a smooth k-scheme, and $\mathcal{D}_X$ the sheaf of differential operators on X. Given a cohomology theory H^*, we have an isomorphism $H_{\mathrm{nc}}^{\frac{*}{2}}(\mathrm{perf}_{\mathrm{dg}}(\mathcal{D}_X)) \simeq H^{\frac{*}{2}}(X)$.*

EXAMPLE 7.4 (Lie algebras). Let G be a connected semisimple algebraic $\mathbb{C}$-group, B a Borel subgroup of G, $\mathfrak{g}$ the Lie algebra of G, and $U_{\mathrm{ev}}(\mathfrak{g})/I$ the quotient of the universal enveloping algebra of $\mathfrak{g}$ by the kernel of the trivial character. Thanks to Beilinson-Bernstein's celebrated "localisation" theorem [**6**], it follows from Theorem 7.3 that $H_{\mathrm{nc}}^{\frac{*}{2}}(U_{\mathrm{ev}}(\mathfrak{g})/I) \simeq H_{\mathrm{nc}}^{\frac{*}{2}}(\mathrm{perf}_{\mathrm{dg}}(\mathcal{D}_{G/B})) \simeq H^{\frac{*}{2}}(G/B)$.

REMARK 7.5. Theorem 7.3 does *not* holds for every additive invariant. For example, in the case of Hochschild homology we have $HH_*(\mathrm{perf}_{\mathrm{dg}}(\mathcal{D}_X)) \simeq H_{dR}^{2d-*}(X)$ for every smooth affine k-scheme X of dimension d; see Wodzicki [**111**, Thm. 2]. Since $H_{dR}^{2d}(X) = 0$, this implies that $HH(\mathrm{perf}_{\mathrm{dg}}(\mathcal{D}_X)) \not\simeq HH(X)$. More generally, we have $HH(\mathrm{perf}_{\mathrm{dg}}(\mathcal{D}_X)) \not\simeq HH(A)$ for every *commutative* k-algebra A.

7.2. Periods. In this subsection we assume that the base field is endowed with an embedding $k \hookrightarrow \mathbb{C}$. Consider the $\mathbb{Z}$-graded $\mathbb{C}$-algebra of Laurent polynomials $\mathbb{C}[t, t^{-1}]$ with t of degree 1. Given a triple $(V, W, \omega) \in \mathrm{Vect}(k, \mathbb{Q})$, let us write $\mathcal{P}(V, W, \omega) \subseteq \mathbb{C}$ for the subset of entries of the matrix representations of ω (with respect to basis of V and W). In the same vein, given an object $\{(V_n, W_n, \omega_n)\}_{n\in\mathbb{Z}}$ of the category $\mathrm{Gr}_{\mathbb{Z}}^b(\mathrm{Vect}(k, \mathbb{Q}))$, let us write $\mathcal{P}(\{(V_n, W_n, \omega_n)\}_{n\in\mathbb{Z}})$ for the $\mathbb{Z}$-graded k-subalgebra of $\mathbb{C}[t, t^{-1}]$ generated in degree n by the elements of the set $P(W_n, W_n, \omega_n)$. In the case of a smooth k-scheme X, $\mathcal{P}(X) := \mathcal{P}(H_{dRB}^*(X))$ is called the ($\mathbb{Z}$-graded) *algebra of periods of* X. This algebra, originally introduced by Grothendieck in the sixties, plays a key role in the study of transcendental numbers; consult, for example, the work of Kontsevich-Zagier [**54**].

Consider the quotient homomorphism $\phi\colon \mathbb{C}[t, t^{-1}] \twoheadrightarrow \mathbb{C}[t, t^{-1}]/\langle 1 - (2\pi i)t^2\rangle$. The next result extends Grothendieck's theory of periods to dg categories:

THEOREM 7.6 ([**83**, Thm. 4.1]). *There exists an assignment $\mathcal{A} \mapsto \mathcal{P}_{\mathrm{nc}}(\mathcal{A})$, with $\mathcal{P}_{\mathrm{nc}}(\mathcal{A})$ a $\mathbb{Z}/2$-graded k-subalgebra of $\mathbb{C}[t, t^{-1}]/\langle 1-(2\pi i)t^2\rangle$, such that $\mathcal{P}_{\mathrm{nc}}(\mathrm{perf}_{\mathrm{dg}}(X))$ is isomorphic to $\phi(\mathcal{P}(X))$ for every smooth k-scheme X.*

Morally speaking, Theorem 7.6 shows that Grothendieck's theory of periods can be extended from schemes to the broad noncommutative setting of dg categories as long as we work modulo $2\pi i$. Among other ingredients, its proof makes use of the noncommutative realization associated to de Rham-Betti cohomology theory.

7.2.1. *Homological projective duality.* Let X and Y be two HP-dual smooth projective k-schemes as in §3.2.1. Recall from *loc. cit.* that the category $\mathrm{perf}(X)$ admits, in particular, a Lefschetz decomposition $\langle \mathbb{A}_0, \mathbb{A}_1(1), \ldots, \mathbb{A}_{i-1}(i-1)\rangle$. Given a linear subspace $L \subset W^*$, consider the linear sections $X_L := X \times_{\mathbb{P}(W^*)} \mathbb{P}(L)$ and $Y_L := Y \times_{\mathbb{P}(W)} \mathbb{P}(L^\perp)$. The next result, proved in [**83**, Thm. 4.6], relates the algebra of periods of X_L with the algebra of periods of Y_L:

THEOREM 7.7 (HPD-invariance). *Let X and Y be as above. Assume that X_L and Y_L are smooth, that $\dim(X_L) = \dim(X) - \dim(L)$, that $\dim(Y_L) = \dim(Y) - \dim(L^\perp)$, and that the category $\mathbb{A}_0$ admits a full exceptional collection. Under these assumptions, the $\mathbb{Z}/2$-graded k-algebras $\phi(\mathcal{P}(X_L))$ and $\phi(\mathcal{P}(Y_L))$ are isomorphic.*

Roughly speaking, Theorem 7.7 shows that (modulo $2\pi i$) Grothendieck's theory of periods is invariant under homological projective duality. This result can be applied, for example, to the Veronese-Clifford duality, to the Spinor duality, to the Grassmannian-Pfaffian duality, to the Determinantal duality, etc.

References

[1] Yves André, *Une introduction aux motifs (motifs purs, motifs mixtes, périodes)* (French, with English and French summaries), Panoramas et Synthèses [Panoramas and Syntheses], vol. 17, Société Mathématique de France, Paris, 2004. MR2115000

[2] P. Ara, M. A. González-Barroso, K. R. Goodearl, and E. Pardo, *Fractional skew monoid rings*, J. Algebra **278** (2004), no. 1, 104–126, DOI 10.1016/j.jalgebra.2004.03.009. MR2068068

[3] M. Artin and J. J. Zhang, *Noncommutative projective schemes*, Adv. Math. **109** (1994), no. 2, 228–287, DOI 10.1006/aima.1994.1087. MR1304753

[4] Vladimir Baranovsky, *Orbifold cohomology as periodic cyclic homology*, Internat. J. Math. **14** (2003), no. 8, 791–812, DOI 10.1142/S0129167X03001946. MR2013145

[5] Hyman Bass, *Some problems in "classical" algebraic K-theory*, Algebraic K-theory, II: "Classical" algebraic K-theory and connections with arithmetic (Proc. Conf., Battelle Memorial Inst., Seattle, Wash., 1972), Springer, Berlin, 1973, pp. 3–73. Lecture Notes in Math., Vol. 342. MR0409606

[6] Alexandre Beĭlinson and Joseph Bernstein, *Localisation de g-modules* (French, with English summary), C. R. Acad. Sci. Paris Sér. I Math. **292** (1981), no. 1, 15–18. MR610137

[7] M. Bernardara, M. Marcolli, and G. Tabuada *Some remarks concerning Voevodsky's nilpotence conjecture*. Available at arXiv:1403.0876. To appear in J. Reine Angew. Math.

[8] R. V. Bezrukavnikov and D. B. Kaledin, *McKay equivalence for symplectic resolutions of quotient singularities* (Russian, with Russian summary), Tr. Mat. Inst. Steklova **246** (2004), no. Algebr. Geom. Metody, Svyazi i Prilozh., 20–42; English transl., Proc. Steklov Inst. Math. **3(246)** (2004), 13–33. MR2101282

[9] A. Białynicki-Birula, *Some theorems on actions of algebraic groups*, Ann. of Math. (2) **98** (1973), 480–497, DOI 10.2307/1970915. MR0366940

[10] S. Bloch, A. Kas, and D. Lieberman, *Zero cycles on surfaces with $p_g = 0$*, Compositio Math. **33** (1976), no. 2, 135–145. MR0435073

[11] A. I. Bondal and M. M. Kapranov, *Representable functors, Serre functors, and reconstructions* (Russian), Izv. Akad. Nauk SSSR Ser. Mat. **53** (1989), no. 6, 1183–1205, 1337; English transl., Math. USSR-Izv. **35** (1989), no. 3, 519–541. MR1039961

[12] Alexey I. Bondal, Michael Larsen, and Valery A. Lunts, *Grothendieck ring of pretriangulated categories*, Int. Math. Res. Not. **29** (2004), 1461–1495, DOI 10.1155/S1073792804140385. MR2051435

[13] A. I. Bondal and D. O. Orlov, *Semiorthogonal decomposition for algebraic varieties*. Available at arXiv:alg-geom/9506012.

[14] Mikhail Bondarko and Gonçalo Tabuada, *Picard groups, weight structures, and (noncommutative) mixed motives*, Doc. Math. **22** (2017), 45–66. MR3609202

[15] Tom Bridgeland, Alastair King, and Miles Reid, *The McKay correspondence as an equivalence of derived categories*, J. Amer. Math. Soc. **14** (2001), no. 3, 535–554, DOI 10.1090/S0894-0347-01-00368-X. MR1824990

[16] Vladimir Chernousov, Stefan Gille, and Alexander Merkurjev, *Motivic decomposition of isotropic projective homogeneous varieties*, Duke Math. J. **126** (2005), no. 1, 137–159, DOI 10.1215/S0012-7094-04-12614-4. MR2110630

[17] P. Deligne, *Catégories tensorielles* (French, with English summary), Mosc. Math. J. **2** (2002), no. 2, 227–248. Dedicated to Yuri I. Manin on the occasion of his 65th birthday. MR1944506

[18] Pierre Deligne, *La conjecture de Weil. I* (French), Inst. Hautes Études Sci. Publ. Math. **43** (1974), 273–307. MR0340258

[19] P. Deligne and J. S. Milne, *Tannakian categories*. In: Hodge Cycles, Motives, and Shimura Varieties. Lecture Notes in Mathematics, vol. **900**, pp. 101–228. Updated online version available at `http://www.jmilne.org/math/xnotes/tc.pdf` (1982).

[20] Bernard Dwork, *On the rationality of the zeta function of an algebraic variety*, Amer. J. Math. **82** (1960), 631–648, DOI 10.2307/2372974. MR0140494

[21] Dan Edidin and William Graham, *Equivariant intersection theory*, Invent. Math. **131** (1998), no. 3, 595–634, DOI 10.1007/s002220050214. MR1614555

[22] A. V. Odesskiĭ and B. L. Feĭgin, *Sklyanin's elliptic algebras* (Russian), Funktsional. Anal. i Prilozhen. **23** (1989), no. 3, 45–54, 96, DOI 10.1007/BF01079526; English transl., Funct. Anal. Appl. **23** (1989), no. 3, 207–214 (1990). MR1026987

[23] Ofer Gabber, *Some theorems on Azumaya algebras*, The Brauer group (Sem., Les Plans-sur-Bex, 1980), Lecture Notes in Math., vol. 844, Springer, Berlin-New York, 1981, pp. 129–209. MR611868

[24] Pierre Gabriel, *Des catégories abéliennes* (French), Bull. Soc. Math. France **90** (1962), 323–448. MR0232821

[25] A. Grothendieck, *Standard conjectures on algebraic cycles*, Algebraic Geometry (Internat. Colloq., Tata Inst. Fund. Res., Bombay, 1968), Oxford Univ. Press, London, 1969, pp. 193–199. MR0268189

[26] Alexander Grothendieck, *Formule de Lefschetz et rationalité des fonctions L* (French), Séminaire Bourbaki, Vol. 9, Soc. Math. France, Paris, 1995, pp. Exp. No. 279, 41–55. MR1608788

[27] Vladimir Guletskiĭ, *Finite-dimensional objects in distinguished triangles*, J. Number Theory **119** (2006), no. 1, 99–127, DOI 10.1016/j.jnt.2005.10.008. MR2228952

[28] V. Guletskiĭ and C. Pedrini, *Finite-dimensional motives and the conjectures of Beilinson and Murre*, K-Theory **30** (2003), no. 3, 243–263, DOI 10.1023/B:KTHE.0000019787.69435.89. Special issue in honor of Hyman Bass on his seventieth birthday. Part III. MR2064241

[29] Daniel R. Grayson, *Finite generation of K-groups of a curve over a finite field (after Daniel Quillen)*, Algebraic K-theory, Part I (Oberwolfach, 1980), Lecture Notes in Math., vol. 966, Springer, Berlin-New York, 1982, pp. 69–90. MR689367

[30] Christian Hasemeyer and Jens Hornbostel, *Motives and etale motives with finite coefficients*, K-Theory **34** (2005), no. 3, 195–207, DOI 10.1007/s10977-005-1563-6. MR2182375

[31] Robin Hartshorne, *On the De Rham cohomology of algebraic varieties*, Inst. Hautes Études Sci. Publ. Math. **45** (1975), 5–99. MR0432647

[32] Michiel Hazewinkel, *Witt vectors. I*, Handbook of algebra. Vol. 6, Handb. Algebr., vol. 6, Elsevier/North-Holland, Amsterdam, 2009, pp. 319–472, DOI 10.1016/S1570-7954(08)00207-6. MR2553661

[33] Annette Huber and Stefan Müller-Stach, *Periods and Nori motives*, Ergebnisse der Mathematik und ihrer Grenzgebiete. 3. Folge. A Series of Modern Surveys in Mathematics [Results in Mathematics and Related Areas. 3rd Series. A Series of Modern Surveys in Mathematics], vol. 65, Springer, Cham, 2017. With contributions by Benjamin Friedrich and Jonas von Wangenheim. MR3618276

[34] Florian Ivorra, *Levine's motivic comparison theorem revisited*, J. Reine Angew. Math. **617** (2008), 67–107, DOI 10.1515/CRELLE.2008.026. MR2400991

[35] Jaya NN Iyer and Stefan Müller-Stach, *Chow-Künneth decomposition for some moduli spaces*, Doc. Math. **14** (2009), 1–18. MR2505057

[36] Uwe Jannsen, *Motives, numerical equivalence, and semi-simplicity*, Invent. Math. **107** (1992), no. 3, 447–452, DOI 10.1007/BF01231898. MR1150598

[37] Uwe Jannsen, *Mixed motives and algebraic K-theory*, Lecture Notes in Mathematics, vol. 1400, Springer-Verlag, Berlin, 1990. With appendices by S. Bloch and C. Schoen. MR1043451

[38] A. de Jong, *A result of Gabber*. Available online at the personal webpage of Aise Johan de Jong: `http://www.math.columbia.edu/dejong/papers/2-gabber.pdf`.

[39] Bruno Kahn, *Zeta functions and motives*, Pure Appl. Math. Q. **5** (2009), no. 1, 507–570, DOI 10.4310/PAMQ.2009.v5.n1.a16. MR2520468

[40] ———, *Motifs et adjoints*. Available at arXiv:1506:08386.

[41] Bruno Kahn and Ronnie Sebastian, *Smash-nilpotent cycles on abelian 3-folds*, Math. Res. Lett. **16** (2009), no. 6, 1007–1010, DOI 10.4310/MRL.2009.v16.n6.a8. MR2576688

[42] M. Kapranov and E. Vasserot, *Kleinian singularities, derived categories and Hall algebras*, Math. Ann. **316** (2000), no. 3, 565–576, DOI 10.1007/s002080050344. MR1752785

[43] N. A. Karpenko, *Cohomology of relative cellular spaces and of isotropic flag varieties* (Russian, with Russian summary), Algebra i Analiz **12** (2000), no. 1, 3–69; English transl., St. Petersburg Math. J. **12** (2000), no. 1, 1–50. MR1758562

[44] Yujiro Kawamata, *D-equivalence and K-equivalence*, J. Differential Geom. **61** (2002), no. 1, 147–171. MR1949787

[45] Bernhard Keller, *On differential graded categories*, International Congress of Mathematicians. Vol. II, Eur. Math. Soc., Zürich, 2006, pp. 151–190. MR2275593

[46] Shun-Ichi Kimura, *Chow groups are finite dimensional, in some sense*, Math. Ann. **331** (2005), no. 1, 173–201, DOI 10.1007/s00208-004-0577-3. MR2107443

[47] Steven L. Kleiman, *The standard conjectures*, Motives (Seattle, WA, 1991), Proc. Sympos. Pure Math., vol. 55, Amer. Math. Soc., Providence, RI, 1994, pp. 3–20, DOI 10.1090/pspum/055.1/1265519. MR1265519

[48] S. L. Kleiman, *Algebraic cycles and the Weil conjectures*, Dix exposés sur la cohomologie des schémas, Adv. Stud. Pure Math., vol. 3, North-Holland, Amsterdam, 1968, pp. 359–386. MR292838

[49] János Kollár, *Conics in the Grothendieck ring*, Adv. Math. **198** (2005), no. 1, 27–35, DOI 10.1016/j.aim.2005.01.004. MR2183248

[50] Maxim Kontsevich, *Mixed noncommutative motives.* Talk at the Workshop on Homological Mirror Symmetry, Miami, 2010. Available at `www-math.mit.edu/auroux/frg/miami10-notes`.

[51] Maxim Kontsevich, *Notes on motives in finite characteristic*, Algebra, arithmetic, and geometry: in honor of Yu. I. Manin. Vol. II, Progr. Math., vol. 270, Birkhäuser Boston, Inc., Boston, MA, 2009, pp. 213–247, DOI 10.1007/978-0-8176-4747-6_7. MR2641191

[52] ———, *Categorification, NC Motives, Geometric Langlands, and Lattice Models.* Talk at the Geometric Langlands Seminar, University of Chicago, 2006. Notes available at the webpage `https://www.ma.utexas.edu/users/benzvi/notes.html`.

[53] ———, *Noncommutative motives.* Talk at the IAS on the occasion of the 61st birthday of Pierre Deligne (2005). Available at `http://video.ias.edu/Geometry-and-Arithmetic`.

[54] Maxim Kontsevich and Don Zagier, *Periods*, Mathematics unlimited—2001 and beyond, Springer, Berlin, 2001, pp. 771–808. MR1852188

[55] A. Kuznetsov, *Semiorthogonal decompositions in algebraic geometry.* Available at 1404.3143. To appear in Proceedings of the ICM 2014.

[56] ———, *Homological projective duality.* Inst. Hautes Études Sci. Publ. Math. (2007), no. **105**, 157–220.

[57] A. G. Kuznetsov, *Hyperplane sections and derived categories* (Russian, with Russian summary), Izv. Ross. Akad. Nauk Ser. Mat. **70** (2006), no. 3, 23–128, DOI 10.1070/IM2006v070n03ABEH002318; English transl., Izv. Math. **70** (2006), no. 3, 447–547. MR2238172

[58] ———, *Homological projective duality for Grassmannians of lines.* Available at arXiv:math/0610957.

[59] Robert Laterveer, *Equivariant motives*, Indag. Math. (N.S.) **9** (1998), no. 2, 255–275, DOI 10.1016/S0019-3577(98)80023-2. MR1691420

[60] W. G. Leavitt, *The module type of a ring*, Trans. Amer. Math. Soc. **103** (1962), 113–130, DOI 10.2307/1993743. MR0132764

[61] Marc Levine, *Mixed motives*, Mathematical Surveys and Monographs, vol. 57, American Mathematical Society, Providence, RI, 1998. MR1623774

[62] David I. Lieberman, *Numerical and homological equivalence of algebraic cycles on Hodge manifolds*, Amer. J. Math. **90** (1968), 366–374, DOI 10.2307/2373533. MR0230336

[63] Yu. I. Manin, *Quantum groups and noncommutative geometry*, Université de Montréal, Centre de Recherches Mathématiques, Montreal, QC, 1988. MR1016381

[64] Yu. I. Manin, *Some remarks on Koszul algebras and quantum groups* (English, with French summary), Ann. Inst. Fourier (Grenoble) **37** (1987), no. 4, 191–205. MR927397

[65] Carlo Mazza, *Schur functors and motives*, K-Theory **33** (2004), no. 2, 89–106, DOI 10.1007/s10977-004-6468-2. MR2131746

[66] Fabien Morel and Vladimir Voevodsky, $\mathbf{A}^1$*-homotopy theory of schemes*, Inst. Hautes Études Sci. Publ. Math. **90** (1999), 45–143 (2001). MR1813224

[67] David Mumford, *The topology of normal singularities of an algebraic surface and a criterion for simplicity*, Inst. Hautes Études Sci. Publ. Math. **9** (1961), 5–22. MR0153682

[68] I. A. Panin, *On the algebraic K-theory of twisted flag varieties*, K-Theory **8** (1994), no. 6, 541–585, DOI 10.1007/BF00961020. MR1326751

[69] Daniel Quillen, *Finite generation of the groups K_i of rings of algebraic integers [MR0349812]*, Cohomology of groups and algebraic K-theory, Adv. Lect. Math. (ALM), vol. 12, Int. Press, Somerville, MA, 2010, pp. 479–488. MR2655185

[70] Daniel Quillen, *Higher algebraic K-theory. I*, Algebraic K-theory, I: Higher K-theories (Proc. Conf., Battelle Memorial Inst., Seattle, Wash., 1972), Springer, Berlin, 1973, pp. 85–147. Lecture Notes in Math., Vol. 341. MR0338129

[71] Daniel Quillen, *On the cohomology and K-theory of the general linear groups over a finite field*, Ann. of Math. (2) **96** (1972), 552–586, DOI 10.2307/1970825. MR0315016

[72] M. Reid, *The complete intersection of two or more quadrics*. Ph. D. thesis. Available at the webpage `https://homepages.warwick.ac.uk/ masda/3folds/qu.pdf`.

[73] Jean-Pierre Serre, *Faisceaux algébriques cohérents* (French), Ann. of Math. (2) **61** (1955), 197–278, DOI 10.2307/1969915. MR0068874

[74] A. M. Šermenev, *Motif of an Abelian variety* (Russian), Funckcional. Anal. i Priložen. **8** (1974), no. 1, 55–61. MR0335523

[75] S. Paul Smith, *Some finite-dimensional algebras related to elliptic curves*, Representation theory of algebras and related topics (Mexico City, 1994), CMS Conf. Proc., vol. 19, Amer. Math. Soc., Providence, RI, 1996, pp. 315–348. MR1388568

[76] J. T. Stafford, *Noncommutative projective geometry*, Proceedings of the International Congress of Mathematicians, Vol. II (Beijing, 2002), Higher Ed. Press, Beijing, 2002, pp. 93–103. MR1957024

[77] J. H. M. Steenbrink, *A summary of mixed Hodge theory*, Motives (Seattle, WA, 1991), Proc. Sympos. Pure Math., vol. 55, Amer. Math. Soc., Providence, RI, 1994, pp. 31–41, DOI 10.1090/pspum/055.1/1265521. MR1265521

[78] Darin R. Stephenson and James J. Zhang, *Noetherian connected graded algebras of global dimension 3*, J. Algebra **230** (2000), no. 2, 474–495, DOI 10.1006/jabr.2000.8323. MR1775800

[79] A. Suslin, *On the K-theory of algebraically closed fields*, Invent. Math. **73** (1983), no. 2, 241–245, DOI 10.1007/BF01394024. MR714090

[80] Gonçalo Tabuada, *A note on secondary K-theory*, Algebra Number Theory **10** (2016), no. 4, 887–906, DOI 10.2140/ant.2016.10.887. MR3519099

[81] ______, *Noncommutative motives*. With a preface by Yuri I. Manin. University Lecture Series **63**, American Mathematical Society, Providence, RI, 2015.

[82] ______, *Equivariant noncommutative motives*, Annals of K-theory **3** (2018), no. 1, 125–156.

[83] ______, *Modified mixed realizations, new additive invariants, and periods of dg categories*, International Mathematics Research Notices **34** (2017), 7607–7638.

[84] ______, *A note on Grothendieck's standard conjectures of type C^+ and D*, Proceedings of the American Mathematical Society **146** (2018), no. 4, 1389–1399.

[85] ______, *A note on the Schur-finiteness of linear sections*. Available at arXiv:1610.06553. To appear in Math. Res. Lett. .

[86] ______, *$\mathbb{A}^1$-homotopy invariants of corner skew Laurent polynomial algebras*, Journal of Noncommutative Geometry **11** (2017), no. 4, 1627–1643.

[87] ______, *Jacques Tits' motivic measure*. Available at arXiv:1604.06407.

[88] ______, *A note on secondary K-theory II*. Available at arXiv:1607.03094.

[89] ______, *Invariants of noncommutative projective schemes*. Available at arXiv:1702.04712. To appear in Moscow Mathematical Journal.

[90] ______, *Noncommutative rigidity*. Available at arXiv:1703.10599. To appear in Mathematische Zeitschrift.

[91] ______, *HPD-invariance of the Tate conjecture*. Available at arXiv:1707.06639.

[92] ______, *Bass and Schur finiteness conjectures for quadric fibrations*. Available at arXiv:1708.05382.

[93] ______, *Noncommutative motives in positive characteristic and their applications*. Available at arXiv:1707.04248.

[94] Gonçalo Tabuada and Michel Van den Bergh, *Noncommutative motives of separable algebras*, Adv. Math. **303** (2016), 1122–1161, DOI 10.1016/j.aim.2016.08.031. MR3552545

[95] ______, *The Gysin triangle via localization and $\mathbb{A}^1$-homotopy invariance*, Transactions of the American Mathematical Society **370** (2018), no. 1, 421–446.

[96] ______, *Additive invariants of orbifolds*. Available at arXiv:1612.03162. To appear in Geometry and Topology.

[97] John Tate, *Conjectures on algebraic cycles in l-adic cohomology*, Motives (Seattle, WA, 1991), Proc. Sympos. Pure Math., vol. 55, Amer. Math. Soc., Providence, RI, 1994, pp. 71–83, DOI 10.1090/pspum/055.1/1265523. MR1265523

[98] John T. Tate, *Algebraic cycles and poles of zeta functions*, Arithmetical Algebraic Geometry (Proc. Conf. Purdue Univ., 1963), Harper & Row, New York, 1965, pp. 93–110. MR0225778

[99] John Tate and Michel van den Bergh, *Homological properties of Sklyanin algebras*, Invent. Math. **124** (1996), no. 1-3, 619–647, DOI 10.1007/s002220050065. MR1369430

[100] R. W. Thomason and Thomas Trobaugh, *Higher algebraic K-theory of schemes and of derived categories*, The Grothendieck Festschrift, Vol. III, Progr. Math., vol. 88, Birkhäuser Boston, Boston, MA, 1990, pp. 247–435, DOI 10.1007/978-0-8176-4576-2_10. MR1106918

[101] B. Toën, *Secondary K-theory.* Talk at the Workshop on Topological Field Theories, Northwestern University, 2009. Available online at David Ben-Zvi's personal webpage `https://www.ma.utexas.edu/users/benzvi/GRASP/lectures/NWTFT/toennw.pdf`.

[102] B. Totaro, *Recent progress on the Tate conjecture.* Available at Burt Totaro's personal webpage `http://www.math.ucla.edu/totaro/papers`.

[103] Angelo Vistoli, *Higher equivariant K-theory for finite group actions*, Duke Math. J. **63** (1991), no. 2, 399–419, DOI 10.1215/S0012-7094-91-06317-9. MR1115114

[104] Vladimir Voevodsky, *Triangulated categories of motives over a field*, Cycles, transfers, and motivic homology theories, Ann. of Math. Stud., vol. 143, Princeton Univ. Press, Princeton, NJ, 2000, pp. 188–238. MR1764202

[105] Vladimir Voevodsky, $\mathbf{A}^1$*-homotopy theory*, Proceedings of the International Congress of Mathematicians, Vol. I (Berlin, 1998), Doc. Math. **Extra Vol. I** (1998), 579–604. MR1648048

[106] V. Voevodsky, *A nilpotence theorem for cycles algebraically equivalent to zero*, Internat. Math. Res. Notices **4** (1995), 187–198, DOI 10.1155/S1073792895000158. MR1326064

[107] Claire Voisin, *Bloch's conjecture for Catanese and Barlow surfaces*, J. Differential Geom. **97** (2014), no. 1, 149–175. MR3229054

[108] Claire Voisin, *Remarks on zero-cycles of self-products of varieties*, Moduli of vector bundles (Sanda, 1994; Kyoto, 1994), Lecture Notes in Pure and Appl. Math., vol. 179, Dekker, New York, 1996, pp. 265–285. MR1397993

[109] Claire Voisin, *Sur les zéro-cycles de certaines hypersurfaces munies d'un automorphisme* (French), Ann. Scuola Norm. Sup. Pisa Cl. Sci. (4) **19** (1992), no. 4, 473–492. MR1205880

[110] André Weil, *Variétés abéliennes et courbes algébriques* (French), Actualités Sci. Ind., no. 1064 = Publ. Inst. Math. Univ. Strasbourg 8 (1946), Hermann & Cie., Paris, 1948. MR0029522

[111] Mariusz Wodzicki, *Cyclic homology of differential operators*, Duke Math. J. **54** (1987), no. 2, 641–647, DOI 10.1215/S0012-7094-87-05426-3. MR899408

Department of Mathematics, MIT, Cambridge, MA 02139, USA; Departamento de Matemática, Faculdade de Ciências e Tecnologia, Universidade Nova de Lisboa, Portugal; Centro de Matemática e Aplicações, Faculdade de Ciências e Tecnologia, Universidade Nova de Lisboa, Portugal.

Email address: `tabuada@math.mit.edu`

Contemporary Mathematics
Volume **707**, 2018
http://dx.doi.org/10.1090/conm/707/14259

The category of Waldhausen categories is a closed multicategory

Inna Zakharevich

Contents

Introduction

The goal of this paper is to develop in detail an example of a closed symmteric multicategory. Closed multicategories were first introduced in [**Man12**]. The literature on closed multicategories has very few examples; in this paper we aim to explain the potentially-useful example of Waldhausen K-theory in enough detail that both the example and the general theory are easier to understand. In addition, we gather in one place some well-known results about K-theory for which it is difficult to find references.

This paper was written because the details were necessary to the future work of Angelica Osorno and Anna Marie Bohmann, but we hope that it will be useful for others as well. Very little in this paper is new, and it is particularly indebted to [**BM11**] for many of the ideas. This paper proves the following theorem:

Theorem 0.1. *The category* **WaldCat** *of Waldhausen categories is a closed symmetric multicategory, in the sense that the hom-sets*

$$\mathbf{WaldCat}_k(A_1, \ldots, A_k; B)$$

all have the structure of Waldhausen categories and composition of morphisms is multiexact. In addition, there is a multiexact

$$\mathrm{ev}\colon \mathcal{C}_1 \times \cdots \times \mathcal{C}_k \times \mathbf{WaldCat}_k(\mathcal{C}_1, \ldots, \mathcal{C}_k; \mathcal{D}) \longrightarrow \mathcal{D}$$

defined by

$$(A_1, \dots, A_k, F) \longmapsto F(A_1, \dots, A_k).$$

In addition, we use this theorem to prove that (after application to the Hom-objects) the K-theory functor can be enriched to be a spectrally-enriched multifunctor $\mathbf{WaldCat_{Sp}} \longrightarrow \mathbf{Sp}$.

The organization of this paper is as follows. In Section 1 we introduce multicategories and closed multicategories and in Section 2 we introduce Waldhausen categories. Sections 3 and 4 discuss k-exactness of functors. The definition of the hom-Waldhausen categories is given in Section 5, and the analysis of the K-theory functor is in Section 6.

1. A quick introduction to symmetric multicategories

A multicategory is a generalization of a symmetric monoidal category where one does not necessarily have a product. The motivation for the definition comes from the notion of tensor product: the tensor product of modules classifies bilinear maps out of the ordinary product of the modules. Thus if one is in a context where the tensor product is difficult to work with directly, one can work with bilinear maps instead. The idea of a multicategory is that we have a notion of "k-linear" map, but we do not necessarily have a representing object, so we must always work with the "k-linear" maps directly.

More formally, we have the following definition [**EM06**]:

DEFINITION 1.1. A *symmetric multicategory* $\mathcal{M}$ is given by the following data:

- A collection of objects $\operatorname{ob}\mathcal{M}$.
- For each $k \geq 0$ and $k+1$-tuple of objects $A_1, \dots, A_k, B \in \operatorname{ob}\mathcal{M}$, a set $\mathcal{M}_k(A_1, \dots, A_k; B)$ of *k-morphisms.*
- A right action of Σ_k on the collection of all k-morphisms such that for $\sigma \in \Sigma_k$,

$$\sigma^* \colon \mathcal{M}_k(A_1, \dots, A_k; B) \longrightarrow \mathcal{M}_k(A_{\sigma(1)}, \dots, A_{\sigma(k)}; B).$$

- A distinguished *unit* $1_A \in \mathcal{M}_1(A; A)$ for every $A \in \operatorname{ob}\mathcal{M}$, and
- A *composition law*

$$\circ \colon \mathcal{M}_\ell(B_1, \dots, B_\ell; C) \times \textstyle\prod_{i=1}^{\ell} M_{k_i}(A_{i1}, \dots, A_{ik_i}; B_i) \\ \longrightarrow \mathcal{M}_{\sum k_i}(A_{11}, \dots, A_{\ell k_\ell}; C)$$

subject to compatibility axioms listed in [**EM06**, p167-168]. We do not state them here as they are quite verbose and unnecessary for the rest of the text.

Any symmetric monoidal category gives rise to a symmetric multicategory, by setting

$$\mathcal{M}_k(A_1, \dots, A_k; B) = \mathcal{M}(((A_1 \otimes A_2) \otimes \cdots \otimes A_k), B).$$

Thus a symmetric multicategory is the "next best thing" to a symmetric monoidal category in many cases. Many proofs that work in a symmetric monoidal category where the product is defined via an appropriate universal property will also work in a symmetric multicategory.

Quite often one wants more than just a symmetric monoidal category structure: one also wants the symmetric monoidal category to be closed, so that there are hom-objects defined in the category itself. In other words, we want $\mathcal{M}$ to be enriched over itself [**Kel82**, Section 1.6] (in a way compatible with the symmetric monoidal structure). This definition is an expansion of the definition of a closed multicategory of [**Man12**, Section 3], as the original did not include the symmetric action. We write $\overline{A} = (A_1, \ldots, A_k)$ in the interest of compactness.

DEFINITION 1.2. A symmetric multicategory $\mathcal{M}$ is called *closed* if for all $k+1$-tuples $A_1, \ldots, A_k, B$ of objects in $\mathcal{M}$ there exists an object $\underline{\mathcal{M}}(\overline{A}; B) \in \mathcal{M}$ with a right Σ_k-action called the *internal hom-object* and an *evaluation morphism*

$$\mathrm{ev}_{\overline{A};B} \in \mathcal{M}_{k+1}(\overline{A}, \underline{\mathcal{M}}(\overline{A}; B); B).$$

These need to satisfy the following axioms:

(CM1) for all ℓ-tuples $C_1, \ldots, C_\ell \in \mathrm{ob}\,\mathcal{M}$ the map

$$\varphi_{\overline{A};\overline{C};B} \colon \mathcal{M}_\ell(\overline{C}; \underline{\mathcal{M}}(\overline{A}; B)) \longrightarrow \mathcal{M}_{k+\ell}(\overline{A}, \overline{C}; B)$$

defined by sending $f \in \mathcal{M}_\ell(\overline{C}; \underline{\mathcal{M}}(\overline{A}; B))$ to the composite

$$\mathrm{ev}_{\overline{A};B} \circ (1_{A_1}, \ldots, 1_{A_k}, f)$$

is a bijection.

(CM2) The evaluation morphism is Σ_k-equivariant, in the sense that the following diagram commutes for all $\sigma \in \Sigma_k$:

$$\begin{array}{ccc} (\overline{A}_\sigma \underline{\mathcal{M}}(\overline{A}; B)) & \xrightarrow{\mathrm{ev}_{\overline{A};B} \cdot (\sigma \times 1)} & B \\ \Big\downarrow{\scriptstyle (1,\ldots,1,\sigma^*)} & \nearrow{\scriptstyle \mathrm{ev}_{\overline{A}_\sigma;B}} & \\ (\overline{A}_\sigma, \underline{\mathcal{M}}(\overline{A}_\sigma; B)) & & \end{array}$$

For a more detailed theory of enriched categories, see [**Lei04**]. We want to thank the anonymous referee for the currect version of (CM2), as their suggestion made it much simpler.

REMARK 1.3. These two axioms are sufficient to give $\mathcal{M}$ the structure of a multicategory enriched over itself [**Man12**, Proposition 3.12]. The underlying multicategory of the enrichment is again $\mathcal{M}$.

For example, if $\mathcal{M}$ is a closed symmetric monoidal category then we can give it the structure of a closed symmetric multicategory by setting $\underline{\mathcal{M}}(A_1, \ldots, A_k; B) = B^{((A_1 \otimes A_2) \otimes \cdots \otimes A_k)}$. Note that if $\mathcal{M}$ is a closed symmetric multicategory then we can think of it as a category enriched over itself.

2. A bit about Waldhausen categories

We begin by recalling the definition of a Waldhausen category. These were first introduced by Waldhausen in [**Wal85**], where they are called "categories with cofibrations and weak equivalences."

DEFINITION 2.1. A *Waldhausen category* is a category $\mathcal{C}$ together with two subcategories, $c\mathcal{C}$ and $w\mathcal{C}$, of cofibrations and weak equivalences, satisfying the following extra axioms:

(W1) All isomorphisms are both weak equivalences and cofibrations.

(W2) $\mathcal{C}$ has a zero object $*$, and the morphism $* \longrightarrow A$ is a cofibration for all $A \in \mathcal{C}$.

(W3) Every diagram $C \longleftarrow A \hookrightarrow B \in \mathcal{C}$ has a pushout, and the morphism $C \hookrightarrow B \cup_A C$ is a cofibration.

(W4) Given any diagram

$$\begin{array}{ccccc} B & \longleftarrow & A & \hookrightarrow & C \\ \downarrow \wr & & \downarrow \wr & & \downarrow \wr \\ B' & \longleftarrow & A' & \hookrightarrow & C' \end{array}$$

the induced morphism $B \cup_A C \xrightarrow{\sim} B' \cup_{A'} C'$ is a weak equivalence.

A functor $F: \mathcal{C} \longrightarrow \mathcal{D}$ between Waldhausen categories $\mathcal{C}$ and $\mathcal{D}$ is 1-*exact* if it preserves weak equivalences, cofibrations, and pushouts of the form described in (W3).

Before we move on to a very quick overview of the $S.$ construction for a Waldhausen category, we introduce a couple of technical definitions which will be of great use to us in the upcoming discussion.

DEFINITION 2.2. Let $\mathcal{I}$ be the category with two objects 0 and 1 and one non-invertible morphism $0 \longrightarrow 1$. Suppose that we are given two functors $F, G: \mathcal{C} \longrightarrow \mathcal{D}$ and a natural transformation $\alpha: F \longrightarrow G$. We write $(F \overset{\alpha}{\Longrightarrow} G)$ for the functor $\mathcal{C} \times \mathcal{I} \longrightarrow \mathcal{D}$ given by

$$(F \overset{\alpha}{\Longrightarrow} G)(A, \epsilon) = \begin{cases} F(A) & \text{if } \epsilon = 0 \\ G(A) & \text{if } \epsilon = 1 \end{cases}$$

and

$$(F \overset{\alpha}{\Longrightarrow} G)(f, g) = \begin{cases} F(f) & \text{if } g = 1_0 \\ G(f) & \text{if } g = 1_1 \\ \alpha_A & \text{if } g: 0 \longrightarrow 1 \end{cases}$$

Note that the existence of such a functor is equivalent to the existence of the natural transformation α.

DEFINITION 2.3. An *n-cube* in $\mathcal{C}$ is a functor $I: \mathcal{I}^n \longrightarrow \mathcal{C}$; a *face* of a cube is a restriction $I\big|_{\mathcal{I}^k \times \{\epsilon\} \times \mathcal{I}^{n-k-1}}$ for $\epsilon = 0, 1$. We will write the objects of $\mathcal{I}^n$ as vectors $\overline{\epsilon} = (\epsilon_1, \ldots, \epsilon_n)$. As shorthand, we will write $I_{k\epsilon}$ for the restriction $I\big|_{\mathcal{I}^{k-1} \times \{\epsilon\} \times \mathcal{I}^{n-k}}$, for $\epsilon = 0, 1$, and $I(1)$ for $I(1, \ldots, 1)$. For any cube I we write I' for the diagram $I\big|_{\overline{\epsilon} \neq (1, \ldots, 1)}$, and define the *southern arrow* of an n-cube I to be the morphism

$$\operatorname{colim} I' \longrightarrow I(1).$$

(The southern arrow may not exist if the colimit does not.) An n-cube I is *good* if its southern arrow exists and is a cofibration and all of the faces of I are good.

In particular, the 0-cubes are the objects of $\mathcal{C}$, and the southern arrow of a 0-cube A is just the morphism $\varnothing \longrightarrow A$, so all 0-cubes are good. The 1-cubes are the morphisms of $\mathcal{C}$, and the southern arrow of a 1-cube is itself. Thus the good 1-cubes are the cofibrations.

The notion of a good cube appears in [**BM11**, Definition 2.1] as a "cubically cofibrant" diagram.

Given a natural transformation $\alpha: I \longrightarrow J$ between n-cubes, we will write $[\alpha]$ for the $n+1$-cube $(I \stackrel{\alpha}{\Longrightarrow} J)$.

Let $\langle n \rangle$ be the ordered set $0 < 1 < \cdots < n$ and let $\mathrm{Ar}\langle n \rangle$ be the arrow category of $\langle n \rangle$; we will denote an object in $\mathrm{Ar}\langle n \rangle$ as $j \leq i$. For a vector $\vec{n} = (n_1, \ldots, n_m)$ we will write $\langle \vec{n} \rangle = \langle n_1 \rangle \times \cdots \times \langle n_m \rangle$.

DEFINITION 2.4. The category $S_n\mathcal{C}$ is defined to have as objects functors $X: \mathrm{Ar}\langle n \rangle \longrightarrow \mathcal{C}$ satisfying the extra conditions

(1) $X(i = i) = *$ for all $i \in \langle n \rangle$, and
(2) $X(i \leq j) \longrightarrow X(i \leq k)$ is a cofibration, and
(3) for all $i < j < k$ the square

$$\begin{array}{ccc} X(i \leq j) & \hookrightarrow & X(i \leq k) \\ \downarrow & & \downarrow \\ X(j = j) & \hookrightarrow & X(j \leq k) \end{array}$$

is a pushout square.

The categories $S_n\mathcal{C}$ form a simplicial category by inheriting the simplicial structure from the simplicial category $[\mathrm{Ar}\langle \cdot \rangle, \mathcal{C}]$.

$S_n\mathcal{C}$ is defined to be a Waldhausen category by setting the weak equivalences to be levelwise, and the cofibrations to be the natural transformations $\alpha: X \longrightarrow Y$ such that for all $i \leq j \leq k$ the square

$$\begin{array}{ccc} X(i \leq k) & \hookrightarrow & X(j \leq k) \\ \downarrow & & \downarrow \\ Y(i \leq k) & \hookrightarrow & Y(j \leq k) \end{array}$$

is good.

As applying $S_\cdot$ to a Waldhausen category produces a simplicial Waldhausen category we can iterate the construction.

LEMMA 2.5 ([**BM11**, Construction 2.2]). *The k-fold iterated construction $S_\cdot^{(k)}\mathcal{C}$ has objects which are functors*

$$X\colon \mathrm{Ar}(\langle n_1 \rangle \times \cdots \times \langle n_k \rangle) \longrightarrow \mathcal{C},$$

such that the following conditions hold:

(i) *For every k-cube $I: \mathcal{I}^k \longrightarrow \mathrm{Ar}\langle \vec{n} \rangle$ such that all objects have 0 as the domain in all coordinates, the k-cube $X \circ I$ is good.*
(ii) *For all $1 \leq j \leq k$, the composition $X \circ (\mathrm{Ar}\langle n_j \rangle \longrightarrow \mathrm{Ar}(\langle n_1 \rangle \times \cdots \times \langle n_k \rangle)$ is an object of $S_n\mathcal{C}$.*

The cofibrations of $S_\cdot^{(k)}\mathcal{C}$ are the natural transformations $\alpha: X \longrightarrow Y$ such that the $k+1$-cube $(X \circ I \stackrel{\alpha}{\Longrightarrow} Y \circ I)$ is good. The weak equivalences are the natural weak equivalences.

Note that $S_\cdot^{(k)}\mathcal{C}$ is a k-fold simplicial category. Using this iterated construction we can define the K-theory of a Waldhausen category.

DEFINITION 2.6. Let **Sp** be the category of symmetric spectra. The functor K: **WaldCat** $\longrightarrow$ **Sp** is defined by taking a Waldhausen category $\mathcal{C}$ to the symmetric spectrum

$$(|w\mathcal{C}|, |wS_.\mathcal{C}|, |wS_.^{(2)}\mathcal{C}|, \ldots).$$

For more details, see [**Wal85**]; for more on an all-at-once construction of $K(\mathcal{C})$, see [**BM11**, Section 2].

DEFINITION 2.7. A functor $F: \mathcal{C} \times \mathcal{D} \longrightarrow \mathcal{E}$ of Waldhausen categories is *biexact* if the following conditions hold.

(1) For any object $A \in \mathcal{C}$, $F(A, -)$ is exact; for any object $B \in \mathcal{D}$, $F(-, B)$ is exact.
(2) For any two cofibrations $f: A \hookrightarrow A' \in \mathcal{C}$ and $g: B \hookrightarrow B' \in \mathcal{D}$, the southern arrow of the square

$$\begin{array}{ccc} F(A,B) & \xrightarrow{F(f,1_B)} & F(A',B) \\ {\scriptstyle F(1_A,g)}\downarrow & & \downarrow{\scriptstyle F(1_{A'},g)} \\ F(A,B') & \xrightarrow{F(f,1_{B'})} & F(A',B') \end{array}$$

is a cofibration.

The definition of biexact functor is meant to be analogous to the definition of bilinear map. If $\mathcal{C}$, $\mathcal{D}$ and $\mathcal{E}$ are all equal then this should correspond to a product structure on $K(\mathcal{C})$. In an ideal situation, **WaldCat** would have a monoidal structure $\otimes$ representing biexact functors, and all we would need to show is that K is symmetric monoidal. Unfortunately, this cannot happen:

PROPOSITION 2.8. *There does not exist a symmetric monoidal product* $\otimes$ *on* **WaldCat** *such that*

$$\left\{ \begin{array}{c} \textit{exact functors} \\ \mathcal{C} \otimes \mathcal{D} \longrightarrow \mathcal{E} \end{array} \right\} \longleftrightarrow \left\{ \begin{array}{c} \textit{biexact functors} \\ \mathcal{C} \times \mathcal{D} \longrightarrow \mathcal{E} \end{array} \right\}.$$

This result is well-known to experts, but as we could not find a reference for it we present a proof here.

PROOF. Let $\mathbf{N}_*$ be the full subcategory of $\mathbf{FinSet}_*$ with objects the finite pointed sets $\underline{n} \stackrel{\text{def}}{=} \{*, 1, \ldots, n\}$ for $n \geq 0$; note that $\mathbf{N}_*$ is equivalent to $\mathbf{FinSet}_*$. We make $\mathbf{N}_*$ into a Waldhausen category by setting the cofibrations to be the injections and the weak equivalences to be the isomorphisms. The set of exact functors $\mathbf{N}_* \longrightarrow \mathbf{FinSet}_*$ is given by a choice of sets $(A_1, \ldots, A_n, \ldots)$ such that $|A_n| = n|A_1|$.

As all Waldhausen categories contain all binary coproducts, any nontrivial Waldhausen category receives at least one exact functor from $\mathbf{N}_*$ such that the image of $\underline{1}$ is nontrivial. Suppose that such a symmetric monoidal structure on **WaldCat** exists, and let $\mathcal{S}$ be the unit. Then by assumption, $\mathcal{S} \otimes \mathbf{N}_* \cong \mathbf{N}_*$ and thus the set of exact functors $\mathcal{S} \otimes \mathbf{N}_*$ is given by a choice of sets $(A_1, \ldots)$ as above. Note that $\mathcal{S}$ must be nontrivial.

Let $\iota: \mathbf{N}_* \longrightarrow \mathcal{S}$ be any exact functor with $\iota(\underline{1}) \neq *$, and let $F: \mathcal{S} \times \mathbf{N}_* \longrightarrow \mathbf{FinSet}_*$ be any biexact functor. By assumption, F must be uniquely determined by a choice of sets $(A_1, \ldots)$. However, unlike in $\mathbf{N}_*$, in $\mathcal{S} \times \mathbf{N}_*$ there are multiple objects whose image under F must have the same cardinality; for example,

$$\big|F(\iota(\underline{1}) \amalg \iota(\underline{1}), \underline{1})\big| = \big|F(\iota(\underline{1}), \underline{1} \amalg \underline{1})\big|.$$

Thus the sequence $(A_1, \ldots)$ cannot determine the value of F, and we see that such a bijection cannot exist. $\square$

We want to show that even though **WaldCat** does not have a symmetric monoidal structure, it does have the next best thing: a symmetric multicategory structure where the 2-ary morphisms are exactly the biexact functors.

3. Cubes

In this section we develop some technical aspects of the theory of cubes. The category $\mathcal{C}$ will always be assumed to be a Waldhausen category.

The general idea of this section is that good n-cubes should behave like objects in a Waldhausen category, and that cofibrations between them should correspond to good $n+1$-cubes. More formally, we have the following proposition, which is designed to be a higher-dimensional analog of Axiom (W3).

PROPOSITION 3.1. *Let I, J and K be good n-cubes in $\mathcal{C}$, and suppose that $\alpha: I \Longrightarrow J$ is a natural transformation. If $[\alpha]$ is a good $n+1$-cube then the diagram*

$$K \Longleftarrow I \overset{\alpha}{\Longrightarrow} J$$

has a pushout $J \cup_I K$, and the natural transformation $\beta: K \Longrightarrow J \cup_I K$ gives a good $n+1$-cube $[\beta]$.

Note that as $J \cup_I K$ is a face of $[\beta]$ it must be good as well.

The proof of this proposition is quite long, so to aid understanding we first spend some time developing a general theory of cubes. The first result we mention is the $n = 1$ case of the proposition, which is proved as lemma 1.1.1 in [**Wal85**]:

LEMMA 3.2. *Consider the diagram*

$$\begin{array}{ccccc} C & \longleftarrow & A & \hookrightarrow & B \\ \big\downarrow & & \big\downarrow & & \big\downarrow \\ C' & \longleftarrow & A' & \hookrightarrow & B' \end{array}$$

in $\mathcal{C}$. If the right-hand square is good (as a 2-cube) then the induced square

$$\begin{array}{ccc} C & \hookrightarrow & B \cup_A C \\ \big\downarrow & & \big\downarrow \\ C' & \hookrightarrow & B' \cup_{A'} C' \end{array}$$

is good.

The next couple of lemmas are general category-theoretic observations whose proofs are simple, but which we will need several times in the forthcoming proofs.

LEMMA 3.3. *If $\mathcal{D}$ is any category with a terminal object $*$ and $F: \mathcal{A} \times \mathcal{D} \longrightarrow \mathcal{C}$ is a functor, then*

$$\operatorname{colim} F \cong \operatorname{colim} F\big|_{\mathcal{A}\times\{*\}}.$$

PROOF. This follows because the functor $\mathcal{A} \times \{*\} \longrightarrow \mathcal{A} \times \mathcal{D}$ is cofinal. □

Many of our proofs rely on computing southern arrows of cubes; luckily, it turns out that these can be deduced from simple pushouts. The following lemma is an n-dimensional generalizataion of a special case of the rigid dual of Proposition 0.2 in [**Goo92**]; we state it here as we will be using it several times in this section. We defer the proof until Appendix A.

LEMMA 3.4. *Suppose that we are given an indexing category $\mathcal{A}$ along with n subcategories $\mathcal{A}_1, \ldots, \mathcal{A}_n$ such that $N\mathcal{A} = \bigcup_{i=1}^n N\mathcal{A}_i$. Let $F: \mathcal{A} \to \mathcal{C}$. Then the southern arrow of the cube I given by*

$$I(\overline{\epsilon}) = \begin{cases} \operatorname{colim} F\big|_{\bigcap_{\epsilon_i=0} \mathcal{A}_i} & \overline{\epsilon} \neq (1, \ldots, 1) \\ \operatorname{colim} F & \overline{\epsilon} = (1, \ldots, 1) \end{cases}$$

is an isomorphism.

The special case that we will use the most often is the following. Let $I: \mathcal{I}^n \longrightarrow \mathcal{C}$ be an n-cube. Set $\mathcal{A} = \mathcal{I}^n \backslash \{(1, \ldots, 1)\}$, $\mathcal{A}_1 = \mathcal{A} \backslash \{(1, \ldots, 1, 0)\}$, and $\mathcal{A}_2 = \mathcal{I}^{n-1} \times \{0\}$. Then

$$\begin{aligned} \operatorname{colim} I|_{\mathcal{A}_1} &= \operatorname{colim}(I_{n1})' \\ \operatorname{colim} I|_{\mathcal{A}_2} &\cong I(1, \ldots, 1, 0) \\ \operatorname{colim} I|_{\mathcal{A}_1 \cap \mathcal{A}_2} &= \operatorname{colim}(I_{n0})'. \end{aligned}$$

Applying Lemma 3.4 we have a pushout square

$$\begin{array}{ccc} \operatorname{colim}(I_{n0})' & \longrightarrow & I(1, \ldots, 1, 0) \\ \downarrow & & \downarrow \\ \operatorname{colim}(I_{n1})' & \longrightarrow & \operatorname{colim} I' \end{array}$$

which will allow us to compute the southern arrow of I using pushouts and induction.

We now turn to the existence of southern arrows.

LEMMA 3.5. *Let I be an n-cube in a Waldhausen category $\mathcal{C}$ such that all of its faces are good. Then $\operatorname{colim} I'_{n0} \longrightarrow \operatorname{colim} I'_{n1}$ is a cofibration, and the southern arrow of I exists.*

PROOF. We prove that $\operatorname{colim} I'_{n0} \longrightarrow \operatorname{colim} I'_{n1}$ is a cofibration by induction on n. The cases $n = 1, 2$ follow directly from the definition of a good n-cube; the case $n = 3$ is a special case of Lemma 3.2. Now suppose that this is true for all cubes of size less than n, and consider the situation for n-cubes.

Let $A = I(0, \ldots, 0)$, $B = I(0, \ldots, 0, 1)$, $X = I(1, \ldots, 1, 0, 0)$ and $Y = I(1, \ldots, 1, 0, 1)$. Then (by Lemma 3.4) we know that

$$\operatorname{colim} I'_{n0} \cong \operatorname{colim}\Big(\operatorname{colim}((I_{n0})_{(n-1)1})' \longleftarrow A \hookrightarrow X\Big)$$

and

$$\operatorname{colim} I'_{n1} \cong \operatorname{colim}\Big((\operatorname{colim}(I_{n1})_{(n-1)1})' \longleftarrow B \hookrightarrow Y\Big).$$

Note that $(I_{n0})_{(n-1)1} = (I_{(n-1)1})_{(n-1)0}$ and $(I_{n1})_{(n-1)1} = (I_{(n-1)1})_{(n-1)1}$, so the inductive hypothesis applies to these and we get a diagram

$$\begin{array}{ccccc} \operatorname{colim}((I_{n0})_{(n-1)1})' & \longleftarrow & A & \hookrightarrow & X \\ \downarrow & & \downarrow & & \downarrow \\ \operatorname{colim}((I_{n1})_{(n-1)1})' & \longleftarrow & B & \hookrightarrow & Y \end{array}$$

As all faces of I are good the right-hand square is also good, and thus we see that Lemma 3.2 applies and the induced morphism between the pushouts is a cofibration, as desired.

In order for the southern arrow to exist we need to show that $\operatorname{colim} I'$ exists. By Lemma 3.4 we know that

$$\begin{aligned} \operatorname{colim} I' &\cong \operatorname{colim}\big(\operatorname{colim} I_{n0} \longleftarrow \operatorname{colim} I'_{n0} \longrightarrow \operatorname{colim} I|_{\overline{\epsilon}\notin\{1\}^{n-1}\times\mathcal{I}}\big) \\ &\cong \operatorname{colim}\big(I_{n0}(1) \longleftarrow \operatorname{colim} I'_{n0} \longrightarrow \operatorname{colim} I'_{n1}\big), \end{aligned}$$

where $\operatorname{colim} I|_{\overline{\epsilon}\notin\{1\}^{n-1}\times\mathcal{I}} \cong \operatorname{colim} I'_{n1}$ by Lemma 3.3. (Recall that $I(1) = I(1,\dots,1)$.) But by the first part of the proof the second morphism in the colimit is a cofibration, so the pushout exists. $\square$

As a consequence we get the following:

LEMMA 3.6. *In a pushout square of n-cubes*

$$\begin{array}{ccc} I & \xrightarrow{\alpha} & J \\ \downarrow & & \downarrow \\ K & \xrightarrow{\beta} & J \cup_I K \end{array}$$

where I, J, K and $[\alpha]$ are good, the southern arrows of $J \cup_I K$ and $[\beta]$ exist.

PROOF. As colimits commute we know that

$$\operatorname{colim}(J \cup_I K)' \cong \operatorname{colim}\Big(\operatorname{colim} K' \longleftarrow \operatorname{colim} I' \overset{i}{\hookrightarrow} \operatorname{colim} J'\Big),$$

where i is a cofibration by Lemma 3.5 because $[\alpha]$ is good. Each term in the pushout on the right exists because I, J and K are good, and the pushout itself exists because i is a cofibration; thus the southern arrow of $J \cup_I K$ exists.

As K is good we know that its southern arrow is a cofibration, and by the special case of Lemma 3.4 discussed on page 182 the southern arrow of $[\beta]$ exists if the southern arrow of $[\beta]_{(n+1)1}$ exists. But $[\beta]_{(n+1)1}$ is exactly $J \cup_I K$ which has a southern arrow, as desired. $\square$

Now we are ready to prove Proposition 3.1.

PROOF OF PROPOSITION 3.1. Note that I gives us a natural transformation of functors from I' to the constant functor at $I(1)$.

We will prove the desired statement by induction on n. Clearly for $n = 0$ we just need to show that cofibrations are preserved under pushout in $\mathcal{C}$, which we know is true because $\mathcal{C}$ is a Waldhausen category. For $n = 1$ this is just Lemma 3.2. Thus assume that $n > 1$ and that we know that the proposition holds for all smaller dimensions. By restricting to faces of I,J,K and using the inductive hypothesis we see that any face of $[\beta]$ that is not equal to K or $J \cup_I K$ is good. K is given to be good, so in order to prove the proposition it suffices to show that the southern arrows of $J \cup_I K$ and $[\beta]$ are cofibrations; we know that they exist by Lemmas 3.5 and 3.6.

The southern arrow of β is the morphism

$$j\colon \operatorname{colim}\Big((J \cup_I K)' \overset{\beta}{\Longleftarrow} K' \Longrightarrow K(1)\Big) \longrightarrow (J \cup_I K)(1);$$

we want to show that it is a cofibration. Note that

$$\operatorname{colim} \begin{array}{ccc} K' & \Longrightarrow & K(1) \\ {\scriptstyle\beta}\Downarrow & & \\ (J \cup_I K)' & & \end{array} \cong \operatorname{colim} \begin{array}{ccccc} I' & \Longrightarrow & K' & \Longrightarrow & K(1) \\ {\scriptstyle\alpha}\Downarrow & & \Downarrow{\scriptstyle\beta} & & \\ J' & \Longrightarrow & (J \cup_I K)' & & \end{array}$$

$$\begin{aligned} &\cong \operatorname{colim}\big(J' \overset{\alpha}{\Longleftarrow} I' \Longrightarrow K(1)\big) \\ &\cong \operatorname{colim}\big(J' \overset{\alpha}{\Longleftarrow} I' \Longrightarrow I(1) \longrightarrow K(1)\big) \end{aligned}$$

where the second line follows because the square on the first line is a pushout square. Therefore we have the folowing diagram, where the three commutative squares are all pushout squares:

$$\begin{array}{ccccc} I' & \Longrightarrow & I(1) & \longrightarrow & K(1) \\ {\scriptstyle\alpha}\Downarrow & & \downarrow & & \downarrow \\ J' & \Longrightarrow & \operatorname{colim}[\alpha]' & \longrightarrow & K(1) \cup_{I(1)} \operatorname{colim}[\alpha]' \\ & \searrow\!\!\searrow & \downarrow{\scriptstyle i} & & \downarrow{\scriptstyle j} \\ & & J(1) & \longrightarrow & (J \cup_I K)(1) \end{array}$$

Since $[\alpha]$ is a good cube it follows that i is a cofibration, and thus j is a cofibration because it is a pushout of i.

It now remains to show that the southern arrow of $J \cup_I K$ is a cofibration. The southern arrow of $J \cup_I K$ factors through the southern arrow of $[\beta]$; thus it suffices to show that the connecting morphism is also a cofibration. By Lemma 3.4 we know that

$$\begin{aligned} \operatorname{colim}[\beta]' &\cong \operatorname{colim}\big(\operatorname{colim} K \longleftarrow \operatorname{colim} K' \longrightarrow \operatorname{colim}[\beta]\big|_{\overline{\epsilon}\notin\{1\}^k\times\mathcal{I}}\big) \\ &\cong \operatorname{colim}\big(K(1) \hookleftarrow \operatorname{colim} K' \longrightarrow \operatorname{colim}(J \cup_I K)'\big) \end{aligned}$$

where the left-hand map on the bottom is a cofibration because K is a good cube, and $\operatorname{colim}[\beta]\big|_{\overline{\epsilon}\notin\{1\}^n\times\mathcal{I}} \cong \operatorname{colim}(J \cup_I K)'$ by Lemma 3.3. The induced morphism $\operatorname{colim}(J \cup_I K)' \longrightarrow \operatorname{colim}[\beta]'$ is the right-hand map in the given pushout square.

Given that it is the pushout of the cofibration $\operatorname{colim} K' \hookrightarrow K(1)$ we know that it is also a cofibration, as desired. Thus $J \cup_I K$ is a good cube, and we are done. □

4. k-exactness

The goal of this section is to show that **WaldCat** is a symmetric multicategory, where $\mathbf{WaldCat}(\mathcal{C}, \mathcal{D})$ is the set of exact functors $\mathcal{C} \longrightarrow \mathcal{D}$ and $\mathbf{WaldCat}_k(\mathcal{C}_1, \ldots, \mathcal{C}_k; \mathcal{D})$ is the set of k-exact functors. Most of this section is devoted to defining k-exactness and working out its properties.

When k is clear from context, we will write $\overline{\mathcal{C}} = \mathcal{C}_1 \times \cdots \times \mathcal{C}_k$, and refer to objects $\overline{A} = (A_1, \ldots, A_k) \in \overline{\mathcal{C}}$, and $\overline{f} = (f_1, \ldots, f_k) \in \overline{\mathcal{C}}$.

DEFINITION 4.1. Given morphisms $f_i: A_{i0} \longrightarrow A_{i1} \in \mathcal{C}_i$ and a functor $F: \overline{\mathcal{C}} \longrightarrow \mathcal{D}$ we define a k-cube $[\overline{f}]_F$ in $\mathcal{D}$ by

$$[\overline{f}]_F(\epsilon_1, \ldots, \epsilon_k) = F(A_{1\epsilon_1}, \ldots, A_{k\epsilon_k})$$

and

$$[\overline{f}]_F(\epsilon_1 \longrightarrow \epsilon_1', \ldots, \epsilon_k \longrightarrow \epsilon_k') = F(h_1, \ldots, h_k),$$

where $h_i = 1_{A_{i\epsilon_i}}$ if $\epsilon_i = \epsilon_i'$ and $h_i = f_i$ otherwise. We define the *box product* of $\overline{f}$ to be the southern arrow of $[\overline{f}]_F$.

DEFINITION 4.2 ([**BM11**, Definition 2.4]). A 0-exact functor with codomain $\mathcal{D}$ is an object of $\mathcal{D}$. A functor $F: \mathcal{C}_1 \times \cdots \times \mathcal{C}_k \longrightarrow \mathcal{D}$ is *k-exact* if

(kE1) $F(\overline{A}) = *$ if $A_i = *$ for any $1 \leq i \leq k$,
(kE2) F is exact in each variable independently, and
(kE4) for all $\overline{f} \in c\overline{\mathcal{C}}$, $[\overline{f}]_F$ is good.

If a functor is k-exact for some k, we will call it *multiexact*.

DEFINITION 4.3. Let **WaldCat** be the following symmetric multicategory:

objects: Waldhausen categories.
k-morphisms: k-exact functors $\mathcal{C}_1 \times \cdots \times \mathcal{C}_k \longrightarrow \mathcal{D}$.
Σ_k-action: permuting the input variables.

To check that this is well-defined it suffices to check that multiexact functors compose properly. This is exactly the result of Proposition 4.7; however, before we can prove that we need to develop a little bit of the theory of k-exact functors. The first lemma we prove shows that in order to prove Axiom (kE4) it suffices to consider a single type of southern arrow.

LEMMA 4.4. *Let $F: \mathcal{C}_1 \times \cdots \times \mathcal{C}_k \longrightarrow \mathcal{D}$ be any functor satisfying Axiom (kE1). If the southern arrow of $[\overline{f}]_F$ is a cofibration for all $\overline{f} \in c\overline{\mathcal{C}}$ then F satisfies Axiom (kE4).*

PROOF. Fix $\overline{f} \in c\mathcal{C}$, writing $f_i: A_{i0} \longrightarrow A_{i1}$. We know that the southern arrow of $[\overline{f}]_F$ is a cofibration, so all that we need to show is that all faces of $[\overline{f}]_F$ are good. Let $J \subseteq \{1, \ldots, k\}$ and let $\eta: \{1, \ldots, k\} \backslash J \longrightarrow \{0, 1\}$. A face of $[\overline{f}]_F$ is determined by J and η by letting J determine which variables are allowed to change, and let η determine the value of the other variables. More formally, we can define the cube $I_{J,\eta}: \mathcal{I}^{|J|} \longrightarrow \mathcal{D}$ by

$$I_{J,\eta}(\epsilon_{j_1}, \ldots, \epsilon_{j_{|J|}}) = F(B_1, \ldots, B_k) \qquad B_j = \begin{cases} A_{j\eta(j)} & \text{if } j \notin J \\ A_{j\epsilon_j} & \text{if } j \in J. \end{cases}$$

Let $\overline{h}$ have $h_j = f_j$ if $j \in J$ and $* \longrightarrow A_{j\eta(j)}$ otherwise. Then $\overline{h} \in c\overline{\mathcal{C}}$, so by assumption we know that the southern arrow of $[\overline{h}]_F$ is a cofibration. But this southern arrow is exactly the southern arrow of $I_{J,\eta}$, so we see that the southern arrow of $I_{J,\eta}$ is a cofibration, as desired. □

We now need to consider composition of multiexact functors. First, a simple observation about southern arrows.

LEMMA 4.5. *A k-exact functor commutes with taking southern arrows in each variable independently.*

PROOF. This follows from Axiom (kE2) and Lemma 3.4, which says that southern arrows can be computed as iterated pushouts. □

The next two results concern the following situtation. Let $j_1, \dots, j_k \in \mathbb{Z}_{\geq 0}$, and write $m_i = \sum_{\ell=1}^{i} j_i$. Given j_i-exact functors

$$G_i \colon \mathcal{C}_{m_{i-1}+1} \times \cdots \times \mathcal{C}_{m_i} \longrightarrow \mathcal{D}_i$$

and a k-exact functor

$$F \colon \mathcal{D}_1 \times \cdots \times \mathcal{D}_k \longrightarrow \mathcal{D},$$

define the composite functor

$$H = F \circ (G_1 \times \cdots \times G_k) \colon \mathcal{C}_1 \times \cdots \times \mathcal{C}_{m_k} \longrightarrow \mathcal{D}.$$

LEMMA 4.6. *Let g_i be the southern arrow of $[(f_{m_{i-1}+1}, \dots, f_{m_i})]_{G_i}$. The southern arrow of $[\overline{f}]_H$ is isomorphic to the southern arrow of $[\overline{g}]_F$.*

PROOF. We will use Lemma 3.4. Let $\mathcal{A}_i = \mathcal{I}^{m_{i-1}} \times (\mathcal{I}^{j_i} \backslash \{(1, \dots, 1)\} \times \mathcal{I}^{m_k - m_i}$, so that for any $J \subseteq \{1, \dots, k\}$,

$$\bigcap_{j \in J} \mathcal{A}_j = \prod_{i=1}^{k} B_i^{(J)}, \qquad \text{where } B_i^{(J)} = \begin{cases} \mathcal{I}^{j_i} & \text{if } i \in J \\ \mathcal{I}^{j_i} \backslash \{(1, \dots, 1)\} & \text{if } i \notin J \end{cases}$$

and

$$\bigcup_{i=1}^{n} \mathcal{A}_i = \mathcal{I}^{m_k} \backslash \{(1, \dots, 1)\}.$$

In addition, by Lemma 4.5

$$\operatorname*{colim}_{\bigcap_{j \in J} \mathcal{A}_j} [\overline{f}]'_H \cong F \left(\operatorname*{colim}_{B_1^{(J)}} [(f_1, \dots, f_{m_1})]'_{G_1}, \dots, \operatorname*{colim}_{B_k^{(J)}} [(f_{m_{k-1}+1}, \dots, f_{m_k})]'_{G_k} \right).$$

Note that these are exactly the entries of the cube $[\overline{g}]_F$. Applying Lemma 3.3 to compute $\operatorname{colim} [\overline{f}]'_H$ in terms of these, we see that the southern arrow of $[\overline{f}]_H$ is exactly the southern arrow of $[\overline{g}]_F$, as desired. □

Using this we can show that H is m_k-exact, and thus that **WaldCat** is a multicategory.

PROPOSITION 4.7. **WaldCat** *is a multicategory.*

PROOF. It suffices to check that H is m_k-exact. Axioms (kE1), (kE2) and (kE3) are direct from the definition, so we just need to check Axiom (kE4).

Let $\overline{f} \in c\overline{\mathcal{C}}$; we want to show that $[\overline{f}]_H$ is good. By Lemma 4.4 it suffices to show that the southern arrow of $[\overline{f}]_H$ is a cofibration. Let g_i be the southern arrow of $[(f_{m_{i-1}+1}, \ldots, f_{m_i})]_{G_i}$. As G_i is j_i-exact we know that g_i is a cofibration; as F is k-exact, the southern arrow of $[\overline{g}]_F$ is also a cofibration. However, by Lemma 4.6 we know that the southern arrow of $[\overline{g}]_F$ is exactly the southern arrow of $[\overline{f}]_H$, so we see that the southern arrow of $[\overline{f}]_H$ must also be a cofibration, as desired. □

5. The closed structure

We would now like to show that **WaldCat** is a closed multicategory.

DEFINITION 5.1. We define the internal hom $\underline{\textbf{WaldCat}}(\mathcal{C}_1, \ldots, \mathcal{C}_k; \mathcal{D})$ in the following manner:

objects: k-exact functors $\mathcal{C}_1 \times \cdots \times \mathcal{C}_k \longrightarrow \mathcal{D}$,
morphisms: natural transformations between functors,
weak equivalences: natural weak equivalences between functors, and
cofibrations: natural transformations $\alpha: F \longrightarrow G$ such that for any $\overline{f} \in c\overline{\mathcal{C}}$, the cube

$$\left([\overline{f}]_F \overset{\alpha}{\Longrightarrow} [\overline{f}]_G\right)$$

is good.

In particular, note that all cofibrations are levelwise cofibrations.

We need to prove that this is well-defined.

LEMMA 5.2. *$\underline{\textbf{WaldCat}}(\mathcal{C}_1, \ldots, \mathcal{C}_k; \mathcal{D})$ is a Waldhausen category.*

PROOF. The only axiom that is nontrivial to check is Axiom (W3); the others follow directly from the fact that $\mathcal{D}$ is a Waldhausen category. Thus we focus our attention on checking Axiom (W3).

Consider a diagram

$$H \longleftarrow F \overset{\alpha}{\hookrightarrow} G \in \underline{\textbf{WaldCat}}(\overline{\mathcal{C}}; \mathcal{D}).$$

We know that α is a levelwise cofibration and that pushouts along cofibrations exist in $\mathcal{D}$, so we get a functor $G \cup_F H: \overline{\mathcal{C}} \longrightarrow \mathcal{D}$. We need to check that this functor is k-exact, and that the induced natural transformation $\beta: H \longrightarrow G \cup_F H$ is a cofibration inside $\underline{\textbf{WaldCat}}(\overline{\mathcal{C}}; \mathcal{D})$.

The functor $G \cup_F H$ satisfies Axiom (kE1) and (kE2) because F, G and H are k-exact, and has Axiom (kE3) because Axiom (W4) holds in $\mathcal{D}$. To prove Axiom (kE4), fix $\overline{f} \in c\overline{\mathcal{C}}$. We know that $[\overline{f}]_F$, $[\overline{f}]_G$ and $[\overline{f}]_H$ are good, so $[\overline{f}]_{G \cup_F H}$ is also good by Proposition 3.1. Thus $G \cup_F H$ is k-exact. Proposition 3.1 also tells us that $([\overline{f}]_H \overset{\beta}{\Longrightarrow} [\overline{f}]_{G \cup_F H})$ is good, which means that β is a cofibration, as desired. □

In order for this definition to make **WaldCat** into a closed multicategory, we need a $k+1$-exact evaluation morphism.

DEFINITION 5.3. The functor

$$\text{ev}_{\mathcal{C}_1, \ldots, \mathcal{C}_k; \mathcal{D}}: \mathcal{C}_1 \times \cdots \times \mathcal{C}_k \times \underline{\textbf{WaldCat}}(\mathcal{C}_1, \ldots, \mathcal{C}_k; \mathcal{D}) \longrightarrow \mathcal{D}$$

is defined by

$$\text{ev}_{\mathcal{C}_1, \ldots, \mathcal{C}_k; \mathcal{D}}(A_1, \ldots, A_k, F) = F(A_1, \ldots, A_k).$$

LEMMA 5.4. *The functor $\text{ev}_{\mathcal{C}_1, \ldots, \mathcal{C}_k; \mathcal{D}}$ is $k+1$-exact.*

PROOF. Axioms (kE1), (kE2) and (kE3) are easily checked from the fact that F is k-exact. Thus we only need to check Axiom (kE4). Given $f_i\colon A_i \hookrightarrow B_i \in \mathcal{C}_i$ and a cofibration $\alpha\colon F \longrightarrow G$ we need to check that $[(\overline{f}, \alpha)]_{\mathrm{ev}}$ is good. However, by definition this cube is the cube $([\overline{f}]_F \overset{\alpha}{\Longrightarrow} [\overline{f}]_G)$, which is good because F and G are k-exact and α is a cofibration. □

It remains to check that this gives a well-defined closed multicategory structure on **WaldCat**, i.e. that for all Waldhausen categories $\mathcal{C}_1, \ldots, \mathcal{C}_k$, $\mathcal{D}$ and $\mathcal{A}_1, \ldots, \mathcal{A}_\ell$ the function

$$\begin{gathered}\mathbf{WaldCat}_\ell(\mathcal{A}_1, \ldots, \mathcal{A}_\ell; \underline{\mathbf{WaldCat}}(\mathcal{C}_1, \ldots, \mathcal{C}_k; \mathcal{D})) \\ \longrightarrow \mathbf{WaldCat}_{k+\ell}(\mathcal{C}_1, \ldots, \mathcal{C}_k, \mathcal{A}_1, \ldots, \mathcal{A}_\ell; \mathcal{D})\end{gathered}$$

given by $F \longmapsto \mathrm{ev}_{\mathcal{C}_1,\ldots,\mathcal{C}_k,\mathcal{D}} \circ (1_{\mathcal{C}_1}, \ldots, 1_{\mathcal{C}_k}, F)$ is a bijection. We can construct an inverse easily by partial application, so assuming that the partial inverse is well-defined we know that this is a bijection. The map ev is also clearly Σ_k-equivariant. Thus we need to show the following:

LEMMA 5.5. *Fix $1 \leq \ell \leq k$. Let $F\colon \mathcal{C}_1 \times \cdots \times \mathcal{C}_k \longrightarrow \mathcal{D}$ be a k-exact functor, and fix $A_i \in \mathcal{C}_i$ for $\ell < i \leq k$. Then the functor*

$$F(-, A_{\ell+1}, \ldots, A_k)\colon \mathcal{C}_1 \times \cdots \times \mathcal{C}_\ell \longrightarrow \mathcal{D}$$

is ℓ-exact.

PROOF. It suffices to prove this for $\ell = k - 1$; the rest will follow by induction. Axioms (kE1), (kE2) and (kE3) hold immediately, so it suffices to consider (kE4). Let $f_i \in c\mathcal{C}_i$ for $1 \leq i < k$ be cofibrations, and let $f_k\colon \varnothing \hookrightarrow A_k$ for the A_k in the statement of the lemma. Since F was k-exact, we know that $[\overline{f}]_F$ is good, which implies that $([\overline{f}]_F)_{k1}$ is good. But this is exactly the cube $[(f_2, \ldots, f_k)]_{F(-,A_k)}$. Thus $F(-, A_k)$ is $k-1$-exact, as desired. □

We have now proved:

PROPOSITION 5.6. **WaldCat** *is a closed multicategory.*

6. K-Theory as an enriched multifunctor

Our goal for this section is to show that the closed multicategory structure on **WaldCat** is compatible with the K-theory functor.

DEFINITION 6.1. A *multifunctor* $F\colon \mathcal{M} \longrightarrow \mathcal{M}'$ between symmetric multicategories is a function $F\colon \mathrm{ob}\,\mathcal{M} \longrightarrow \mathrm{ob}\,\mathcal{M}'$, and a function

$$\mathcal{M}(A_1, \ldots, A_k; B) \longrightarrow \mathcal{M}'(F(A_1), \ldots, F(A_k); F(B))$$

for all tuples $B, A_1, \ldots, A_k$ of objects. These must preserve the units and be compatible with composition and the Σ_k-action. In the case when $\mathcal{M}$ and $\mathcal{M}'$ are enriched over a symmetric monoidal $\mathcal{V}$, we just need

$$\mathcal{M}(A_1, \ldots, A_k; B) \longrightarrow \mathcal{M}'(F(A_1), \ldots, F(A_k); F(B))$$

to be a $\mathcal{V}$-morphism.

First we consider the unenriched setting.

PROPOSITION 6.2. *The functor K:* **WaldCat** $\longrightarrow$ **Sp** *is a multifunctor.*

This statement is well-known to specialists, and has been mentioned in many papers, including [**Wal85**], [**BM11**, Theorem 2.6], and [**GH99**]. However, we could not find a reference that explicitly checked that the multifunctor structure was compatible with the structure maps of the symmetric spectra produced by K-theory, so in the interest of completeness we present the proof here, as well.

Before starting the proof of this proposition we will first make an auxillary construction. In order to show that K is a multifunctor we will need to show that any k-exact functor $F\colon \mathcal{C}_1 \times \cdots \times \mathcal{C}_k \longrightarrow \mathcal{C}$ gives rise to a morphism $K(\mathcal{C}_1) \wedge \cdots \wedge K(\mathcal{C}_k) \longrightarrow K(\mathcal{C})$. When $k = 0$ this says that a choice of object $A \in \mathcal{C}$ gives a morphism $\mathbb{S} \longrightarrow K(\mathcal{C})$. To construct this, choose an equivalence $\alpha\colon \mathbb{S} \longrightarrow K(\mathbf{FinSet}_*)$, and then take the exact functor $p_A\colon \mathbf{FinSet}_* \longrightarrow \mathcal{C}$ given by $I \longmapsto \coprod_I A$. Then $K(p_A)\alpha$ is the desired morphism. In the case $k = 1$ this is just a definition check to see that the definition of K-theory in Definition 2.6 is functorial in $\mathcal{C}$.

Now consider $k > 1$. In the interest of simplifying the following analysis, we will restrict our attention to the case when $k = 2$; the higher cases follow analogously. The data of a 2-morphism is, for every pair m_1, m_2, a map of spaces

$$\mu_{m_1,m_2}\colon K(\mathcal{C}_1)_{m_1} \wedge K(\mathcal{C}_2)_{m_2} \longrightarrow K(\mathcal{C})_{m_1+m_2}.$$

These maps need to be coherent with respect to the spectral structure maps; in particular, we need the following diagram to commute:

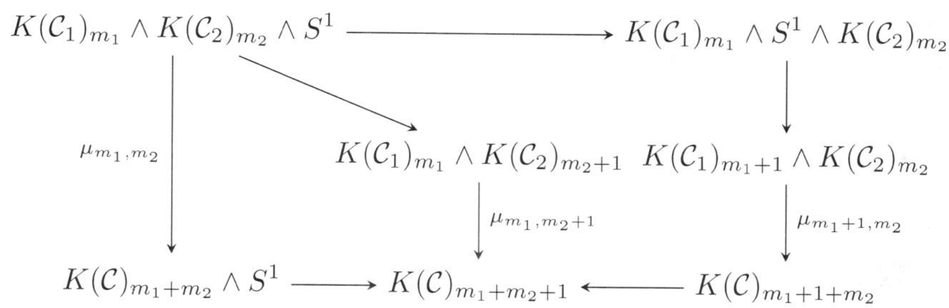

The right-hand map along the bottom of the diagram comes from the $\Sigma_{m_1+m_2+1}$-action from the symmetric spectrum structure of $K(\mathcal{C})$.

For a Waldhausen category $\mathcal{C}$ and $0 \le i \le n$ we define a functor $\rho_{ni}\colon \mathcal{C} \longrightarrow S_n\mathcal{C}$, which is defined on objects by

$$\rho_{ni}(A)(j \le k) = \begin{cases} * & \text{if } k < i \text{ or } j \ge i, \\ A & \text{otherwise} \end{cases}$$

and extends in the analogous manner to morphisms. Let S^1 be the pointed simplicial set which at level n is equal to the set $\{0, 1, \dots, n\}$; we can also consider S^1 to be a pointed category with only trivial morphisms. Then we have a morphism of simplicial categories $P\colon \mathcal{C} \times S^1 \longrightarrow S.\mathcal{C}$ $(A, i) \longmapsto \rho_{ni}(A)$.

LEMMA 6.3. *P is a well-defined functor of simplicial categories.*

PROOF. In order for P to be well-defined we need to show that the image of P is in $S.\mathcal{C}$, and that P is compatible with the simplicial maps. The first part of this is true by definition, since ρ_{ni} is constructed to be a valid element of $S_n\mathcal{C}$. For the second part, note that we have

$$\partial_j \rho_{ni}(A) = \begin{cases} \rho_{(n-1)i}(A) & \text{if } j \geq i \text{ and } i \neq n \\ \rho_{(n-1)(i-1)}(A) & \text{if } j < i \\ \rho_{(n-1)0} & \text{if } j = i = n \end{cases}$$
$$= \rho_{(n-1)(\partial_j(i))}(A),$$

where in the right-hand side of the above, $i \in S^1_n$. Analogously,

$$s_j \rho_{ni}(A) = \begin{cases} \rho_{(n+1)i} & \text{if } j \geq i \\ \rho_{(n+1)(i+1)} & \text{if } j < i \end{cases} = \rho_{(n-1)(s_j(i))}(A),$$

so we are done. □

We thus have functors

$$P \colon S_\cdot^{(m)}\mathcal{C} \times S^1 \longrightarrow S_\cdot^{(m+1)}\mathcal{C}.$$

By definition, if either $i = 0$ or $A = *$ then $P(A, i) = *$, so P lifts to a map

$$P \colon NwS_\cdot^{(m)}\mathcal{C} \wedge S^1 \longrightarrow NwS_\cdot^{(m+1)}\mathcal{C}.$$

This is the spectral structure map of the K-theory of a symmetric spectrum.

Proof of Proposition 6.2. Consider a biexact functor $F \colon \mathcal{C}_1 \times \mathcal{C}_2 \longrightarrow \mathcal{C}$. We want to use F to construct morphisms $\mu_{m_1,m_2} \colon K(\mathcal{C}_1)_{m_1} \wedge K(\mathcal{C}_2)_{m_2} \longrightarrow K(\mathcal{C})_{m_1+m_2}$.

The key fact we need about the objects of $S_{n_1} \cdots S_{n_m}\mathcal{C}$ is that they will be preserved by biexact functors in the following manner. Consider the composition

$$S_{\vec{n}_1}^{(m_1)}\mathcal{C}_1 \times S_{\vec{n}_2}^{(m_2)}\mathcal{C}_2 \longrightarrow \big[\mathrm{Ar}[\vec{n}_1], \mathcal{C}_1\big] \times \big[\mathrm{Ar}[\vec{n}_2], \mathcal{C}_2\big] \longrightarrow$$
$$\big[\mathrm{Ar}([\vec{n}_1] \times [\vec{n}_2]), \mathcal{C}_2 \times \mathcal{C}_2\big] \xrightarrow{F\circ} \big[\mathrm{Ar}([\vec{n}_1] \times [\vec{n}_2]), \mathcal{C}\big].$$

The key extra condition on the objects of $S_{\vec{n}_1\vec{n}_2}^{(m_1+m_2)}\mathcal{C}$ is that this functor lands in $S_{\vec{n}_1\vec{n}_2}^{(m_1+m_2)}\mathcal{C}$. By varying the coordinates of $\vec{n}_1$ and $\vec{n}_2$ these assemble into exact functors

$$S_\cdot^{(m_1)}\mathcal{C}_1 \times S_\cdot^{(m_2)}\mathcal{C}_2 \longrightarrow S_\cdot^{(m_1+m_2)}\mathcal{C}.$$

Applying $|Nw \cdot|$ to these and noting that any point with the basepoint as one of the coordinates gets mapped to the basepoint, we get maps

$$\mu_{m_1,m_2} \colon K(\mathcal{C}_1)_{m_1} \wedge K(\mathcal{C}_2)_{m_2} \longrightarrow K(\mathcal{C})_{m_1+m_2}.$$

In order to check these assemble into a map $K(\mathcal{C}_1) \wedge K(\mathcal{C}_2) \longrightarrow K(\mathcal{C})$ we that these satisfy the coherence conditions stated earlier. In order to show this, we will show that the following diagram commutes:

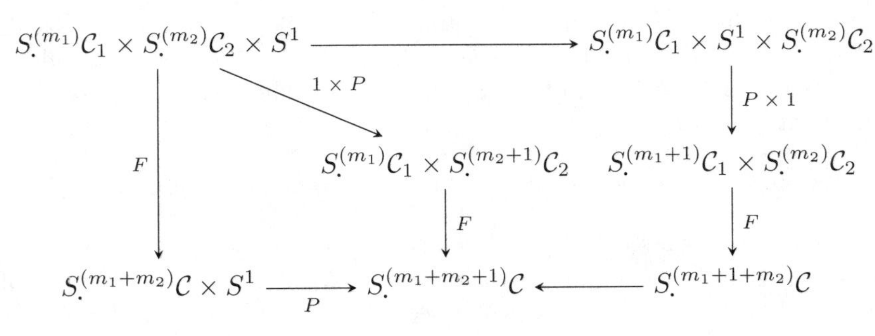

In fact, all of the morphisms except for the two horizontal morphisms are obtained by postcomposing functors $\mathrm{Ar}([\vec{n}_1] \times [\vec{n}_2]) \longrightarrow s\mathbf{Cat}$ with P or F. The horizontal morphisms, on the other hand, permute both source and target categories, and then permute the source categories back; everything in between is, once again, postcomposing with P or F. Thus in order for this diagram to commute it suffices to show that the diagram

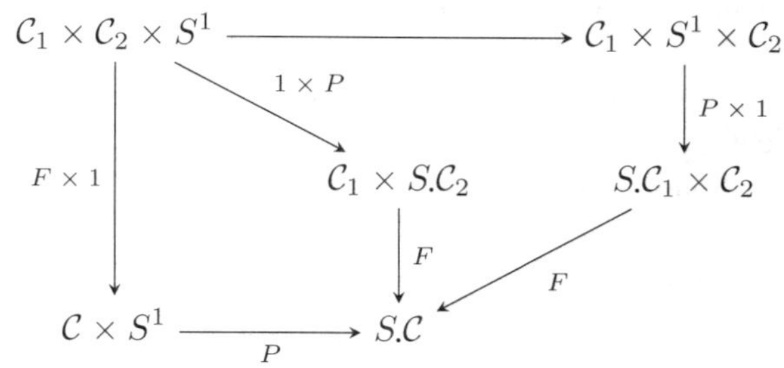

commutes. Consider a triple $(A_1, A_2, i) \in W_1 \times W_2 \times S^1_n$. To check that the diagram commutes, we need to show that

$$\rho_{ni}(F(A_1, A_2)) = F(A_1, \rho_{ni}(A_2)) = F(\rho_{ni}(A_1), A_2).$$

Looking at each of these at spot $j \leq k$ we have that if $j < i$ or $k \geq i$, the first is $*$, the second is $F(A_1, *)$ and the third is $F(*, A_2)$, which are all equal because F is biexact. Otherwise, these are all equal to $F(A_1, A_2)$, so are again all equal. So these diagrams commute on objects. Analogously, they commute on all morphisms.

This completes the proof of Proposition 6.2. □

We still have not shown that K is compatible with the closed structure on **WaldCat**. However, as we have just shown that K is in fact a multifunctor, we can consider **WaldCat** to be enriched over **Sp** just by applying K to each of the internal hom-objects to produce a spectrally-enriched category $\mathbf{WaldCat}_{\mathbf{Sp}}$. We then have the following:

PROPOSITION 6.4. $K\colon \mathbf{WaldCat}_{\mathbf{Sp}} \longrightarrow \mathbf{Sp}$ *is a spectrally-enriched multifunctor.*

PROOF. We need to show that K gives a morphism

$$K(\underline{\mathbf{WaldCat}}(\mathcal{C}_1, \ldots, \mathcal{C}_k; \mathcal{D})) \longrightarrow K(\mathcal{D})^{K(\mathcal{C}_1)\wedge\cdots\wedge K(\mathcal{C}_k)}.$$

As **Sp** is closed symmetric monoidal, it suffices to show that we get a morphism $K(\mathcal{C}_1)\wedge\cdots\wedge K(\mathcal{C}_k)\wedge K(\underline{\mathbf{WaldCat}}(\mathcal{C}_1, \ldots, \mathcal{C}_k; \mathcal{D})) \longrightarrow K(\mathcal{D})$. The evaluation functor defined in Definition 5.3 gives us such a morphism, and it follows directly that this produces the enrichment on K. □

For more on multifunctors between closed multicategories see [**Man12**, Section 3].

Appendix A. Proof of Lemma 3.4

For ease of reading we restate Lemma 3.4 here.

LEMMA 3.4. *Suppose that we are given an indexing category $\mathcal{A}$ along with n subcategories $\mathcal{A}_1, \ldots, \mathcal{A}_n$ such that $\mathcal{A} = \bigcup_{i=1}^n \mathcal{A}_i$. Let $F\colon \mathcal{A} \to \mathcal{C}$. Then the southern arrow of the cube I given by*

$$I(\overline{\epsilon}) = \operatorname{colim} F\big|_{\bigcap_{\epsilon_i=0} \mathcal{A}_i} \qquad I(1) = \operatorname{colim} F$$

is an isomorphism.

PROOF. We want to show that $\operatorname{colim} I' \cong \operatorname{colim} F$; we will prove this by induction on n.

We begin by proving the inductive step. Suppose that we know that this lemma is true for values lower than n. Let $\mathcal{A}' = \bigcup_{i=1}^{n-1} \mathcal{A}_i$. Let $\mathcal{B} = \mathcal{I}^n \backslash \{(1, 1, \ldots, 0), (1, \ldots, 1)\}$. By the $n = 2$ case of the lemma applied to I',

$$\begin{array}{ccc} \operatorname{colim} I'\big|_{\mathcal{B}\cap(\mathcal{I}^{n-1}\times\{0\})} & \longrightarrow & \operatorname{colim} I'\big|_{\mathcal{I}^{n-1}\times\{0\}} \\ \downarrow & & \downarrow \\ \operatorname{colim} I'\big|_{\mathcal{B}} & \longrightarrow & \operatorname{colim} I' \end{array}$$

is a pushout square. By the $n-1$ case of the lemma, the upper-left corner is $\operatorname{colim} F\big|_{\mathcal{A}'\cap\mathcal{A}_n}$, and the lower-left corner is $\operatorname{colim} F\big|_{\mathcal{A}'}$. The upper-right corner is just $I'(1, \ldots, 1, 0) = \operatorname{colim} F\big|_{\mathcal{A}_n}$. Rewriting this, we see that

$$\begin{array}{ccc} \operatorname{colim} F\big|_{\mathcal{A}'\cap\mathcal{A}_n} & \longrightarrow & \operatorname{colim} F\big|_{\mathcal{A}_n} \\ \downarrow & & \downarrow \\ \operatorname{colim} F\big|_{\mathcal{A}'} & \longrightarrow & \operatorname{colim} I' \end{array}$$

is a pushout square. On the other hand, by the $n = 2$ case of the lemma applied directly to F,

$$\begin{array}{ccc} \operatorname{colim} F\big|_{\mathcal{A}'\cap\mathcal{A}_n} & \longrightarrow & \operatorname{colim} F\big|_{\mathcal{A}_n} \\ \downarrow & & \downarrow \\ \operatorname{colim} F\big|_{\mathcal{A}'} & \longrightarrow & \operatorname{colim} F \end{array}$$

is also a pushout square. Thus $\operatorname{colim} I' \cong \operatorname{colim} F$, as desired.

The base case is $n = 2$. In particular, we want to show that

$$\begin{array}{ccc} \operatorname{colim} F\big|_{\mathcal{A}_1\cap\mathcal{A}_2} & \longrightarrow & \operatorname{colim} F\big|_{\mathcal{A}_1} \\ \downarrow & & \downarrow \\ \operatorname{colim} F\big|_{\mathcal{A}_2} & \longrightarrow & \operatorname{colim} F \end{array}$$

is a pushout square. Let X be the pushout of the upper-left part of the square. We have

$$
\begin{aligned}
\operatorname{colim} F &\cong \operatorname{coeq}\Bigl(\coprod_{f\in\mathcal{A}} \operatorname{dom} F(f) \rightrightarrows \coprod_{A\in\mathcal{A}} F(A)\Bigr)\\
&\cong \operatorname{colim}\left(\begin{array}{ccc}
\displaystyle\coprod_{f\in\mathcal{A}_1\cap\mathcal{A}_2} \operatorname{dom} F(f) & \overset{t}{\underset{s}{\rightrightarrows}} & \displaystyle\coprod_{A\in\mathcal{A}_1\cap\mathcal{A}_2} F(A)\\
{\scriptstyle L}\downarrow\downarrow{\scriptstyle R} & & {\scriptstyle L}\downarrow\downarrow{\scriptstyle R}\\
\displaystyle\coprod_{f\in\mathcal{A}_1\amalg\mathcal{A}_2} \operatorname{dom} F(f) & \overset{t}{\underset{s}{\rightrightarrows}} & \displaystyle\coprod_{A\in\mathcal{A}_1\amalg\mathcal{A}_2} F(A)
\end{array}\right)\\
&\cong \operatorname{coeq}\left(\begin{array}{c}
\operatorname{coeq}\Bigl(\displaystyle\coprod_{f\in\mathcal{A}_1\cap\mathcal{A}_2} \operatorname{dom} F(f) \overset{t}{\underset{s}{\rightrightarrows}} \coprod_{A\in\mathcal{A}_1\cap\mathcal{A}_2} F(A)\Bigr)\\
{\scriptstyle L}\downarrow \qquad\qquad\qquad \downarrow{\scriptstyle R}\\
\operatorname{coeq}\Bigl(\displaystyle\coprod_{f\in\mathcal{A}_1\amalg\mathcal{A}_2} \operatorname{dom} F(f) \overset{t}{\underset{s}{\rightrightarrows}} \coprod_{A\in\mathcal{A}_1\amalg\mathcal{A}_2} F(A)\Bigr)
\end{array}\right)\\
&\cong \operatorname{coeq}\Bigl(\operatorname{colim} F\big|_{\mathcal{A}_1\cap\mathcal{A}_2} \rightrightarrows \operatorname{colim} F\big|_{\mathcal{A}_1} \amalg \operatorname{colim} F\big|_{\mathcal{A}_2}\Bigr)\\
&\cong X,
\end{aligned}
$$

as claimed. Here, s is the morphism which takes $\operatorname{dom} f$ to itself and t is the morphism given by f on the component indexed by f. L includes $\mathcal{A}_1 \cap \mathcal{A}_2$ into $\mathcal{A}_1$, and R includes $\mathcal{A}_1 \cap \mathcal{A}_2$ into $\mathcal{A}_2$. □

Acknowledgments

The author would like to thank the organizers of the mid-Atlantic Topology Seminar for organizing these proceedings, as well as the anonymous referee whose comments vastly improved this paper. The author was supported by the Institute for Advanced Studies and by an NSF MSRFP grant.

References

[BM11] Andrew J. Blumberg and Michael A. Mandell, *Derived Koszul duality and involutions in the algebraic K-theory of spaces*, J. Topol. **4** (2011), no. 2, 327–342, DOI 10.1112/jtopol/jtr003. MR2805994

[EM06] A. D. Elmendorf and M. A. Mandell, *Rings, modules, and algebras in infinite loop space theory*, Adv. Math. **205** (2006), no. 1, 163–228, DOI 10.1016/j.aim.2005.07.007. MR2254311

[GH99] Thomas Geisser and Lars Hesselholt, *Topological cyclic homology of schemes*, Algebraic K-theory (Seattle, WA, 1997), Proc. Sympos. Pure Math., vol. 67, Amer. Math. Soc., Providence, RI, 1999, pp. 41–87, DOI 10.1090/pspum/067/1743237. MR1743237

[Goo92] Thomas G. Goodwillie, *Calculus. II. Analytic functors*, K-Theory **5** (1991/92), no. 4, 295–332, DOI 10.1007/BF00535644. MR1162445

[Kel82] Gregory Maxwell Kelly, *Basic concepts of enriched category theory*, London Mathematical Society Lecture Note Series, vol. 64, Cambridge University Press, Cambridge-New York, 1982. MR651714

[Lei04] Tom Leinster, *Higher operads, higher categories*, London Mathematical Society Lecture Note Series, vol. 298, Cambridge University Press, Cambridge, 2004. MR2094071

[Man12] Oleksandr Manzyuk, *Closed categories vs. closed multicategories*, Theory Appl. Categ. **26** (2012), No. 5, 132–175. MR2909641

[Wal85] Friedhelm Waldhausen, *Algebraic K-theory of spaces*, Algebraic and geometric topology (New Brunswick, N.J., 1983), Lecture Notes in Math., vol. 1126, Springer, Berlin, 1985, pp. 318–419, DOI 10.1007/BFb0074449. MR802796

Department of Mathematics, Cornell University, Ithaca, NY 14853-4201
Email address: zakh@math.cornell.edu

Selected Published Titles in This Series

707 **Nitya Kitchloo, Mona Merling, Jack Morava, Emily Riehl, and W. Stephen Wilson, Editors,** New Directions in Homotopy Theory, 2018

704 **Alain Escassut, Cristina Perez-Garcia, and Khodr Shamseddine, Editors,** Advances in Ultrametric Analysis, 2018

703 **Andreas Malmendier and Tony Shaska, Editors,** Higher Genus Curves in Mathematical Physics and Arithmetic Geometry, 2018

702 **Mark Grant, Gregory Lupton, and Lucile Vandembroucq, Editors,** Topological Complexity and Related Topics, 2018

701 **Joan-Carles Lario and V. Kumar Murty, Editors,** Number Theory Related to Modular Curves, 2018

700 **Alexandre Girouard, Dmitry Jakobson, Michael Levitin, Nilima Nigam, Iosif Polterovich, and Frédéric Rochon, Editors,** Geometric and Computational Spectral Theory, 2017

699 **Mark L. Agranovsky, Matania Ben-Artzi, Catherine Bénéteau, Lavi Karp, Dmitry Khavinson, Simeon Reich, David Shoikhet, Gilbert Weinstein, and Lawrence Zalcman, Editors,** Complex Analysis and Dynamical Systems VII, 2017

698 **Alexander M. Blokh, Leonid A. Bunimovich, Paul H. Jung, Lex G. Oversteegen, and Yakov G. Sinai, Editors,** Dynamical Systems, Ergodic Theory, and Probability: in Memory of Kolya Chernov, 2017

697 **Fabrizio Broglia, Françoise Delon, Max Dickmann, Danielle Gondard-Cozette, and Victoria Ann Powers, Editors,** Ordered Algebraic Structures and Related Topics, 2017

696 **Ara S. Basmajian, Yair N. Minsky, and Alan W. Reid, Editors,** In the Tradition of Ahlfors–Bers, VII, 2017

695 **Katrina Barron, Elizabeth Jurisich, Antun Milas, and Kailash Misra, Editors,** Lie Algebras, Vertex Operator Algebras, and Related Topics, 2017

694 **Manjul Bhargava, Robert Guralnick, Gerhard Hiss, Klaus Lux, and Pham Huu Tiep, Editors,** Finite Simple Groups: Thirty Years of the Atlas and Beyond, 2017

693 **Michael Cwikel and Mario Milman, Editors,** Functional Analysis, Harmonic Analysis, and Image Processing, 2017

692 **Anatole Katok, Yakov Pesin, and Federico Rodriguez Hertz, Editors,** Modern Theory of Dynamical Systems, 2017

691 **Farrell Brumley, Maria Paula Gómez Aparicio, and Alberto Mínguez, Editors,** Around Langlands Correspondences, 2017

690 **Andrés Eduardo Caicedo, James Cummings, Peter Koellner, and Paul B. Larson, Editors,** Foundations of Mathematics, 2017

689 **Erica Flapan, Allison Henrich, Aaron Kaestner, and Sam Nelson, Editors,** Knots, Links, Spatial Graphs, and Algebraic Invariants, 2017

688 **Jeffrey Bergen, Stefan Catoiu, and William Chin, Editors,** Groups, Rings, Group Rings, and Hopf Algebras, 2017

687 **Fernanda Botelho, Raena King, and T. S. S. R. K. Rao, Editors,** Problems and Recent Methods in Operator Theory, 2017

686 **Alp Bassa, Alain Couvreur, and David Kohel, Editors,** Arithmetic, Geometry, Cryptography and Coding Theory, 2017

685 **Heather A. Harrington, Mohamed Omar, and Matthew Wright, Editors,** Algebraic and Geometric Methods in Discrete Mathematics, 2017

684 **Anna Beliakova and Aaron D. Lauda, Editors,** Categorification in Geometry, Topology, and Physics, 2017